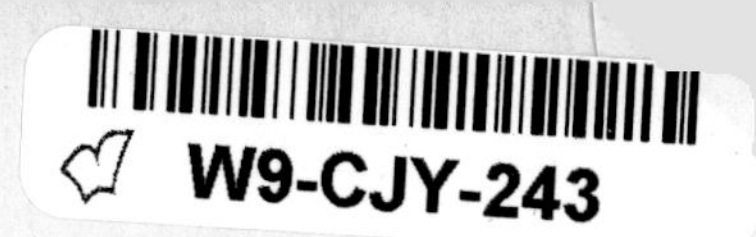

PRACTICE BOOK
for
Conceptual Integrated Science

Paul G. Hewitt

Suzanne Lyons

John Suchocki

Jennifer Yeh

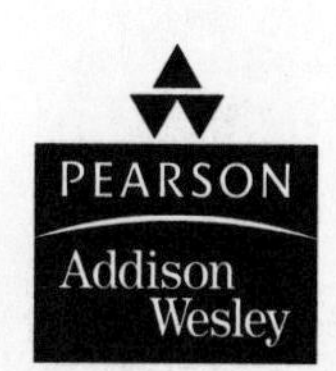

San Francisco Boston New York
Capetown Hong Kong London Madrid Mexico City
Montreal Munich Paris Singapore Sydney Tokyo Toronto

Editor-in-Chief: Adam R.S. Black, Ph.D.
Senior Acquisitions Editor: Lothlórien Homet
Editorial Assistant: Ashley Taylor Anderson
Managing Editor: Corinne Benson
Production Supervisor: Lori Newman
Manufacturing Buyer: Pam Augspurger
Cover Designer: Richard Whitaker, Seventeenth Street Studios
Production Services: Progressive Publishing Alternatives
Composition: Progressive Information Technologies
Illustrations: Paul G. Hewitt
Cover and Text Printer: Courier Stoughton

ISBN: 0-8053-9039-1

5 6 7 8 9 10 –CRS– 10 09 08
www.aw-bc.com

Table of Contents

PART 3 BIOLOGY

PART 4 EARTH SCIENCE

PART 5 ASTRONOMY

Name ______________________________ Date __________

Chapter 2: Describing Motion

Vectors and Equilibrium

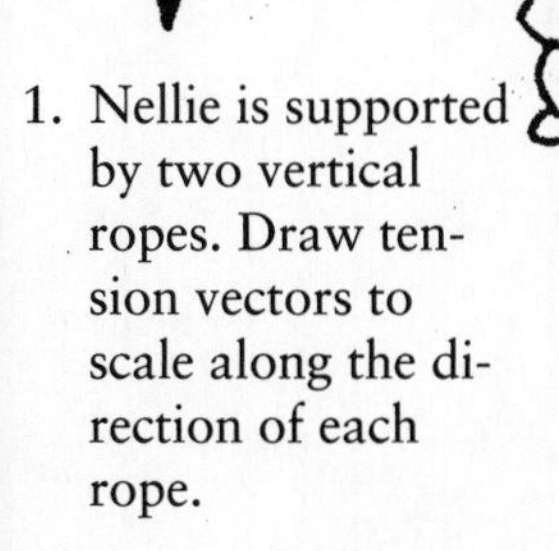

Nellie Newton dangles from a vertical rope in equilibrium: $\Sigma F = 0$. The tension in the rope (upward vector) has the same magnitude as the downward pull of gravity (downward vector).

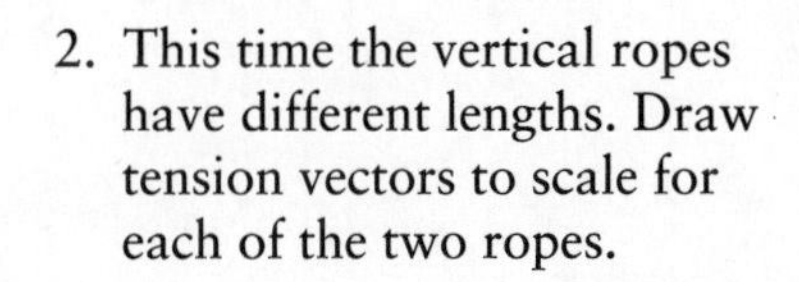

1. Nellie is supported by two vertical ropes. Draw tension vectors to scale along the direction of each rope.

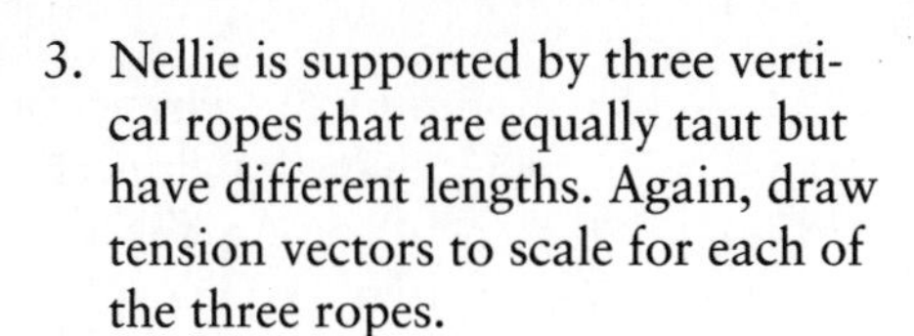

2. This time the vertical ropes have different lengths. Draw tension vectors to scale for each of the two ropes.

3. Nellie is supported by three vertical ropes that are equally taut but have different lengths. Again, draw tension vectors to scale for each of the three ropes.

Circle the correct answer:

4. We see that tension in a rope is [dependent on] [independent of] the length of the rope. So the length of a vector representing rope tension is [dependent on] [independent of] the length of the rope.

Rope tension depends on the angle the rope makes with the vertical, as Practice Pages for Chapter 3 will show!

Chapter 2: Describing Motion

Free Fall Speed

1. Aunt Minnie gives you $10 per second for 4 seconds. How much money do you have after 4 seconds? ____________

2. A ball dropped from rest picks up speed at 10 m/s per second. After it falls for 4 seconds, how fast is it going? ____________
3. You have $20, and Uncle Harry gives you $10 each second for 3 seconds. How much money do you have after 3 seconds? ____________
4. A ball is thrown straight down with an initial speed of 20 m/s. After 3 seconds, how fast is it going? ____________
5. You have $50 and you pay Aunt Minnie $10/second. When will your money run out? ____________
6. You shoot an arrow straight up at 50 m/s. When will it run out of speed? ____________
7. What will be the arrow's speed 5 seconds after you shoot it? ____________
8. What will its speed be 6 seconds after you shoot it? 7 seconds? ____________ ____________

Free Fall Distance

1. Speed is one thing; distance another. *Where* is the arrow you shoot up at 50 m/s when it runs out of speed? ____________
2. How high will the arrow be 7 seconds after being shot up at 50 m/s? ____________

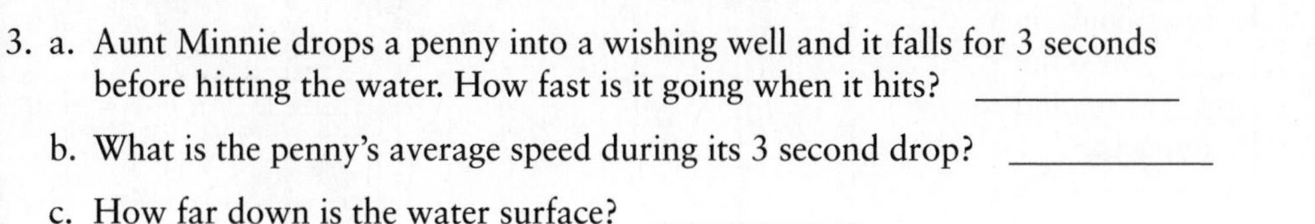

3. a. Aunt Minnie drops a penny into a wishing well and it falls for 3 seconds before hitting the water. How fast is it going when it hits? ____________
   b. What is the penny's average speed during its 3 second drop? ____________
   c. How far down is the water surface? ____________

4. Aunt Minnie didn't get her wish, so she goes to a deeper wishing well and throws a penny straight down into it at 10 m/s. How far does this penny go in 3 seconds? ____________

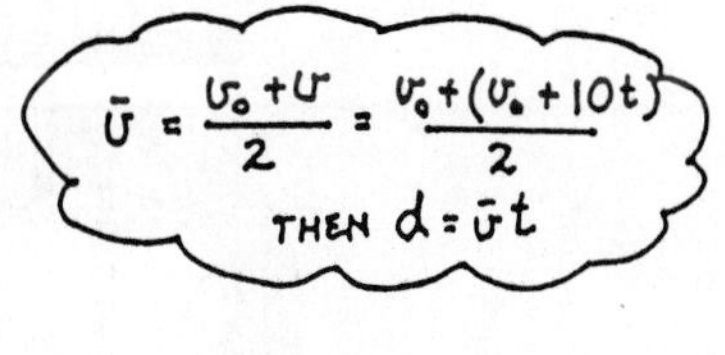

$\bar{v} = \frac{v_0 + v}{2} = \frac{v_0 + (v_0 + 10t)}{2}$
THEN $d = \bar{v}t$

Distinguish between "how fast," "how far," and "how long"!

Name ______________________________ Date __________

Acceleration of Free Fall

A rock dropped from the top of a cliff picks up speed as it falls. Pretend that a speedometer and odometer are attached to the rock to show readings of speed and distance at 1 second intervals. Both speed and distance are zero at time = zero (see sketch). Note that after falling 1 second the speed reading is 10 m/s and the distance fallen is 5 m. The readings for succeeding seconds of fall are not shown and are left for you to complete. Draw the position of the speedometer pointer and write in the correct odometer reading for each time. Use $g = 10\ m/s^2$ and neglect air resistance.

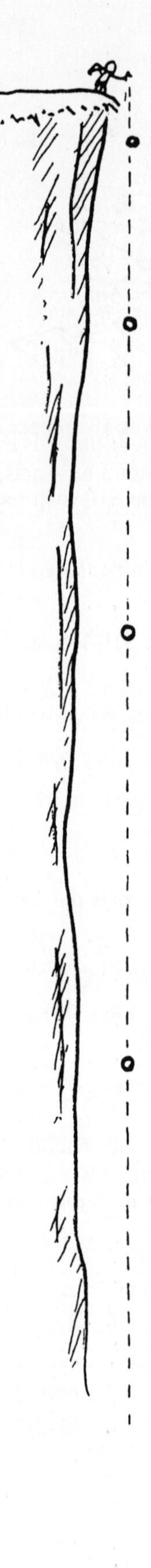

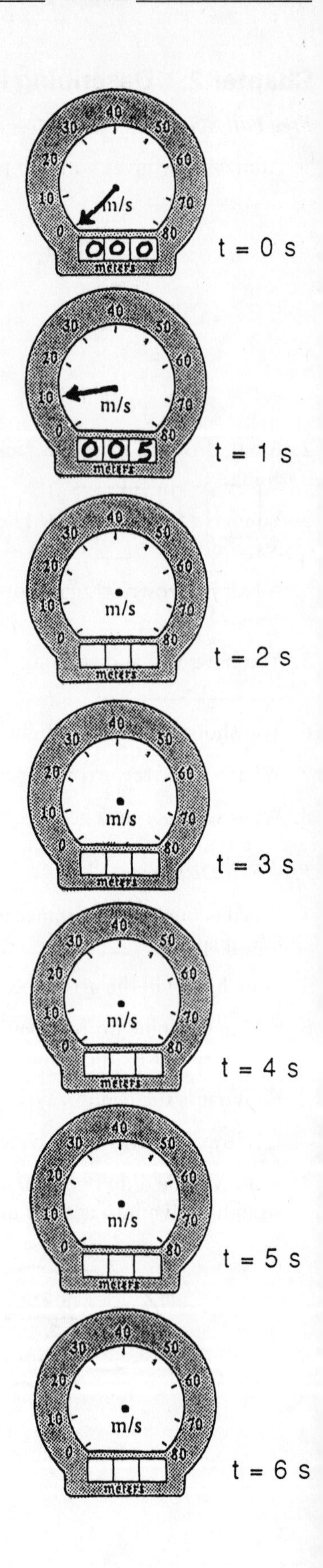

YOU NEED TO KNOW:
Instantaneous speed of fall from rest:

$$v = gt$$

Distance fallen from rest:

$$d = \frac{1}{2}gt^2$$

1. The speedometer reading increases by the same amount, ___________ m/s, each second. This increase in speed per second is called ________________________.

2. The distance fallen increases as the square of the ___________.

3. If it takes 7 seconds to reach the ground, then its speed at impact is ___________ m/s, the total distance fallen is ___________ m, and its acceleration of fall just before impact is ___________ m/s^2.

Name ______________________________ Date ____________

Chapter 3: Newton's Laws of Motion

Newton's First Law and Friction

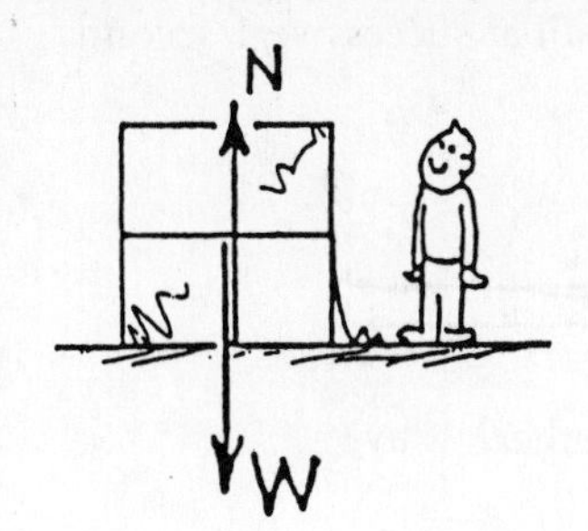

1. A crate filled with video games rests on a horizontal floor. Only gravity and the support force of the floor act on it, as shown by the vectors for weight **W** and normal force **N**.

   a. The net force on the crate is (zero) (greater than zero).

   b. Evidence for this is ______________________________.

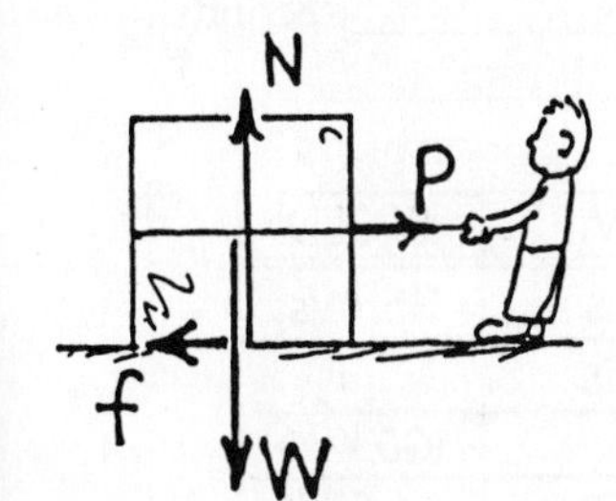

2. A slight pull **P** is exerted on the crate, not enough to move it. A force of friction **f** now acts,

   a. which is (less than) (equal to) (greater than) **P**.

   b. Net force on the crate is (zero) (greater than zero).

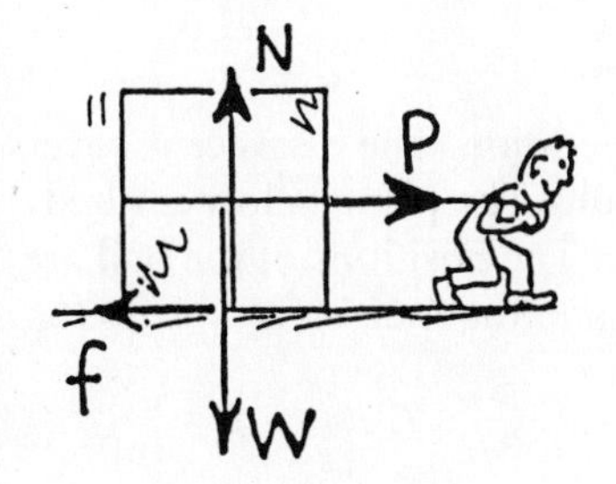

3. Pull **P** is increased until the crate begins to move. It is pulled so that it moves with constant velocity across the floor.

   a. Friction **f** is (less than) (equal to) (greater than) **P**.

   b. Constant velocity means acceleration is (zero) (greater than zero).

   c. Net force on the crate is (less than) (equal to) (greater than) zero.

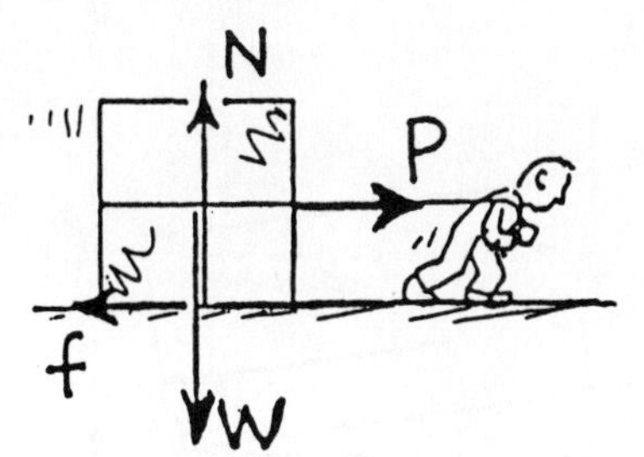

4. Pull **P** is further increased and is now greater than friction **f**.

   a. Net force on the crate is (less than) (equal to) (greater than) zero.

   b. The net force acts toward the right, so acceleration acts toward the (left) (right).

5. If the pulling force **P** is 150 N and the crate doesn't move, what is the magnitude of **f**? __________

6. If the pulling force **P** is 200 N and the crate doesn't move, what is the magnitude of **f**? __________

7. If the force of sliding friction is 250 N, what force is necessary to keep the crate sliding at constant velocity?

   ____

8. If the mass of the crate is 50 kg and sliding friction is 250 N, what is the acceleration of the crate when the pulling force is 250 N? __________ 300 N? __________ 500 N? __________

Name ______________________________ Date __________

Nonaccelerated Motion

1. The sketch shows a ball rolling at constant velocity along a level floor. The ball rolls from the first position shown to the second in 1 second. The two positions are 1 meter apart. Sketch the ball at successive 1-second intervals all the way to the wall (Neglect resistance.)

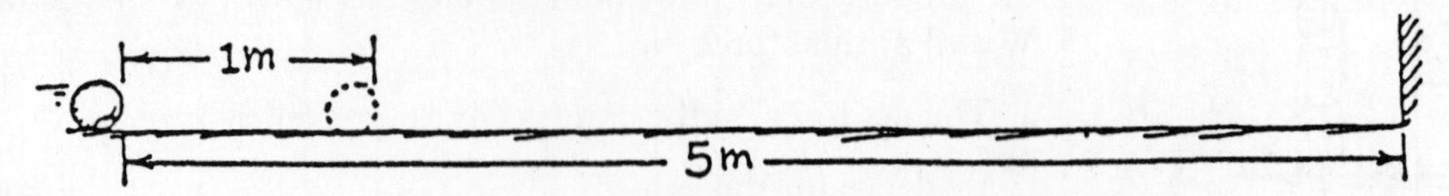

a. Did you draw successive ball positions evenly spaced, farther apart, or close together? Why?

____

b. The ball reaches the wall with a speed of __________ m/s and takes a time of __________ seconds.

2. Table 1 shows data of sprinting speeds of some animals. Make whatever computations are necessary to complete the table.

Table 1

ANIMAL	DISTANCE	TIME	SPEED
CHEETAH	75 m	3 s	25 m/s
GREYHOUND	160 m	10 s	
GAZELLE	1 km		100 km/h
TURTLE		30 s	1 cm/s

Accelerated Motion

3. An object starting from rest gains a speed $v = at$ when it undergoes uniform acceleration. The distance it covers is $d = 1/2\ at^2$. Uniform acceleration occurs for a ball rolling down an inclined plane. The plane below is tilted so a ball picks up a speed of 2 m/s each second; then its acceleration is $a = 2\ m/s^2$. The positions of the ball are shown at 1-second intervals. Complete the six blank spaces for distance covered, and the four blank spaces for speeds.

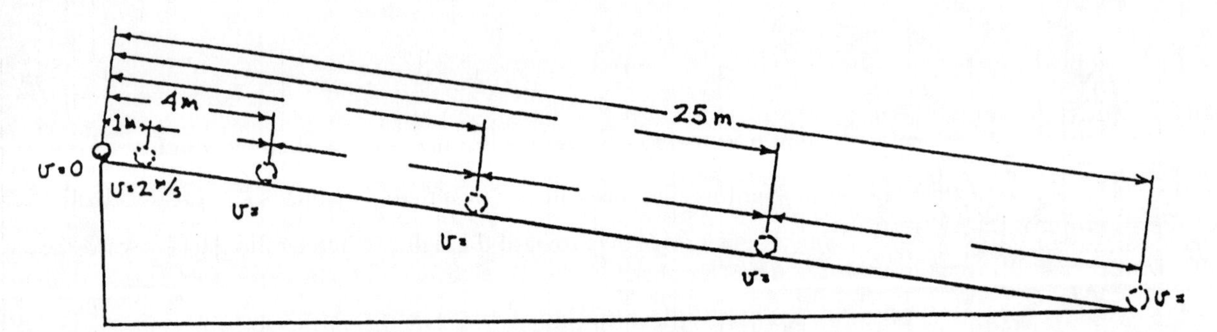

a. Do you see that the total distance from the starting point increases as the square of the time? This was discovered by Galileo. If the incline were to continue, predict the ball's distance from the starting point for the next 3 seconds.

____

b. Note the increase of distance between ball positions with time. Do you see an odd-integer pattern (also discovered by Galileo) for the increase? If the incline were to continue, predict the successive distances between ball positions for the next 3 seconds.

____

Name ______________________________ Date ____________

Chapter 3: Newton's Laws of Motion

A Day at the Races with Newton's Second Law: $a = \frac{F}{m}$

In each situation below, Cart A has a mass of **1 kg.** The mass of Cart B varies as indicated. Circle the correct answer (A, B, or Same for both).

1. Cart A is pulled with a force of **1 N.** Cart B also has a mass of **1 kg** and is pulled with a force of **2 N.** Which undergoes the greater acceleration?

A B Same for both

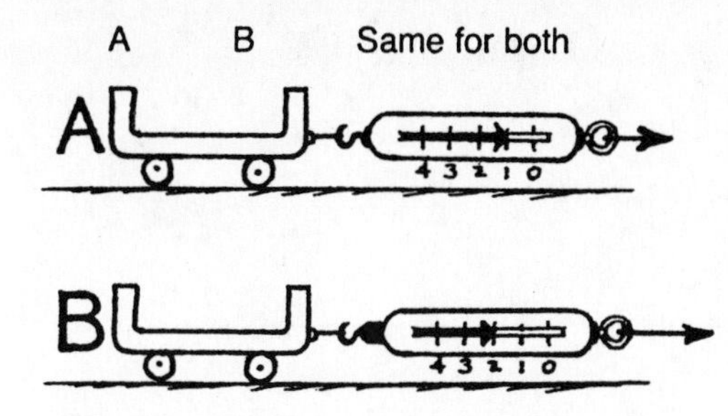

2. Cart A is pulled with a force of **1 N.** Cart B has a mass of **2 kg** and is also pulled with a force of **1 N.** Which undergoes the greater acceleration?

A B Same for both

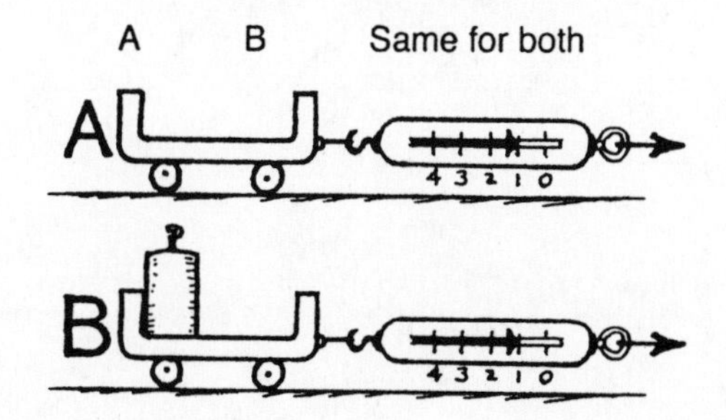

3. Cart A is pulled with a force of **1 N.** Cart B has a mass of **2 kg** and is pulled with a force of **2 N.** Which undergoes the greater acceleration?

A B Same for both

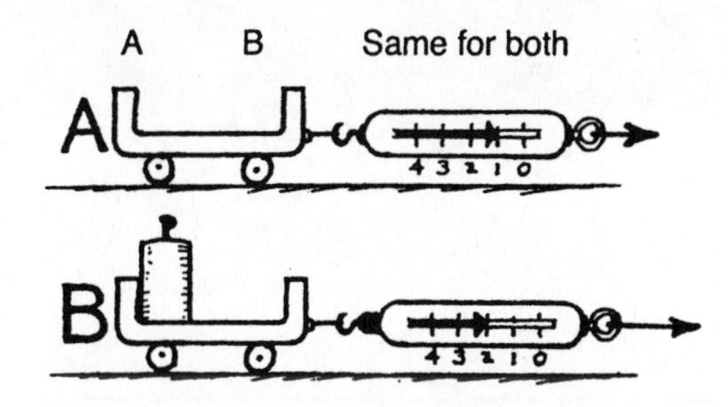

4. Cart A is pulled with a force of **1 N.** Cart B has a mass of **3 kg** and is pulled with a force of **3 N.** Which undergoes the greater acceleration?

A B Same for both

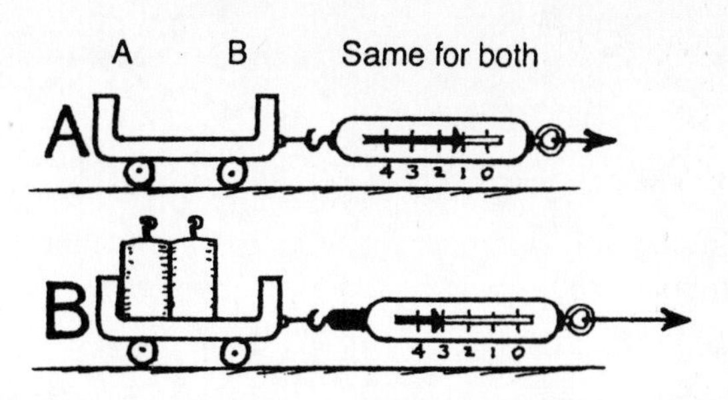

5. This time Cart A is pulled with a force of **4 N.** Cart B has a mass of **4 kg** and is pulled with a force of **4 N.** Which undergoes the greater acceleration?

A B Same for both

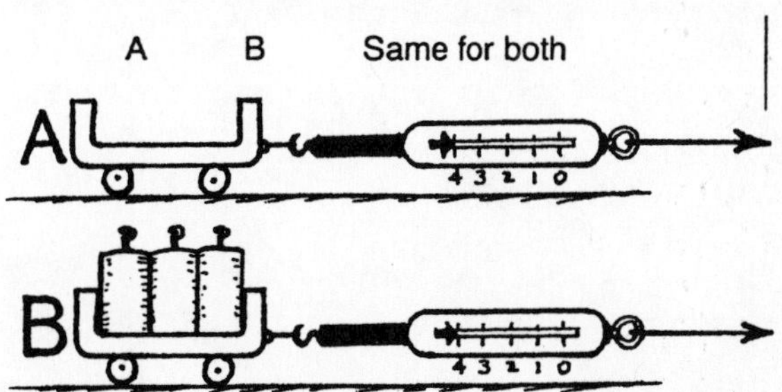

6. Cart A is pulled with a force of **2 N.** Cart B has a mass of **4 kg** and is pulled with a force of **3 N.** Which undergoes the greater acceleration?

A B Same for both

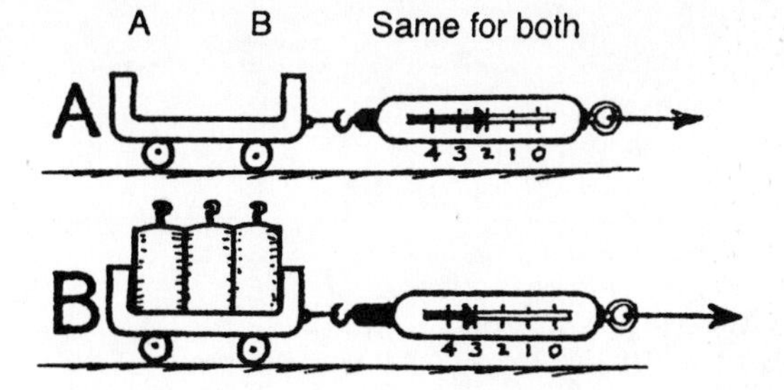

Name ______________________________ Date __________

Chapter 3: Newton's Laws of Motion

Dropping Masses and Accelerating Cart

1. Consider the simple case of a 1-kg cart being pulled by a 10 N applied force. According to Newton's Second Law, acceleration of the cart is

$$a = \frac{F}{m} = \frac{10\text{ N}}{1\text{ kg}} = 10\text{ m/s}^2$$

This is the same as the acceleration of free fall, *g*—because a force equal to the cart's weight accelerates it.

2. Now consider the acceleration of the cart when a second mass is also accelerated. This time the applied force is due to a 10-N iron weight attached to a string draped over a pulley. Will the cart accelerate as before, at 10 m/s²? The answer is no, because the mass being accelerated is the mass of the cart *plus* the mass of the piece of iron that pulls it. Both masses accelerate. The mass of the 10-N iron weight is 1 kg—so the total mass being accelerated (cart + iron) is 2 kg. Then,

The pulley changes only the direction of the force.

$$a = \frac{F}{m} = \frac{10\text{ N}}{2\text{ kg}} = 5\text{ m/s}^2$$

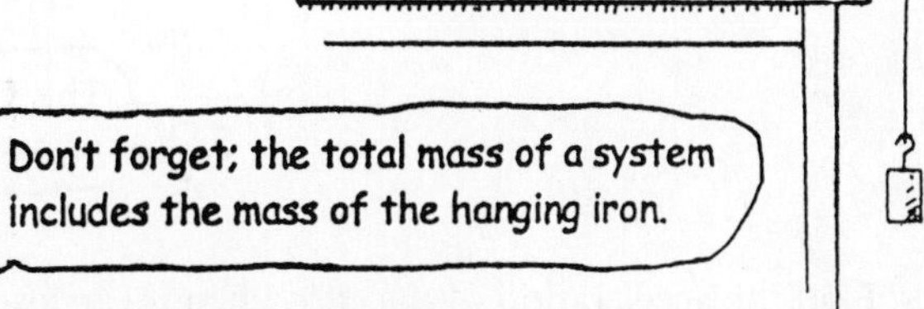

Don't forget; the total mass of a system includes the mass of the hanging iron.

Note this is half the acceleration due to gravity alone, *g*. So the acceleration of 2 kg produced by the weight of 1 kg is *g*/2.

a. Find the acceleration of the 1-kg cart when two identical 10-N weights are attached to the string.

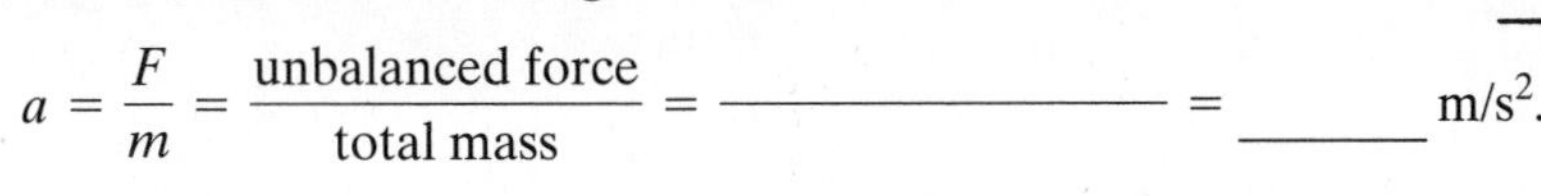

$$a = \frac{F}{m} = \frac{\text{unbalanced force}}{\text{total mass}} = \underline{\qquad\qquad\qquad} = \underline{\qquad}\text{ m/s}^2.$$

Note that the mass being accelerated is 1 kg for the cart + 1 kg each for the weights = 3 kg.

Name ______________________________ Date __________

Dropping Masses and Accelerating Cart—continued

b. Find the acceleration of the 1-kg cart when three identical 10-N weights are attached to the string.

$$a = \frac{F}{m} = \frac{\text{unbalanced force}}{\text{total mass}} = \underline{\hspace{6em}} = \underline{\hspace{3em}}\ \text{m/s}^2.$$

c. Find the acceleration of the 1-kg cart when four identical 10-N weights (not shown) are attached to the string.

$$a = \frac{F}{m} = \frac{\text{unbalanced force}}{\text{total mass}} = \underline{\hspace{6em}} = \underline{\hspace{3em}}\ \text{m/s}^2.$$

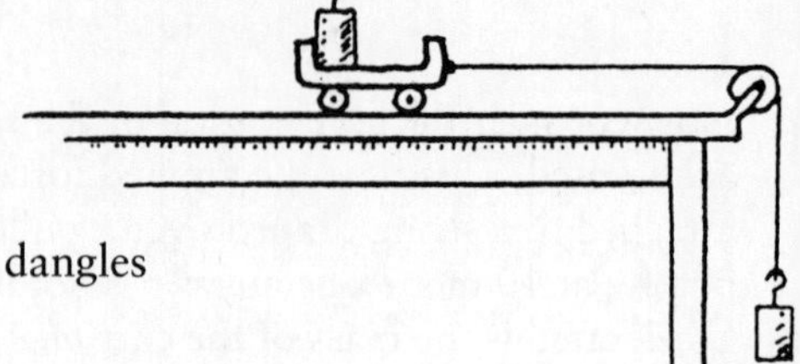

d. This time, 1 kg of iron is added to the cart, and only one iron piece dangles from the pulley. Find the acceleration of the cart.

$$a = \frac{F}{m} = \frac{\text{unbalanced force}}{\text{total mass}} = \underline{\hspace{6em}} = \underline{\hspace{3em}}\ \text{m/s}^2.$$

The force due to gravity on a mass *m* is *mg*. So gravitational force on 1 kg is (1 kg)(10 m/s²) = 10 N.

e. Find the acceleration of the cart when it carries two pieces of iron and only one iron piece dangles from the pulley.

$$a = \frac{F}{m} = \frac{\text{unbalanced force}}{\text{total mass}} = \underline{\hspace{6em}} = \underline{\hspace{3em}}\ \text{m/s}^2.$$

Name ______________________________ Date ____________

Dropping Masses and Accelerating Cart—continued

f. Find the acceleration of the cart when it carries three pieces of iron and only one iron piece dangles from the pulley.

$$a = \frac{F}{m} = \frac{\text{unbalanced force}}{\text{total mass}} = \frac{\qquad\qquad}{\qquad\qquad} = \underline{\qquad}\ \text{m/s}^2.$$

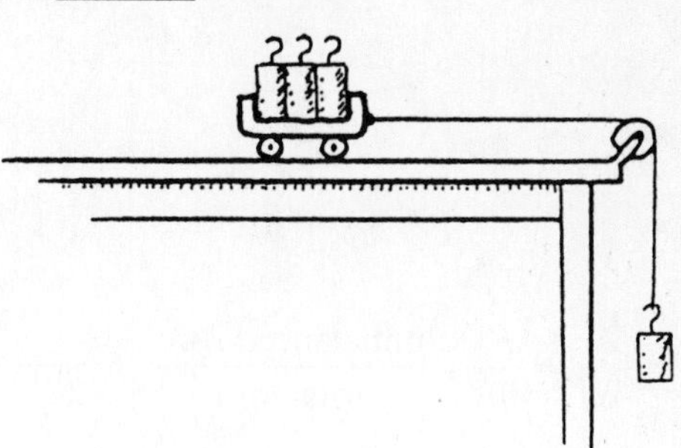

g. Find the acceleration of the cart when it carries three pieces of iron and four iron pieces dangle from the pulley.

$$a = \frac{F}{m} = \frac{\text{unbalanced force}}{\text{total mass}} = \frac{\qquad\qquad}{\qquad\qquad} = \underline{\qquad}\ \text{m/s}^2.$$

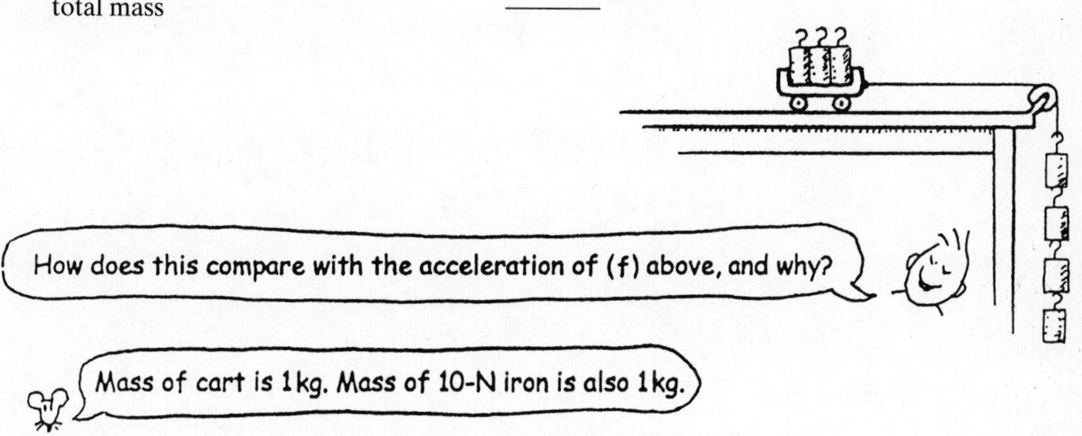

How does this compare with the acceleration of (f) above, and why?
Mass of cart is 1 kg. Mass of 10-N iron is also 1 kg.

h. Draw your own combination of masses and find the acceleration.

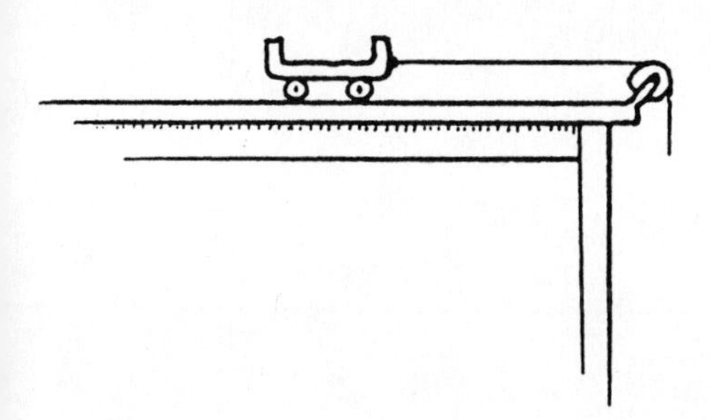

$$a = \frac{F}{m} = \frac{\text{unbalanced force}}{\text{total mass}} = \frac{\qquad\qquad}{\qquad\qquad} = \underline{\qquad}\ \text{m/s}^2.$$

How does this compare with the acceleration of [illegible] above, and why?

Name ______________________________ Date __________

Chapter 3: Newton's Laws of Motion

Mass and Weight

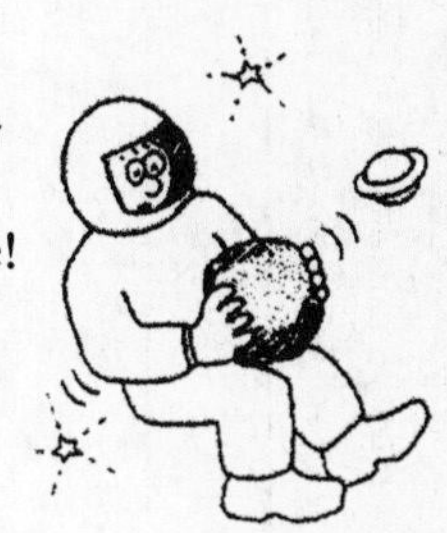

Learning physics is learning the connections among concepts in nature, and also learning to distinguish between closely related concepts. Velocity and acceleration are often confused. Similarly, in this chapter, we find that mass and weight are often confused. They aren't the same! Please review the distinction between mass and weight in your textbook. To reinforce your understanding of this distinction, circle the correct answers below.

Comparing the concepts of mass and weight, one is basic—fundamental—depending only on the internal makeup of an object and the number and kind of atoms that compose it. The concept that is fundamental is [mass] [weight].

The concept that additionally depends on location in a gravitational field is [mass] [weight].

[Mass] [Weight] is a measure of the amount of matter in an object and only depends on the number and kind of atoms that compose it.

We can correctly say that [mass] [weight] is a measure of an object's "laziness."

[Mass] [Weight] is related to the gravitational force acting on the object.

[Mass] [Weight] depends on an object's location, whereas [mass] [weight] does not.

In other words, a stone would have the same [mass] [weight] whether it is on the surface of Earth or on the surface of the Moon. However, its [mass] [weight] depends on its location.

On the Moon's surface, where gravity is only about 1/16 of Earth's gravity, [mass] [weight] [both the mass and the weight] of the stone would be the same as on Earth.

While mass and weight are not the same, they are [directly proportional] [inversely proportional] to each other. In the same location, twice the mass has [twice] [half] the weight.

The Standard International (SI) unit of mass is the [kilogram] [newton], and the SI unit of force is the [kilogram] [newton].

In the United States, it is common to measure the mass of something by measuring its gravitational pull to Earth, its weight. The common unit of weight in the U.S. is the [pound] [kilogram] [newton].

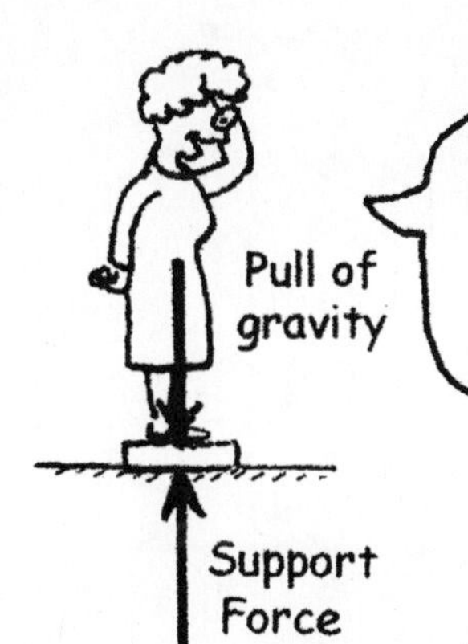
Pull of gravity
Support Force
When I step on a scale, two forces act on it: a downward pull of gravity, and an upward support force. These equal and opposite forces effectively compress a spring inside the scale that is calibrated to show weight. When in equilibrium, my weight = *mg*.

Name ______________________________ Date __________

Chapter 3: Newton's Laws of Motion

Converting Mass to Weight

Objects with mass also have weight (although they can be weightless under special conditions). If you know the mass of something in **kilograms** and want its weight in **newtons,** at Earth's surface, you can take advantage of the formula that relates weight and mass:

$$\text{Weight} = \text{mass} \times \text{acceleration due to gravity}$$

$$W = mg.$$

This is in accord with Newton's Second Law, written as $F = ma$. When the force of gravity is the only force, the acceleration of any object of mass m will be g, the acceleration of free fall. Importantly, g acts as a proportionality constant, 9.8 N/kg, which is equivalent to 9.8 m/s^2.

Sample Question:

How much does a 1-kg bag of nails weigh on Earth?

$W = mg = (1\text{ kg})(9.8\text{ m/s}^2) = 9.8\text{ m/s}^2 = 9.8\text{ N}.$

or simply, $W = mg = (1\text{ kg})(9.8\text{ N/kg}) = 9.8\text{ N}.$

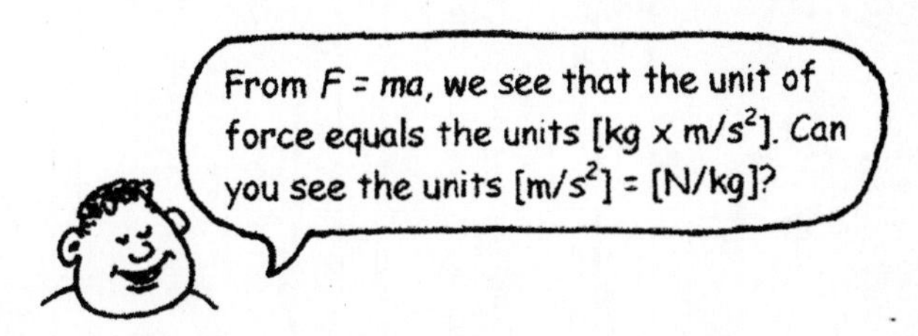

From F = ma, we see that the unit of force equals the units [kg x m/s²]. Can you see the units [m/s²] = [N/kg]?

Answer the following questions:

Felicia the ballet dancer has a mass of 45.0 kg.

1. What is Felicia's weight in newtons at Earth's surface? __________
2. Given that 1 kilogram of mass corresponds to 2.2 pounds at Earth's surface, what is Felicia's weight in pounds on Earth? __________
3. What would be Felicia's mass on the surface of Jupiter? __________
4. What would be Felcia's weight on Jupiter's surface, where the acceleration due to gravity is 25.0 m/s^2? __________

Different masses are hung on a spring scale calibrated in newtons. The force exerted by gravity on 1 kg = 9.8 N.

5. The force exerted by gravity on 5 kg = __________ N.
6. The force exerted by gravity on __________ kg = 98 N.

Make up your own mass and show the corresponding weight:

The force exerted by gravity on __________ kg = __________ N.

By whatever means (spring scales, measuring balances, etc.), find the mass of your integrated science book. Then complete the table.

OBJECT	MASS	WEIGHT
MELON	1 kg	
APPLE		1 N
BOOK		
A FRIEND	60 kg	

Name ______________________________ Date ____________

Chapter 3: Newton's Laws of Motion

Bronco and Newton's Second Law

Bronco skydives and parachutes from a stationary helicopter. Various stages of fall are shown in positions *a* through *f*. Using Newton's Second Law:

$$a = \frac{F_{NET}}{m} = \frac{W - R}{m}$$

Find Bronco's acceleration at each position (answer in the blanks to the right). You need to know that Bronco's mass m is 100 kg so his weight is a constant 1000 N. Air resistance R varies with speed and cross-sectional area as shown.

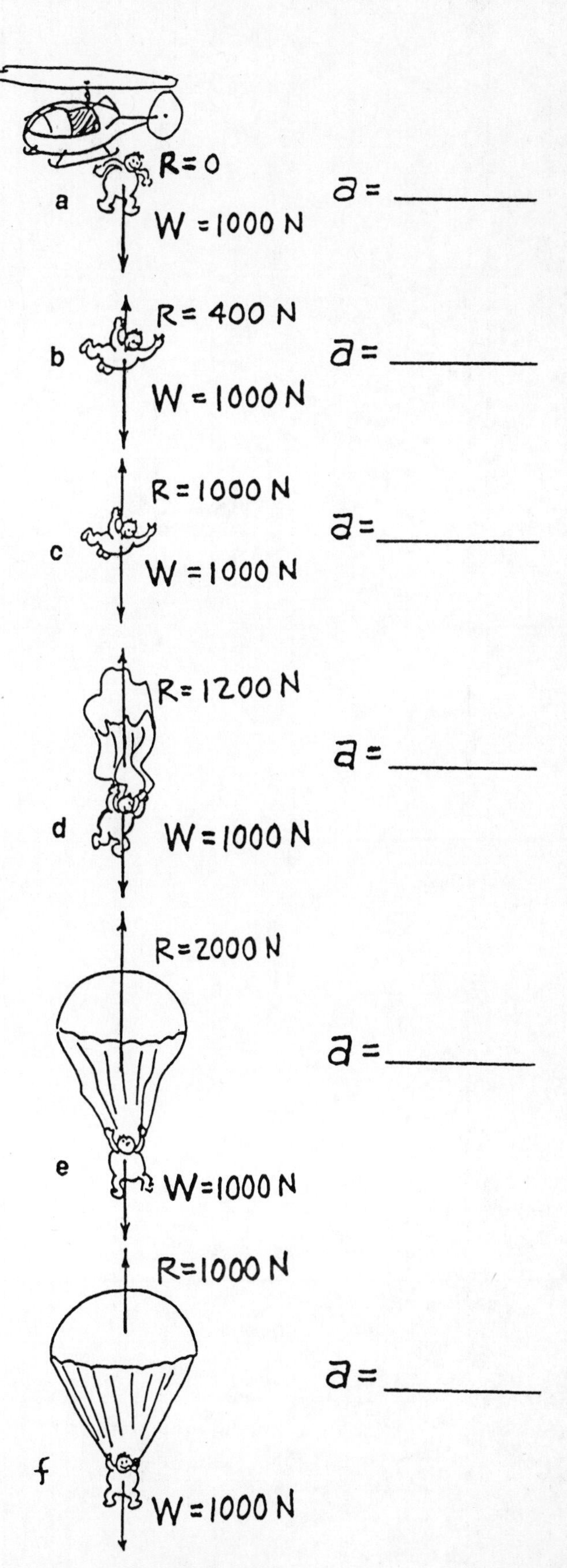

Circle the correct answers:

1. When Bronco's speed is least, his acceleration is

   (least) (most).

2. In which position(s) does Bronco experience a downward acceleration?

   (a) (b) (c) (d) (e) (f)

3. In which position(s) does Bronco experience an upward acceleration?

   (a) (b) (c) (d) (e) (f)

4. When Bronco experiences an upward acceleration, his velocity is

   (still downward) (upward also).

5. In which position(s) is Bronco's velocity constant?

   (a) (b) (c) (d) (e) (f)

6. In which position(s) does Bronco experience terminal velocity?

   (a) (b) (c) (d) (e) (f)

7. In which position(s) is terminal velocity greatest?

   (a) (b) (c) (d) (e) (f)

8. If Bronco were heavier, his terminal velocity would be

   (greater) (less) (the same).

Name ______________________________ Date ____________

Chapter 3: Newton's Laws of Motion

Newton's Third Law

Your thumb and finger pull on each other when you stretch a rubber band between them. This pair of forces, thumb on finger and finger on thumb, make up an action-reaction pair of forces, both of which are equal in magnitude and oppositely directed. Draw the reaction vector and state in words the reaction force for each of the examples **a** through **g**. Then make up your own example in **h**.

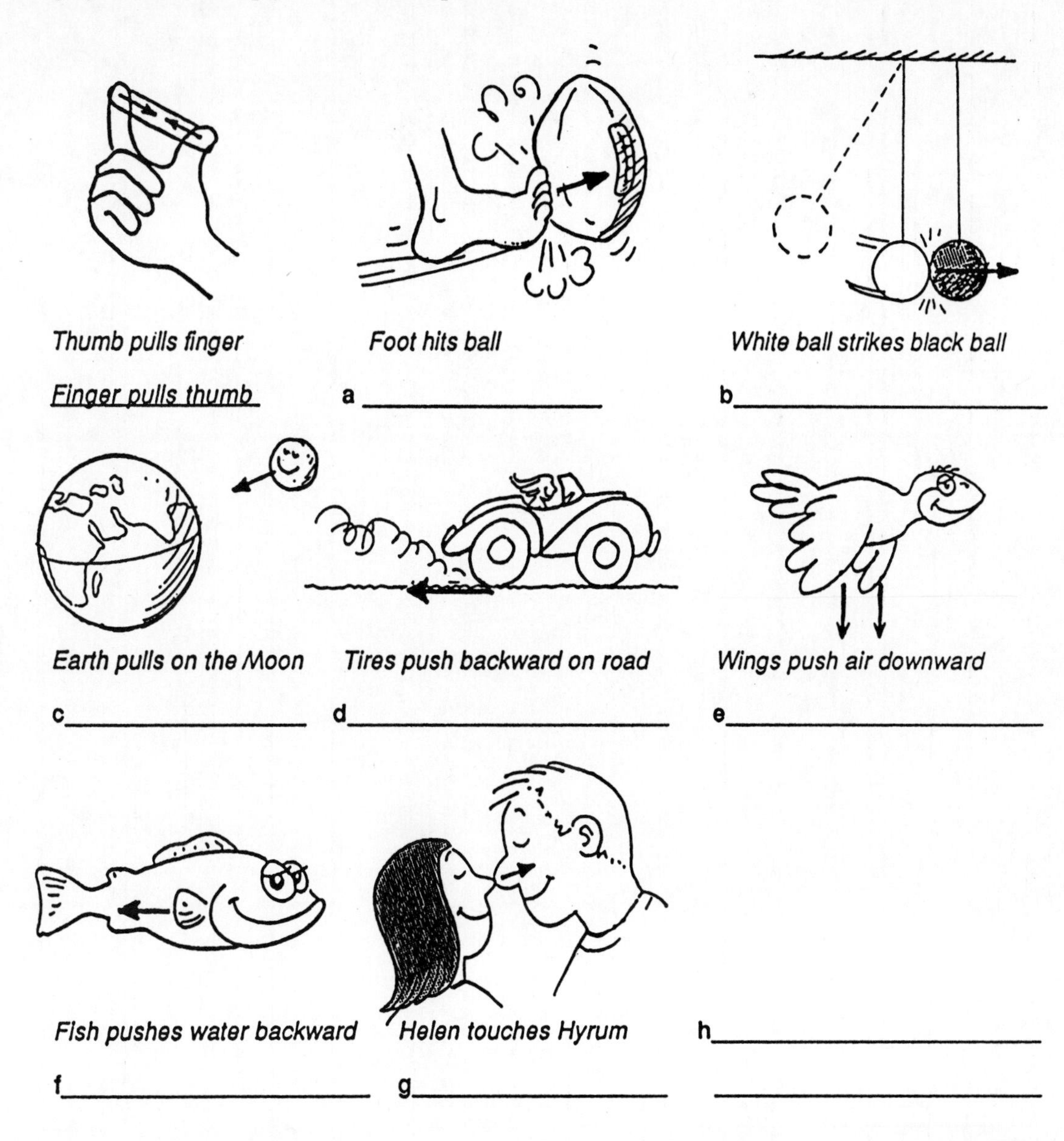

YOU CAN'T TOUCH WITHOUT BEING TOUCHED—*NEWTON'S THIRD LAW*

Name ______________________________ Date __________

Chapter 3: Newton's Laws of Motion

Nellie and Newton's Third Law

Nellie Newton holds an apple weighing 1 newton at rest on the palm of her hand. *Circle the correct answers.*

1. To say the weight (W) of the apple is 1 N is to say that a downward gravitational force of 1 N is exerted on the apple by

   (Earth) (her hand).

2. Nellie's hand supports the apple with normal force N, which acts in a direction opposite to W. We can say N

   (equals W) (has the same magnitude as W).

3. Since the apple is at rest, the net force on the apple is

   (zero) (nonzero).

4. Since N is equal and opposite to W, we (can) (cannot) say that N and W comprise an action-reaction pair. The reason is because action and reaction (act on the same object) (act on different objects), and here we see N and W (both acting on the apple) (acting on different objects).

5. In accord with the rule, "If ACTION is A acting on B, then REACTION is B acting on A," if we say action is Earth pulling down on the apple, reaction is

   (the apple pulling up on Earth) (N, Nellie's hand pushing up on the apple.)

6. To repeat for emphasis, we see that N and W are equal and opposite to each other

   (and comprise an action-reaction pair) (but do *not* comprise an action-reaction pair).

To identify a pair of action-reaction forces in any situation, first identify the pair of interacting objects involved. Something is interacting with something else. In this case, the whole Earth is interacting (gravitationally) with the apple. So, Earth pulls downward on the apple (call it action), while the apple pulls upward on Earth (reaction).

Simply put, Earth pulls on apple (action); apple pulls on Earth (reaction).

Better put, apple and Earth *pull on each other* with equal and opposite forces that comprise a *single* interaction.

7. Another pair of forces is N [shown] and the downward force of the apple against Nellie's hand [not shown]. This pair of forces (is) (isn't) an action-reaction pair.

8. Suppose Nellie now pushes upward on the apple with the force of 2 N. The apple (is still in equilibrium) (accelerates upward), and compared with W, the magnitude of N is (the same) (twice) (not the same, and not twice).

9. Once the apple leaves Nellie's hand, N is (zero) (still twice the magnitude of W), and the net force on the apple is (zero) (only W) (still W − N, which is a negative force).

Name ______________________________ Date __________

Chapter 3: Newton's Laws of Motion

Vectors and the Parallelogram Rule

1. When vectors **A** and **B** are at an angle to each other, they add to produce the resultant **C** by the *parallelogram rule.* Note that C is the diagonal of a parallelogram where **A** and **B** are adjacent sides. Resultant **C** is shown in the first two diagrams, *a* and *b*. Construct the resultant **C** in diagrams *c* and *d*. Note that in diagram *d* you form a rectangle (a special case of a parallelogram).

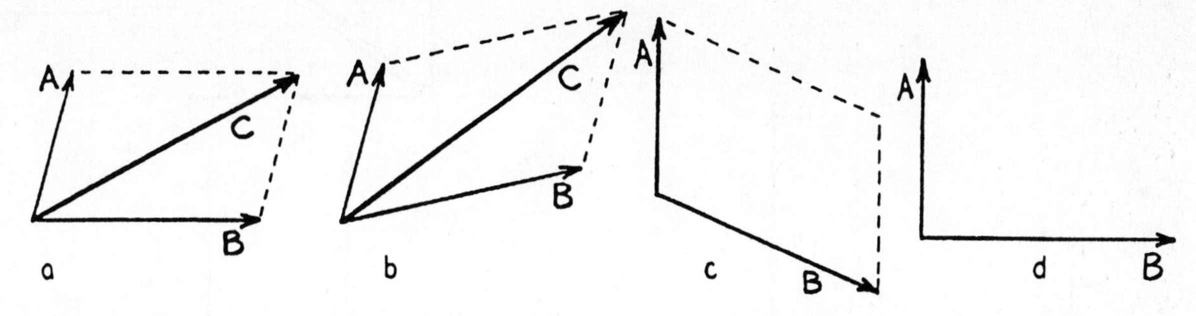

2. Below we see a top view of an airplane being blown off course by wind in various directions. Use the parallelogram rule to show the resulting speed and direction of travel for each case. In which case does the airplane travel fastest across the ground? __________ Slowest? __________

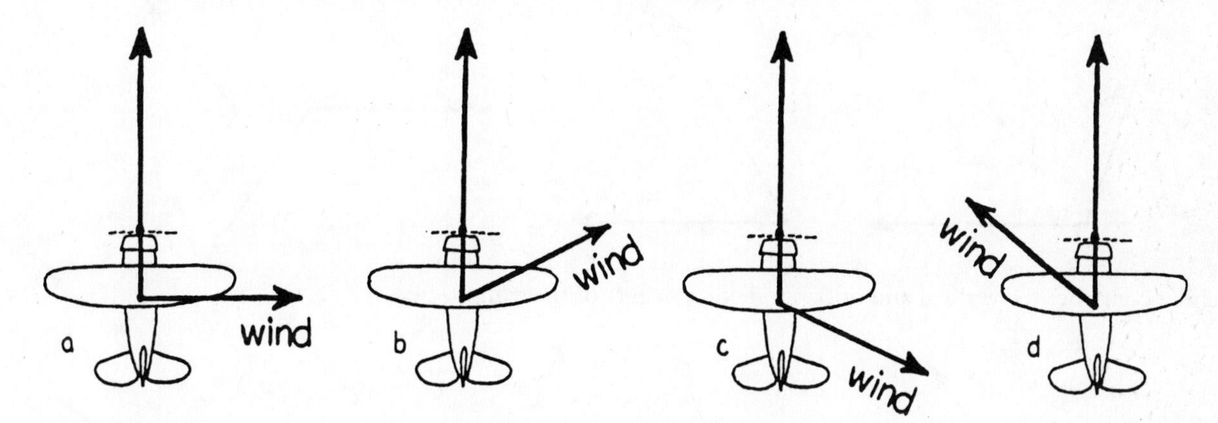

3. To the right we see top views of three motorboats crossing a river. All have the same speed relative to the water, and all experience the same water flow.

   Construct resultant vectors showing the speed and direction of the boats.

   a. Which boat takes the shortest path to the opposite shore? __________

   b. Which boat reaches the opposite shore first? __________

   c. Which boat provides the fastest ride? __________

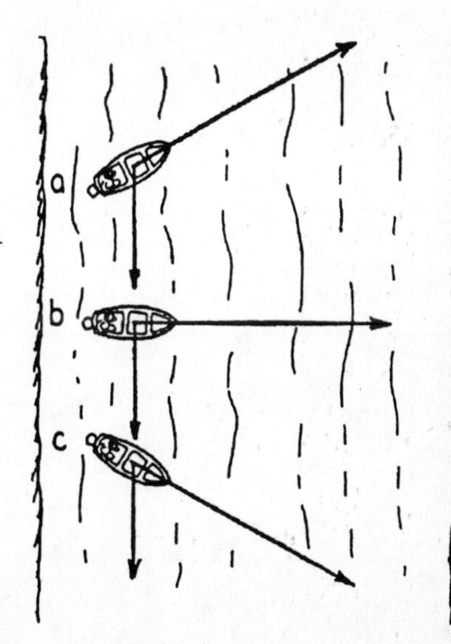

Name ______________________________ Date ____________

Vectors

Use the parallelogram rule to carefully construct the resultants for the eight pairs of vectors.

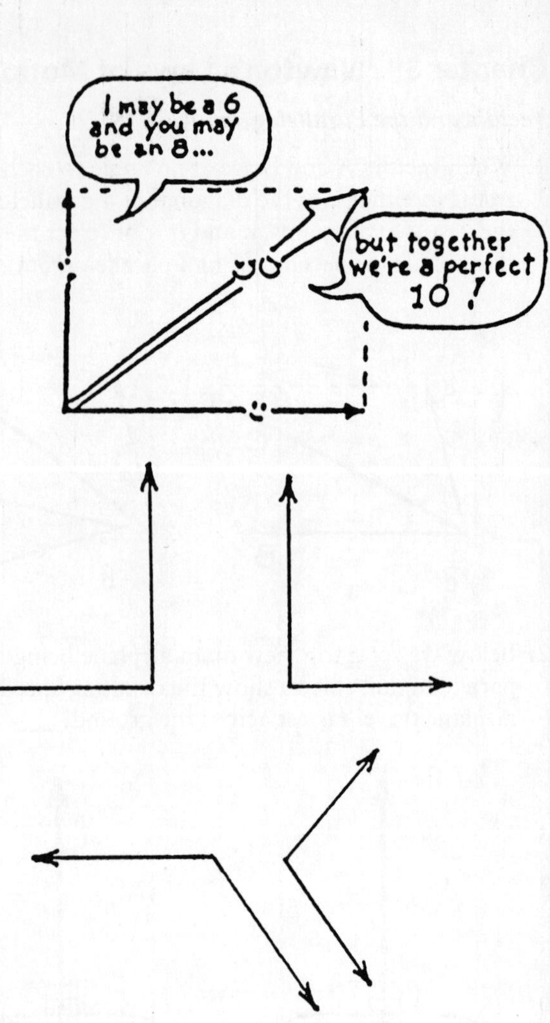
I may be a 6 and you may be an 8...
but together we're a perfect 10!

Carefully construct the vertical and horizontal components of the eight vectors.

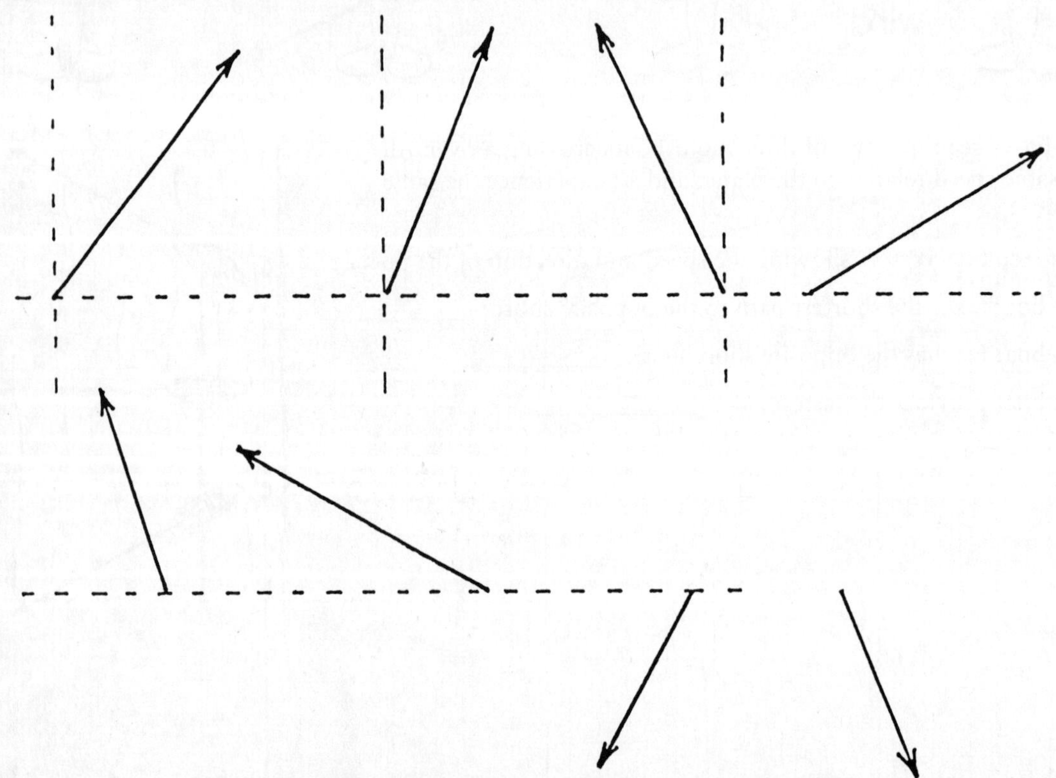

Name ______________________________ Date __________

Chapter 3: Newton's Laws of Motion

Force Vectors and the Parallelogram Rule

1. The heavy ball is supported in each case by two strands of rope. The tension in each strand is shown by the vectors. Use the parallelogram rule to find the resultant of each vector pair.

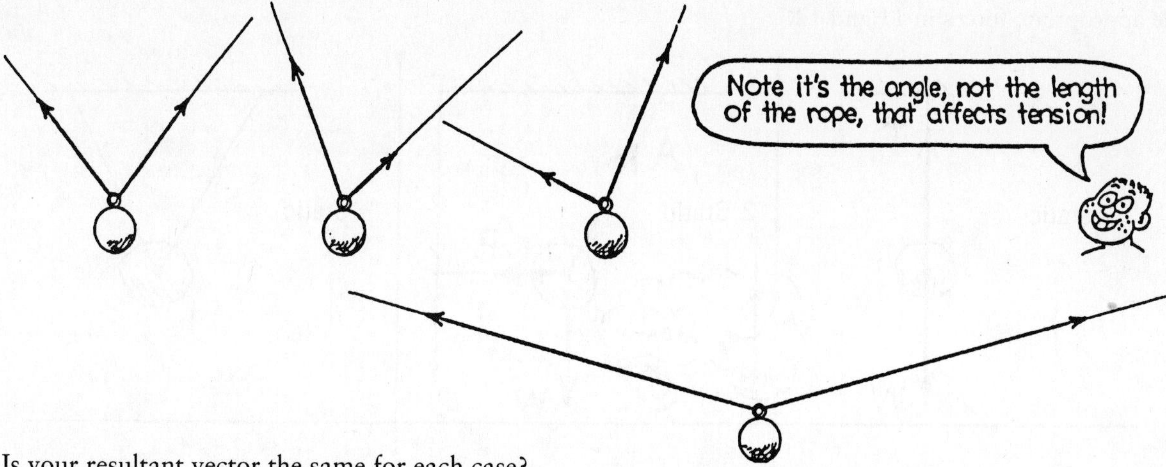

Note it's the angle, not the length of the rope, that affects tension!

   a. Is your resultant vector the same for each case? __________

   b. How do you think the resultant vector compares to the weight of the ball?

   ____

2. Now let's do the opposite of what we've done above. More often, we know the weight of the suspended object, but we don't know the rope tensions. In each case below, the weight of the ball is shown by the vector W. Each dashed vector represents the resultant of the pair of rope tensions. Note that each is equal and opposite to vectors W (they must be; otherwise the ball wouldn't be at rest).

   a. Construct parallelograms where the ropes define adjacent sides and the dashed vectors are the diagonals.

   b. How do the relative lengths of the sides of each parallelogram compare to rope tensions?

   c. Draw rope-tension vectors, clearly showing their relative magnitudes.

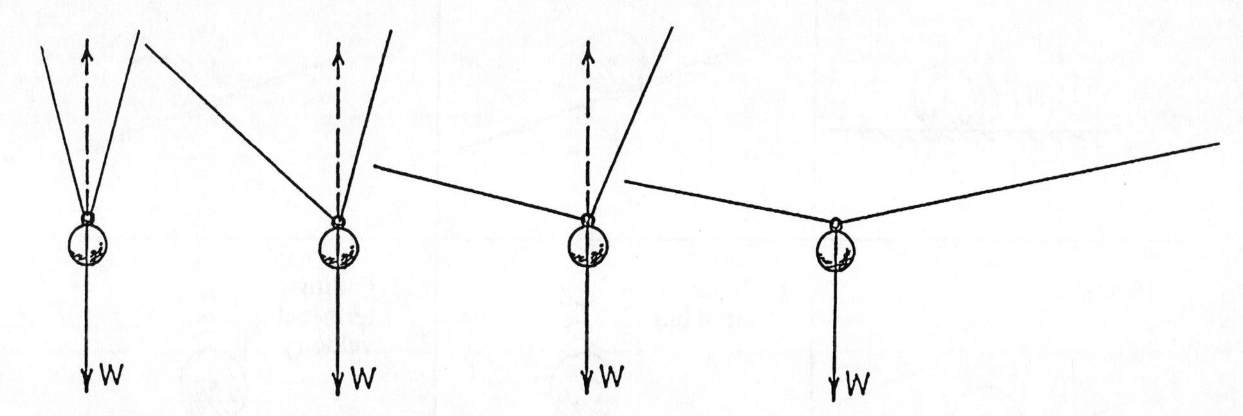

3. A lantern is suspended as shown. Draw vectors to show the relative tensions in ropes **A**, **B**, and **C**. Do you see a relationship between your vectors **A** + **B** and vector **C**? Between vectors **A** + **C** and vector **B**?

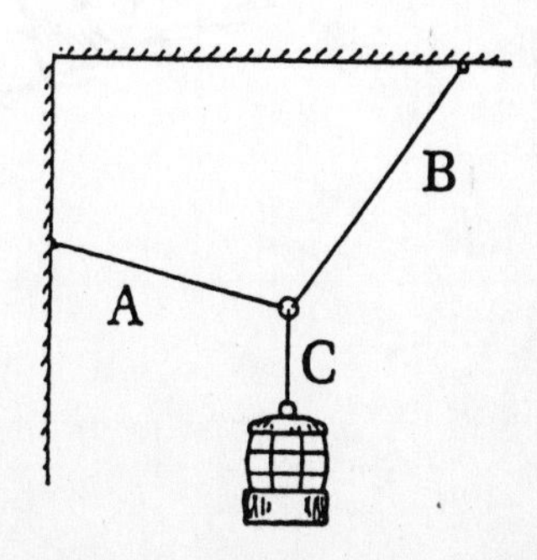

Name ______________________________ Date __________

Force-Vector Diagrams

In each case, a rock is acted on by one or more forces. Draw an accurate vector diagram showing all forces acting on the rock, and no other forces. Use a ruler, and do it in pencil so you can correct mistakes. The first two are done as examples. Show by the parallelogram rule in 2 that the vector sum of **A** + **B** is equal and opposite to **W** (that is, **A** + **B** = −**W**). Do the same for 3 and 4. Draw and label vectors for the weight and normal forces in 5 to 10, and for the appropriate forces in 11 and 12.

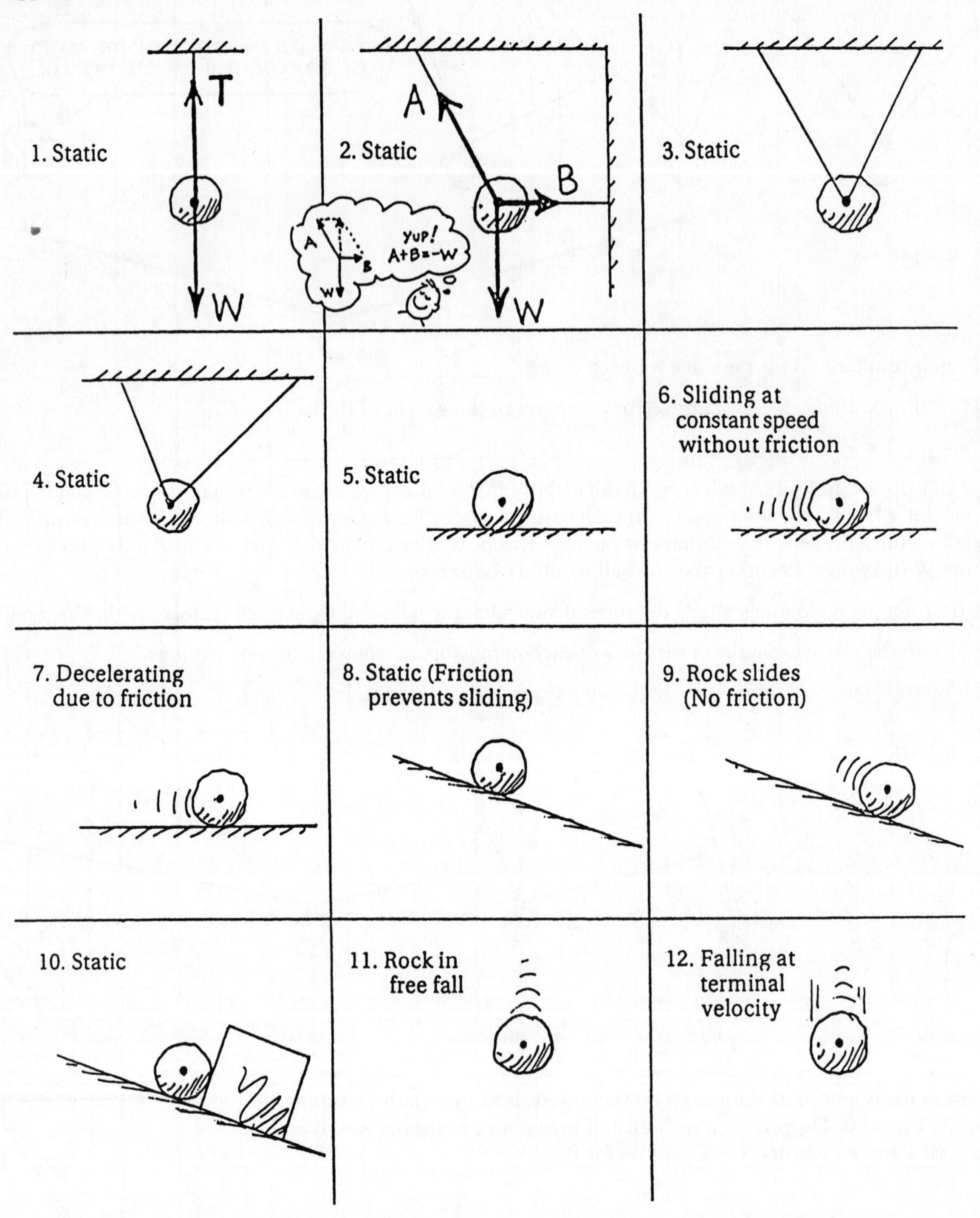

Name ______________________________ Date __________

Chapter 4: Momentum and Energy

Momentum

1. A moving car has momentum. If it moves twice as fast, its momentum is __________ as much.

2. Two cars, one twice as heavy as the other, move down a hill at the same speed. Compared to the lighter car, the momentum of the heavier car is __________ as much.

3. The recoil momentum of a gun that kicks is

   (more than) (less than) (the same as)

   the momentum of the gases and bullet it fires.

4. If a man firmly holds a gun when fired, then the momentum of the bullet and expelled gases is equal to the recoil momentum of the

   (gun alone) (gun-man system) (man alone).

5. Suppose you are traveling in a bus at highway speed on a nice summer day and the momentum of an unlucky bug is suddenly changed as it splatters onto the front window.

   a. Compared to the force that acts on the bug, how much force acts on the bus?

   (more) (the same) (less)

   b. The time of impact is the same for both the bug and the bus. Compared to the impulse on the bug, this means the impulse on the bus is

   (more) (the same) (less).

   c. Although the momentum of the bus is very large compared to the momentum of the bug, the change in momentum of the bus, compared to the *change* of momentum of the bug is

   (more) (the same) (less).

   d. Which undergoes the greater acceleration?

   (bus) (both the same) (bug)

   e. Which, therefore, suffers the greater damage?

   (bus) (both the same) (the bug of course!)

Name ______________________________ Date __________

Chapter 4: Momentum and Energy

Systems

Momentum conservation (and Newton's Third Law) apply to *systems* of bodies. Here we identify some systems.

1. When the compressed spring is released, Blocks A and B will slide apart. There are three systems to consider here, indicated by the closed dashed lines below—System A, System B, and System A+B. Ignore the vertical forces of gravity and the support force of the table.

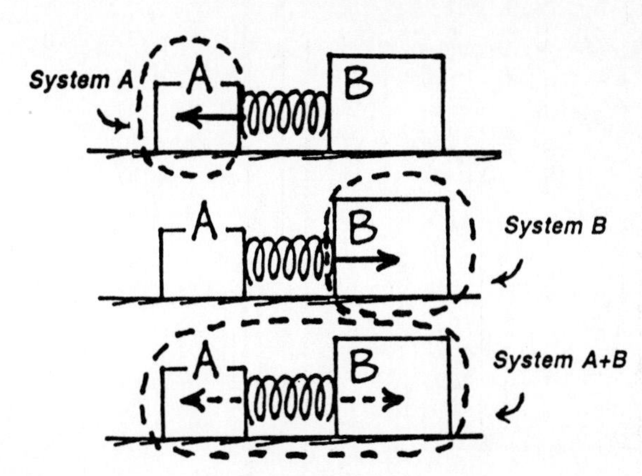

   a. Does an external force act on System A? (yes) (no)

   Will the momentum of System A change? (yes) (no)

   b. Does an external force act on System B? (yes) (no)

   Will the momentum of System B change? (yes) (no)

   c. Does an external force act on System A+B? (yes) (no)

   Will the momentum of System A+B change? (yes) (no)

2. Billiard ball A collides with billiard ball B at rest. Isolate each system with a closed dashed line. Draw only the external force vectors that act on each system.

System A

System B

System A+B

   a. Upon collision, the momentum of System A (increases) (decreases) (remains unchanged).
   b. Upon collision, the momentum of System B (increases) (decreases) (remains unchanged).
   c. Upon collision, the momentum of System A+B (increases) (decreases) (remains unchanged).

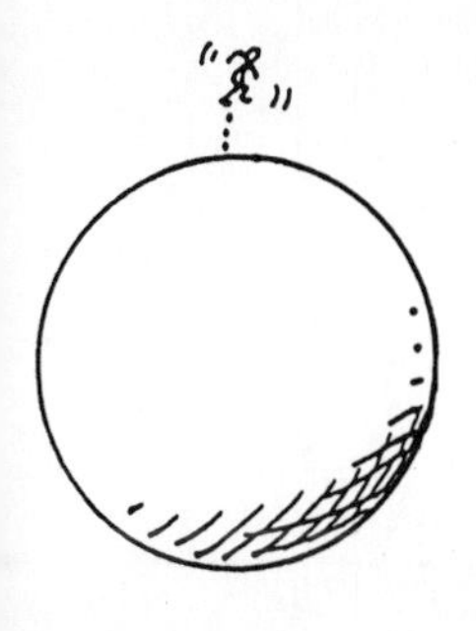

3. A girl jumps upward from Earth's surface. In the sketch to the left, draw a closed dashed line to indicate the system of the girl.

   a. Is there an external force acting on her? (yes) (no)

   Does her momentum change? (yes) (no)

   Is the girl's momentum conserved? (yes) (no)

   b. In the sketch to the right, draw a closed dashed line to indicate the system [girl + Earth]. Is there an external force due to the interaction between the girl and Earth that acts on the system? (yes) (no)

   Is the momentum of the system conserved? (yes) (no)

4. A block strikes a blob of jelly. Isolate three systems with a closed dashed line and show the external force on each. In which system is momentum conserved?

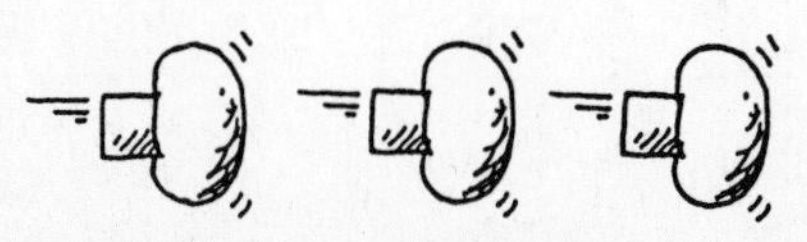

5. A truck crashes into a wall. Isolate three systems with a closed dashed line and show the external force on each. In which system is momentum conserved?

Name ____ Date __________

Chapter 4: Momentum and Energy

Impulse–Momentum

Bronco Brown wants to put $Ft = \Delta mv$ to the test and try bungee jumping. Bronco leaps from a high cliff and experiences free fall for 3 seconds. Then the bungee cord begins to stretch, reducing his speed to zero in 2 seconds. Fortunately, the cord stretches to its maximum length just short of the ground below.

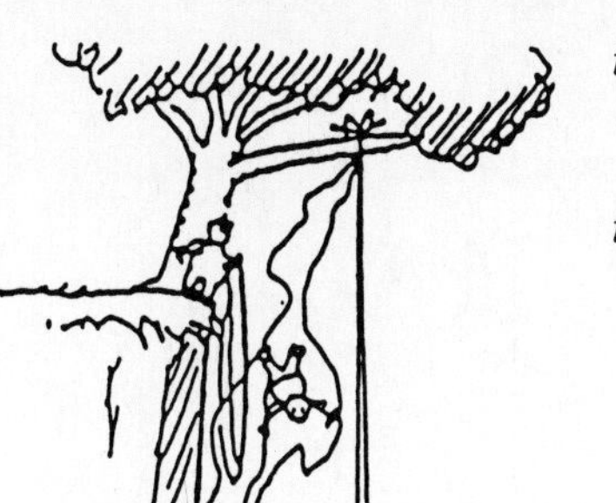

$t = 0$ s $v =$ ______
momentum = __________

$t = 1$ s $v =$ ______
momentum = __________

$t = 2$ s $v =$ ______
momentum = __________

$t = 3$ s $v =$ ______
momentum = __________

$t = 5$ s $v =$ ______
momentum = __________

Fill in the blanks. Bronco's mass is 100 kg. Acceleration of free fall is 10 m/s^2.

Express values in SI units (distance in m, velocity in m/s, momentum in kg·m/s, impulse in N·s, and deceleration in m/s^2*).*

1. The 3-s free-fall distance of Bronco just before the bungee cord begins to stretch

   = __________

2. Δmv during the 3-s interval of free fall

   = __________

3. Δmv during the 2-s interval of slowing down

   = __________

4. *Impulse* during the 2-s interval of slowing down

   = __________

5. *Average force* exerted by the cord during the 2-s interval of slowing down

   = __________

6. How about *work* and *energy*? How much KE does Bronco have 3 s after his jump?

   ____

7. How much does gravitational PE decrease during this 3 s?

   ____

8. What two kinds of PE are changing during the slowing-down interval?

____

Name ____ Date ____________

Chapter 4: Momentum and Energy

Conservation of Momentum

Granny whizzes around the rink and is suddenly confronted with Ambrose at rest directly in her path. Rather than knock him over, she picks him up and continues in motion without "braking." Consider both Granny and Ambrose as two parts of one system. Since no outside forces act on the system, the momentum of the system before collision equals the momentum of the system after collision.

a. Complete the before-collision data in the table below.

BEFORE COLLISION	
Granny's mass	80 kg
Granny's speed	3 m/s
Granny's momentum	________
Ambrose's mass	40 kg
Ambrose's speed	0 m/s
Ambrose's momentum	________
Total momentum	____________

b. After collision, does Granny's speed increase or decrease?

____

c. After collision, does Ambrose's speed increase or decrease?

____

d. After collision, what is the total mass of Granny + Ambrose?

____

e. After collision, what is the total momentum of Granny + Ambrose?

____

f. Use the conservation of momentum law to find the speed of Granny and Ambrose together after collision. (Show your work in the space below.)

New speed = ___________

Name ______________________________ Date ____________

Chapter 4: Momentum and Energy

Work and Energy

1. How much work (energy) is needed to lift an object that weighs 200 N to a height of 4 m?

____

2. How much power is needed to lift the 200-N object to a height of 4 m in 4 s?

____

3. What is the power output of an engine that does 60,000 J of work in 10 s?

____

4. The block of ice weighs 500 newtons.

   a. Neglecting friction, how much force is needed to push it up the incline?

   ____

   b. How much work is required to push it up the incline compared with lifting the block vertically 3 m?

   ____

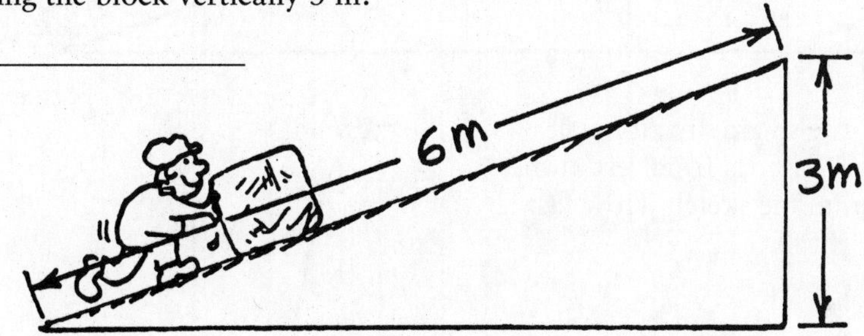

5. All the ramps are 5 m high. We know that the KE of the block at the bottom of the ramp will be equal to the loss of PE (conservation of energy). Find the speed of the block at ground level in each case. [Hint: Do you recall from earlier chapters how long it takes something to fall a vertical distance of 5 m from a position of rest (assume $g = 10\ m/s^2$)? And how much speed a falling object acquires in this time? This gives you the answer to Case 1. Discuss with your classmates how energy conservation gives you the answers to Cases 2 and 3.]

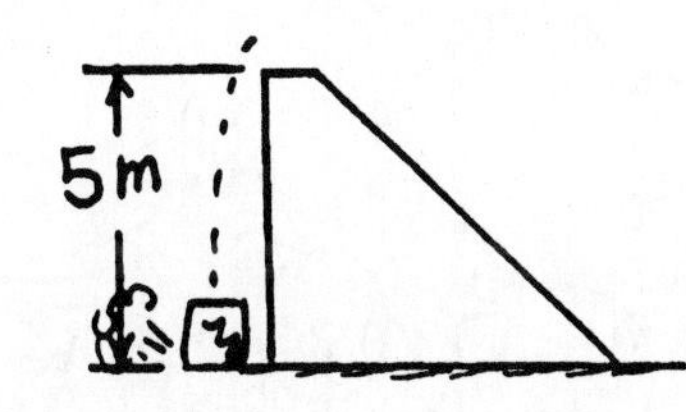

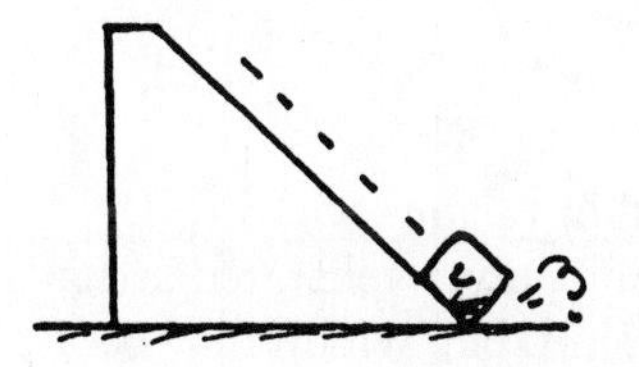

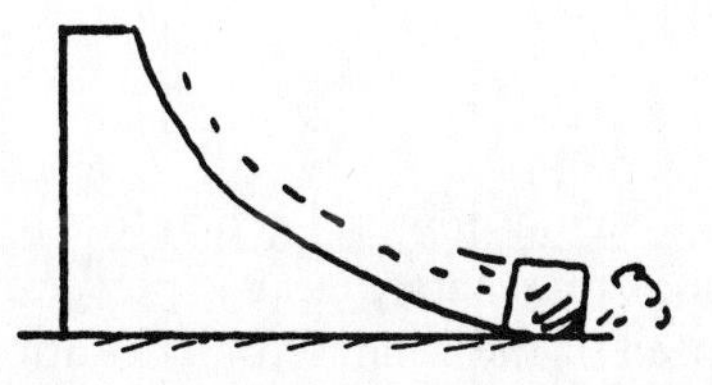

Case 1: Speed = __________ m/s Case 2: Speed = __________ m/s Case 3: Speed = __________ m/s

Name ______________________________ Date __________

Work and Energy—continued

6. Which block gets to the bottom of the incline first? Assume there is no friction. (Be careful!) Explain your answer.

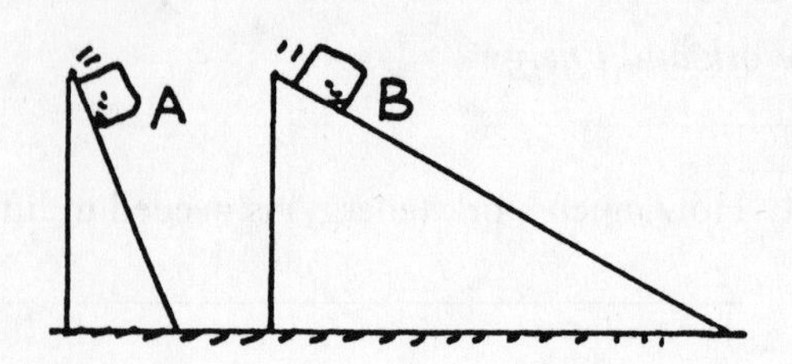

____

____

____

7. The KE and PE of a block freely sliding down a ramp are shown in only one place in the sketch. Fill in the missing information.

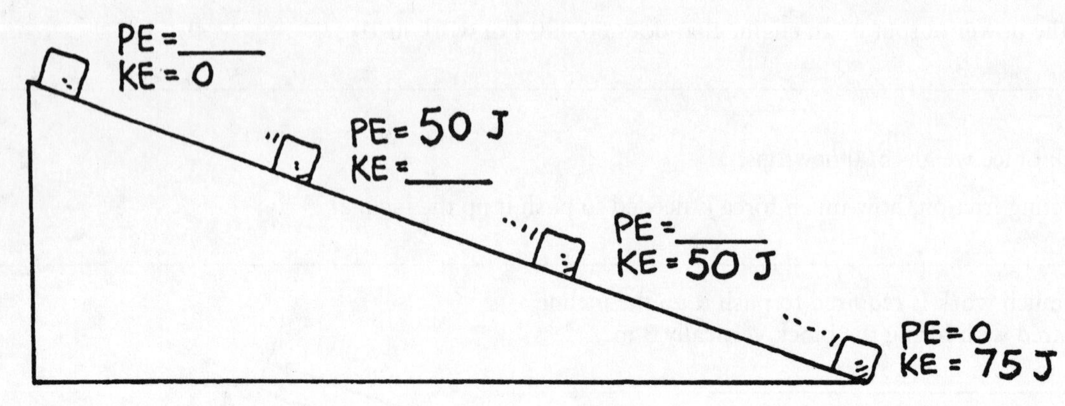

8. A big metal bead slides due to gravity along an upright friction-free wire. It starts from rest at the top of the wire as shown in the sketch. How fast is it traveling as it passes

Point B? __________

Point D? __________

Point E? __________

At what point does it have the maximum speed? __________

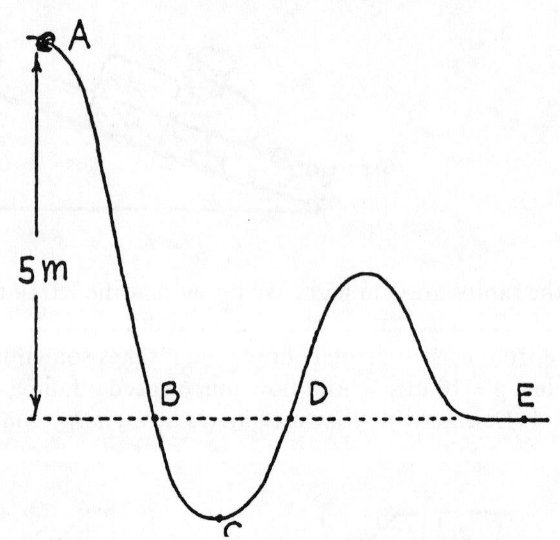

9. Rows of wind-powered generators are used in various windy locations to generate electric power. Does the power generated affect the speed of the wind? Would locations behind the "windmills" be windier if they weren't there? Discuss this in terms of energy conservation with your classmates.

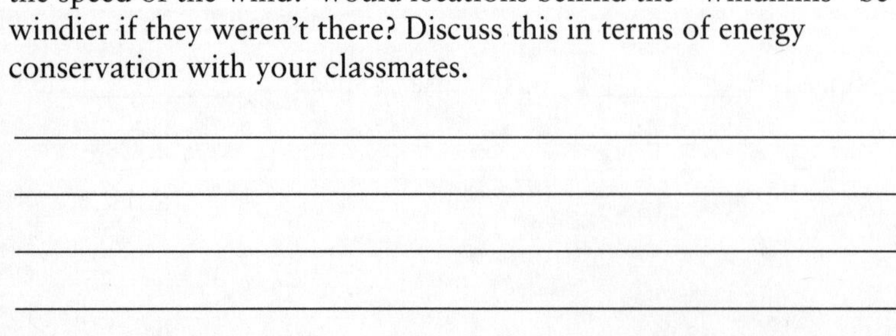

____

____

____

____

Name ______________________________ Date __________

Chapter 4: Momentum and Energy

Conservation of Energy

Fill in the blanks for the six systems shown:

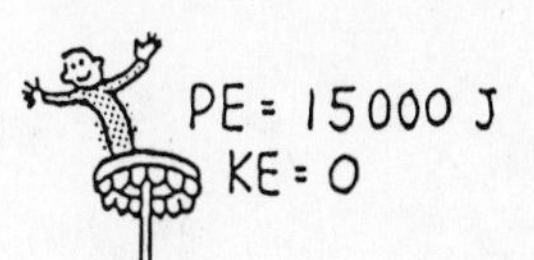

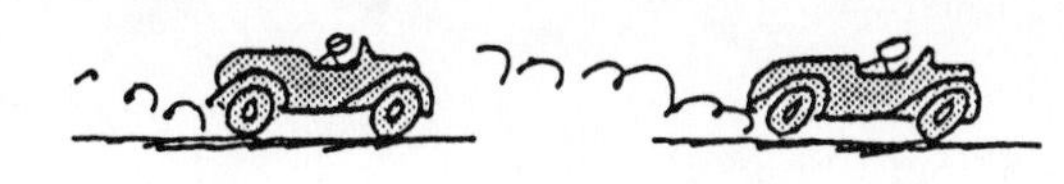

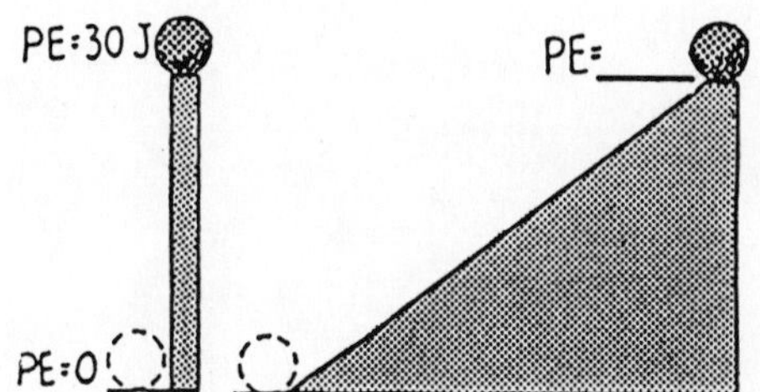

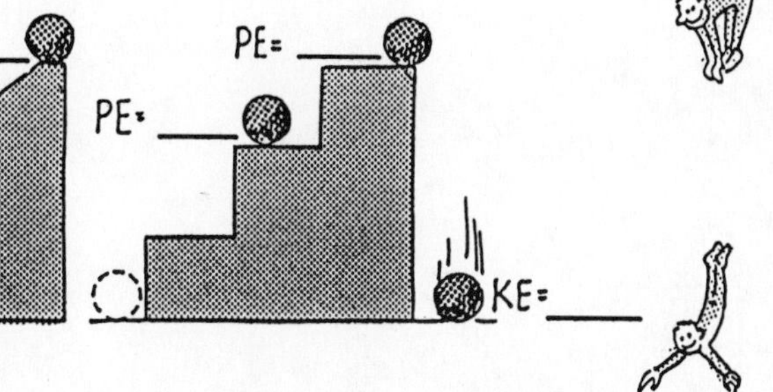

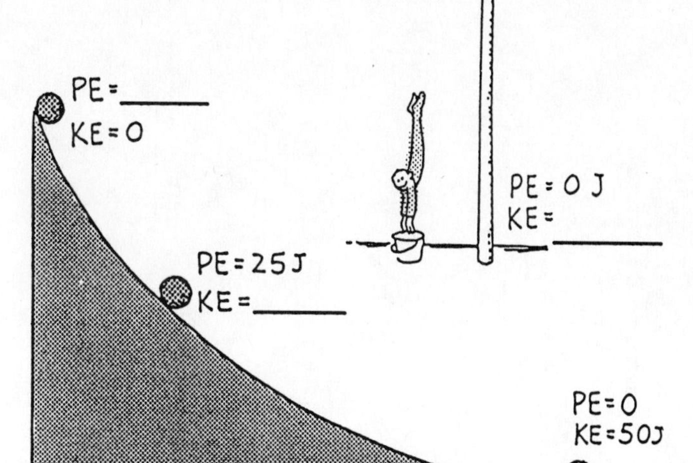

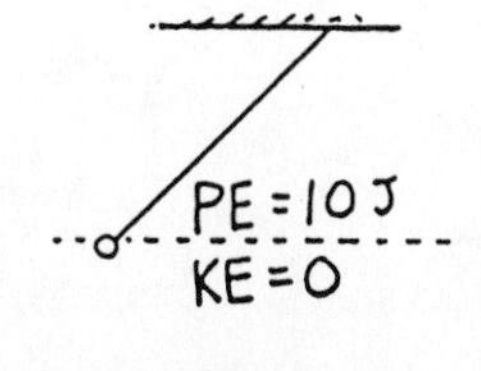

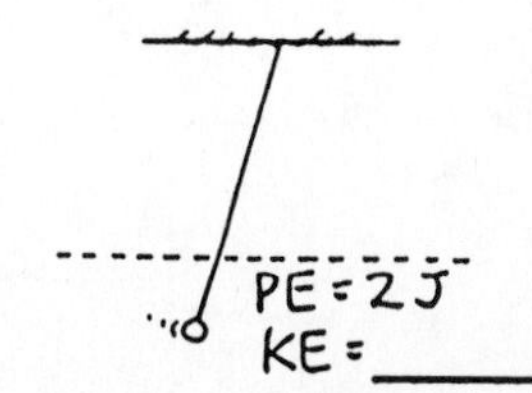

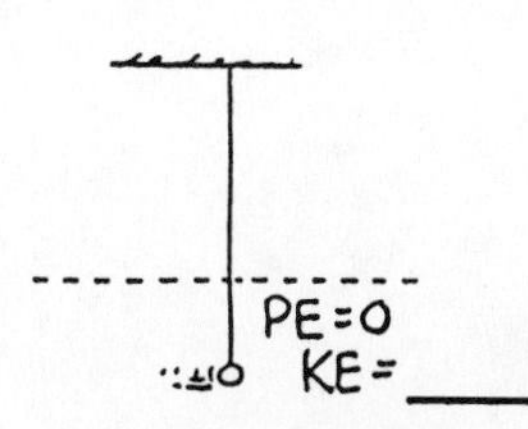

PE = ______
KE = ______

Name ______________________________ Date __________

Chapter 4: Momentum and Energy

Momentum, Impulse, and Kinetic Energy

A Honda Civic and a Lincoln Town Car are initially at rest on a horizontal parking lot at the edge of a steep cliff. For simplicity, we assume that the Town Car has twice as much mass as the Civic. Equal constant forces are applied to each car and they accelerate across equal distances (we ignore the effects of friction). When they reach the far end of the lot the force is suddenly removed, whereupon they sail through the air and crash to the ground below. (The cars are beat up to begin with, and this is a scientific experiment!)

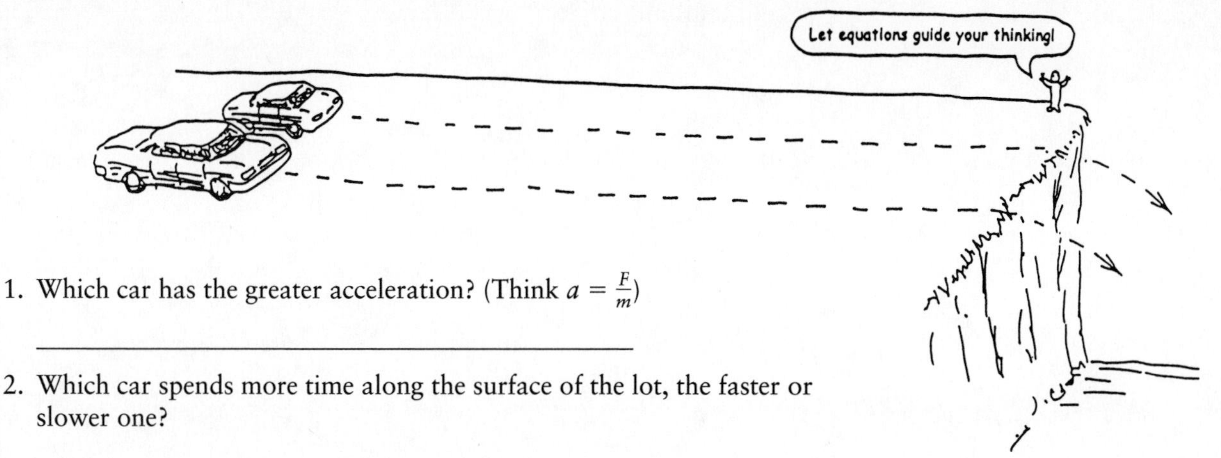

Let equations guide your thinking!

1. Which car has the greater acceleration? (Think $a = \frac{F}{m}$)

____

2. Which car spends more time along the surface of the lot, the faster or slower one?

____

3. Which car has the larger impulse imparted to it by the applied force? (Think Impulse = Ft.) Defend your answer.

____

4. Which car has the greater momentum at the cliff's edge? (Think $Ft = \Delta mv$.) Defend your answer.

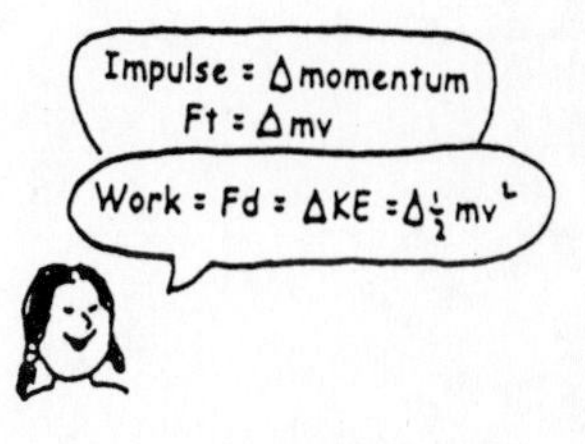

____

5. Which car has the greater work done on it by the applied force? (Think $W = Fd$) Defend your answer in terms of the distance traveled.

____

6. Which car has the greater kinetic energy at the edge of the cliff? (Think $W = \Delta KE$) Does your answer follow from your explanation of 5? Does it contradict your answer to 3? Why or why not?

____

7. Which car spends more time in the air, from the edge of the cliff to the ground below?

____

8. Which car lands farthest horizontally from the edge of the cliff onto the ground below?

Making the distinction between momentum and kinetic energy is high-level physics.

____

Challenge: Suppose the slower car crashes a horizontal distance of 10 m from the ledge. At what horizontal distance does the faster car hit?

Name ______________________________ Date ____________

Chapter 5: Gravity

The Inverse-Square Law—Weight

1. Paint spray travels radially away from the nozzle of the can in straight lines. Like gravity, the strength (intensity) of the spray obeys an inverse-square law. Complete the diagram by filling in the blank spaces.

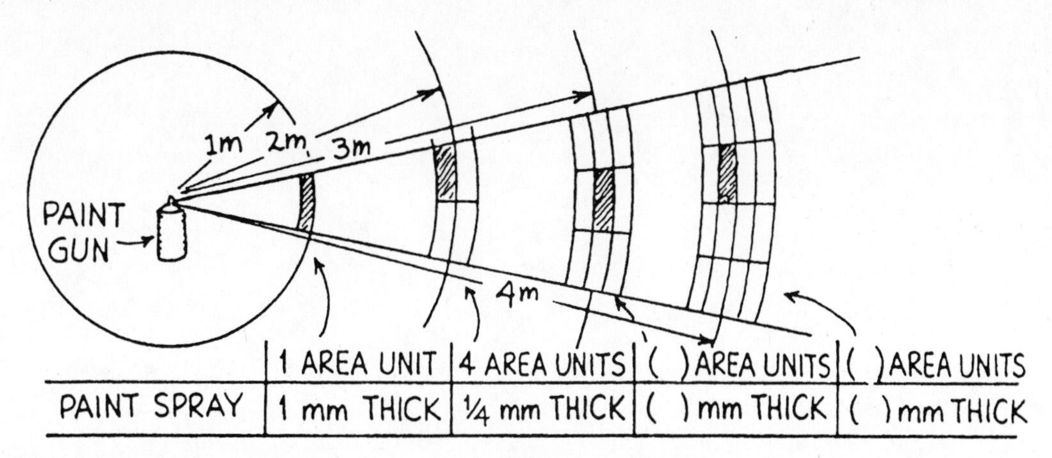

2. A small light source located 1 m in front of an opening of area 1 m^2 illuminates a wall behind. If the wall is 1 m behind the opening (2 m from the light source), the illuminated area covers 4 m^2. How many square meters will be illuminated if the wall is

   5 m from the source? ____________

   10 m from the source? ____________

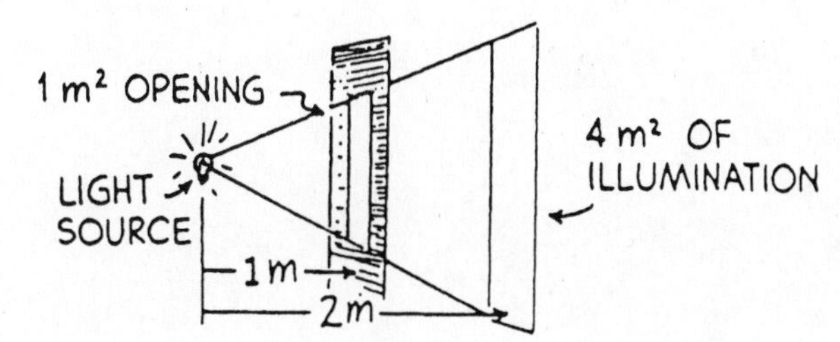

3. If we stand on a weighing scale and find that we are pulled toward Earth with a force of 500 N, then we weigh ____________ N. Strictly speaking, we weigh ____________ N relative to Earth. How much does Earth weigh? If we tip the scale upside down and repeat the weighing process, we can say that we and Earth are still pulled together with a force of ____________ N, and, therefore, relative to us, the whole 6,000,000,000,000,000,000,000,000-kg Earth weighs ____________ N! Weight, unlike mass, is a relative quantity.

We are pulled to Earth with a force of 500 N, so we weigh 500 N.

Earth is pulled toward us with a force of 500 N, so it weighs 500 N.

Name ______________________________ Date ____________

Chapter 5: Gravity

Ocean Tides

1. Consider two equal-mass blobs of water, A and B, initially at rest in the Moon's gravitational field. The vector shows the gravitational force of the Moon on A.

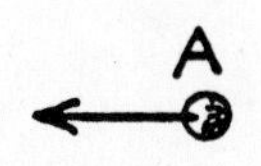

   a. Draw a force vector on B due to the Moon's gravity.

   b. Is the force on B more or less than the force on A? __________

   c. Why? ______________________________

   d. The blobs accelerate toward the Moon. Which has the greater acceleration? (A) (B)

   e. Because of the different accelerations, with time

      (A gets farther ahead of B) (A and B gain identical speeds) and the distance between A and B

      (increases) (stays the same) (decreases).

   f. If A and B were connected by a rubber band, with time the rubber band would

      (stretch) (not stretch).

   g. This (stretching) (nonstretching) is due to the (difference) (nondifference) in the Moon's gravitational pulls.

   h. The two blobs will eventually crash into the Moon. To orbit around the Moon instead of crashing into it, the blobs should move (away from the Moon) (tangentially). Then their accelerations will consist of changes in (speed) (direction).

2. Now consider the same two blobs located on opposite sides of Earth.

   a. Because of differences in the Moon's pull on the blobs, they tend to

      (spread away from each other) (approach each other). This produces ocean tides!

   b. If Earth and the Moon were closer, gravitational force between them would be

      (more) (the same) (less), and the difference in gravitational forces on the near and far parts of the ocean would be (more) (the same) (less).

   c. Because Earth's orbit about the Sun is slightly elliptical, Earth and the Sun are closer in December than in June. Taking the Sun's tidal force into account, on a world average, ocean tides are greater in

      (December) (June) (no difference).

Name ____________________ Date __________

Chapter 5: Gravity

Projectile Motion

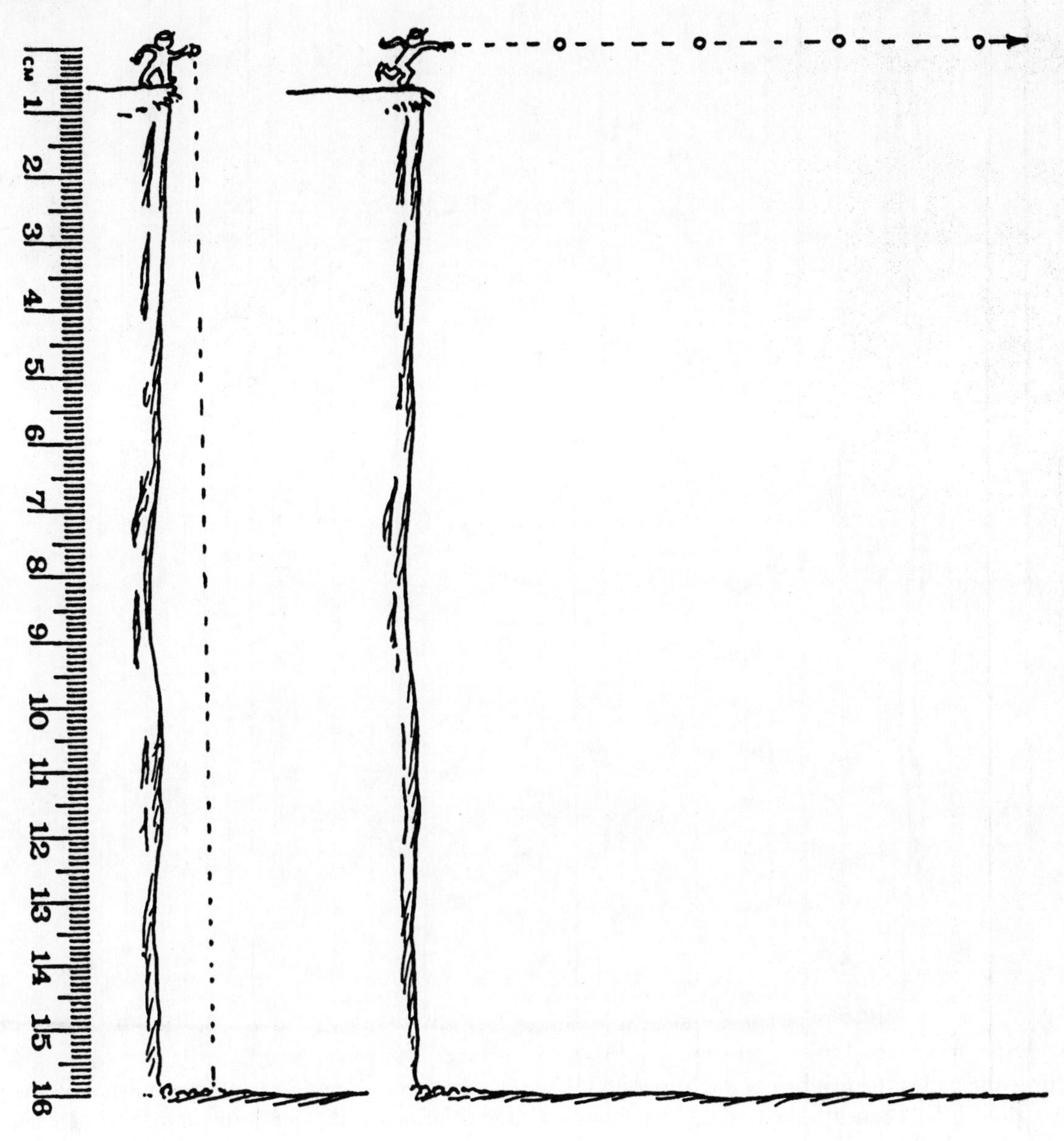

1. Above left: Use the scale 1 cm : 5 m and draw the positions of the dropped ball at 1-second intervals. Neglect air drag and assume $g = 10\ m/s^2$. Estimate the number of seconds the ball is in the air.

   __________ seconds.

2. Above right: The four positions of the thrown ball with *no gravity* are at 1-second intervals. At 1 cm : 5 m, carefully draw the positions of the ball *with* gravity. Neglect air drag and assume $g = 10\ m/s^2$. Connect your positions with a smooth curve to show the path of the ball. How is the motion in the vertical direction affected by motion in the horizontal direction?

   ____

Name ______________________________ Date __________

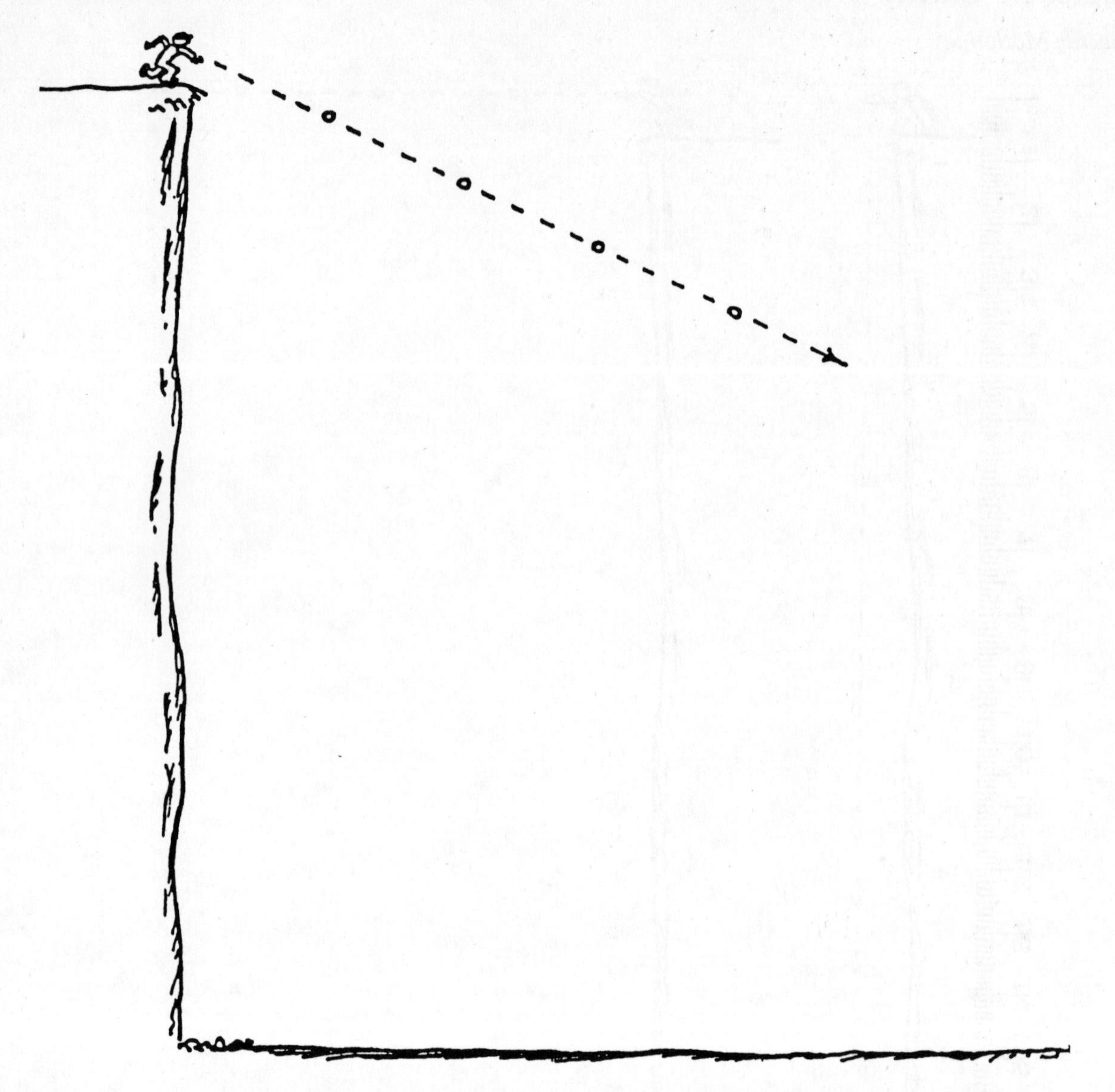

3. This time the ball is thrown below the horizontal. Use the same scale 1 cm : 5 m and carefully draw the positions of the ball as it falls beneath the dashed line. Connect your positions with a smooth curve. Estimate the number of seconds the ball remains in the air. ______________ seconds

4. Suppose that you are an accident investigator and you are asked to figure out whether or not the car was speeding before it crashed through the rail of the bridge and into the mudbank as shown. The speed limit on the bridge is 55 mph = 24 m/s. What is your conclusion?

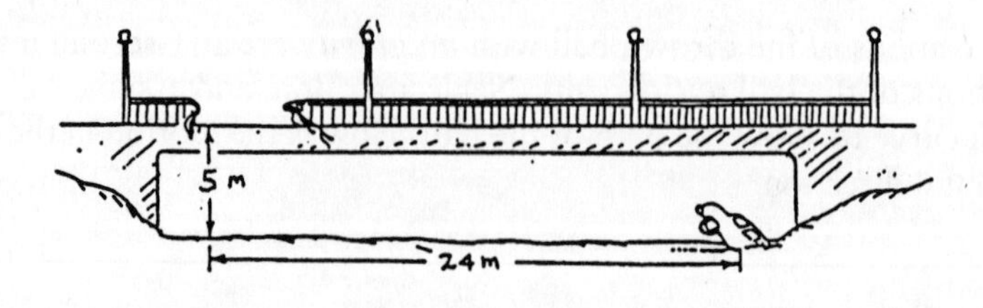

____

____

____

Name ______________________________ Date ____________

Chapter 5: Gravity

Tossed-Ball Vectors

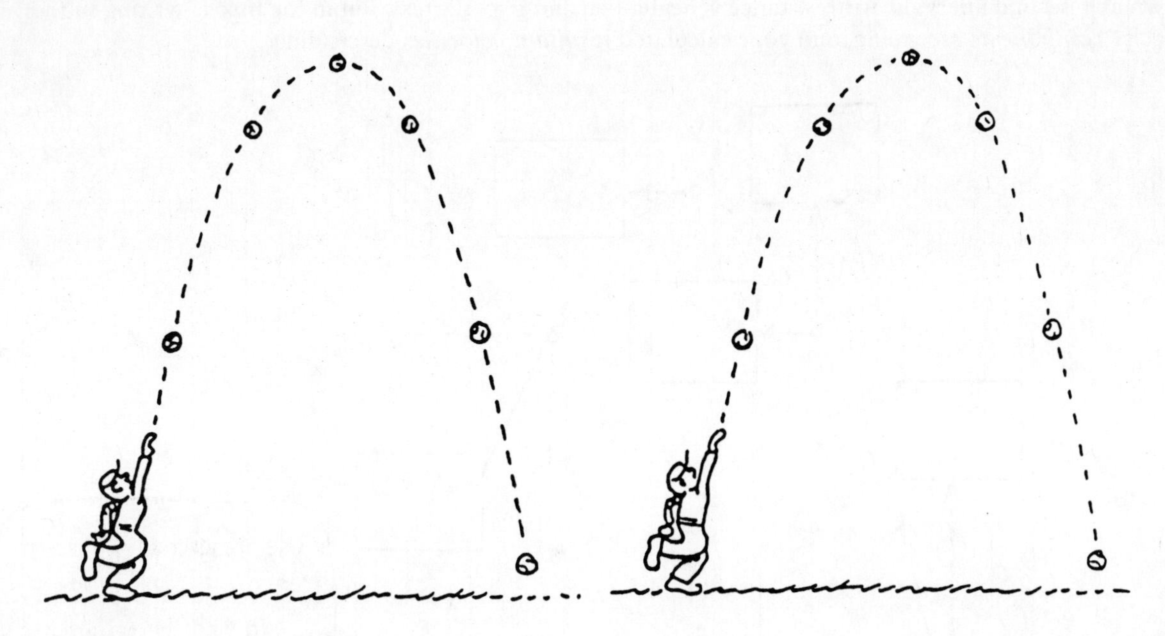

1. Draw sample vectors to represent the force of gravity on the ball in the positions shown above (after it leaves the thrower's hand). Neglect air drag.

2. Draw sample bold vectors to represent the velocity of the ball in the positions shown above. With lighter vectors, show the horizontal and vertical components of velocity for each position.

3. a. Which velocity component in the previous question remains constant? Why?

   ____

   b. Which velocity component changes along the path? Why?

   ____

4. It is important to distinguish between force and velocity vectors. Force vectors combine with other force vectors, and velocity vectors combine with other velocity vectors. Do velocity vectors combine with force vectors? __________

Name ______________________________ Date __________

Tossed-Ball Vectors—continued

A ball tossed upward has initial velocity components 30 m/s vertical and 5 m/s horizontal. The position of the ball is shown at 1-second intervals. Air resistance is negligible, and $g = 10\ m/s^2$. Fill in the boxes, writing in the values of velocity *components* ascending, and your calculated *resultant velocities* descending.

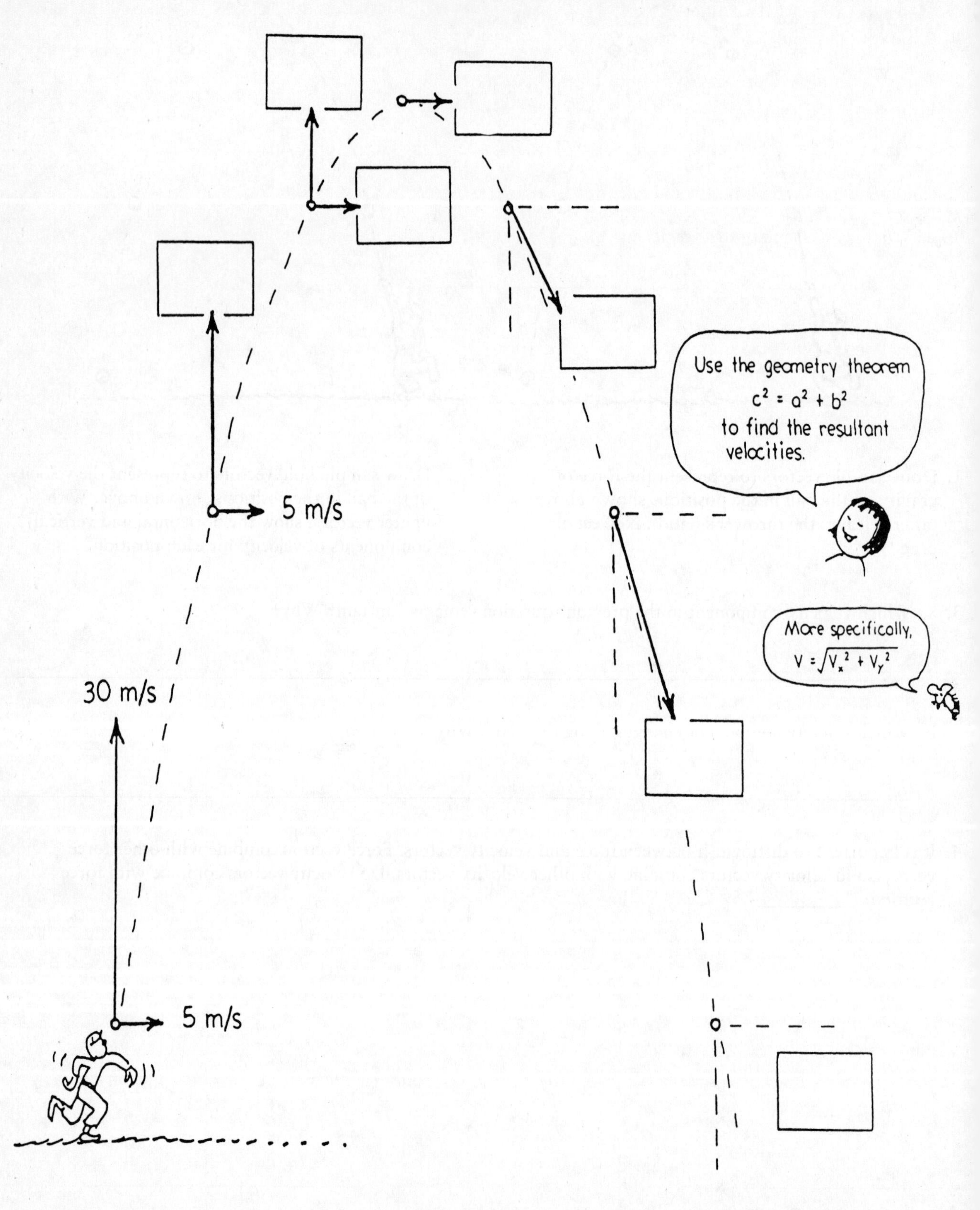
30 m/s
5 m/s
5 m/s
Use the geometry theorem $c^2 = a^2 + b^2$ to find the resultant velocities.
More specifically, $V = \sqrt{V_x^2 + V_y^2}$

Name ________________________________ Date __________

Chapter 5: Gravity

Circular and Elliptical Orbits

I. *Circular Orbits*

1. Figure 1 shows "Newton's Mountain," so high that its top is above the drag of the atmosphere. The cannonball is fired and hits the ground as shown.

   a. Draw the path the cannonball might take if it were fired a little bit faster.

   b. Repeat for a still greater speed, but still less than 8 km/s.

   c. Draw the orbital path it would take if its speed were 8 km/s.

   d. What is the shape of the 8-km/s curve?

   ____

   e. What would be the shape of the orbital path if the cannonball were fired at a speed of about 9 km/s?

   ____

Figure 1

2. Figure 2 shows a satellite in circular orbit.

   a. At each of the four positions draw a vector that represents the gravitational *force* exerted on the satellite.

   b. Label the force vectors ***F***.

   c. Draw at each position a vector to represent the *velocity* of the satellite at that position and label it ***V***.

   d. Are all four ***F*** vectors the same length? Why or why not?

   ____

   ____

   e. Are all four ***V*** vectors the same length? Why or why not?

   ____

   f. What is the angle between your ***F*** and ***V*** vectors?

   ____

   g. Is there any component of ***F*** along ***V***? ______________________________

   h. What does this tell you about the work the force of gravity does on the satellite?

   ____

   i. Does the KE of the satellite in Figure 2 remain constant, or does it vary? ______________

   j. Does the PE of the satellite remain constant, or does it vary?

   ____

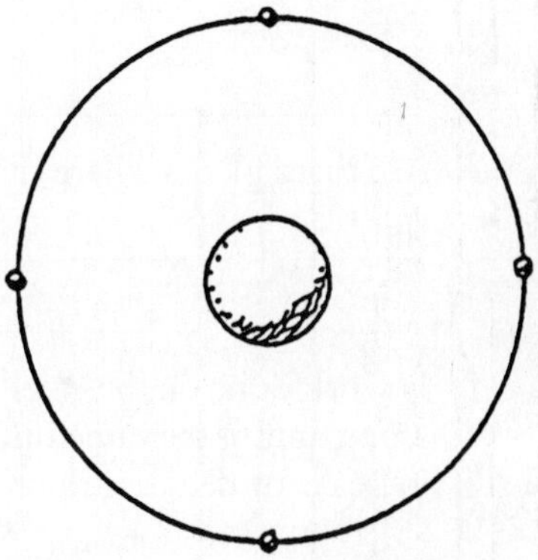

Figure 2

Name ______________________________ Date __________

Circular and Elliptical Orbits—continued

II. *Elliptical Orbits*

3. Figure 3 shows a satellite in elliptical orbit.

   a. Repeat the procedure you used for the circular orbit, drawing vectors ***F*** and ***V*** for each position, including proper labeling. Show equal magnitudes with equal lengths, and greater magnitudes with greater lengths, but don't bother making the scale accurate.

   b. Are your vectors ***F*** all the same magnitude? Why or why not?

   ____

   ____

   c. Are your vectors ***V*** all the same magnitude? Why or why not?

   ____

   ____

   ____

   ____

   d. Is the angle between vectors ***F*** and ***V*** everywhere the same, or does it vary?

   ____

   e. Are there places where there is a component of ***F*** along ***V***?

   ____

   ____

   f. Is work done on the satellite when there is a component of ***F*** along and in the same direction of ***V***, and if so, does this increase or decrease the KE of the satellite?

   ____

   g. When there is a component of ***F*** along and opposite to the direction of ***V***, does this increase or decrease the KE of the satellite?

   ____

   ____

   ____

   h. What can you say about the sum KE + PE along the orbit?

   ____

   ____

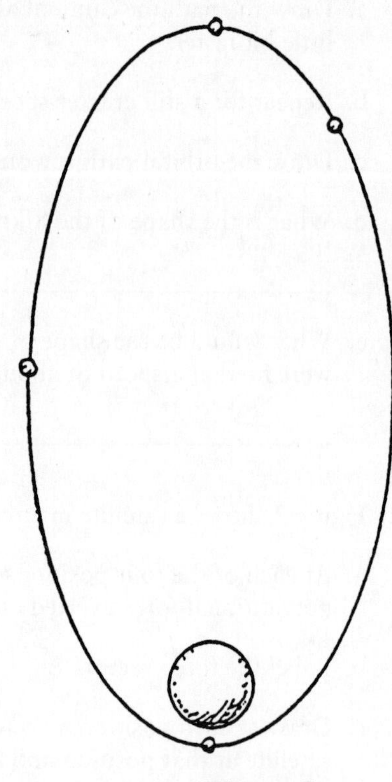

Figure 3

Be very, very careful when placing both velocity and force vectors on the same diagram. Not a good practice, for one may construct the resultant of the vectors—ouch!

Name ______________________________ Date ____________

Chapter 5: Gravity

Mechanics Overview

1. The sketch shows the elliptical path described by a satellite about Earth. In which of the marked positions, A–D, (put S for "same everywhere") does the satellite experience the maximum

   a. gravitational force? ____________

   b. speed? ____________

   c. velocity? ____________

   d. momentum? ____________

   e. kinetic energy? ____________

   f. gravitational potential energy? ____________

   g. total energy (KE + PE)? ____________

   h. acceleration? ____________

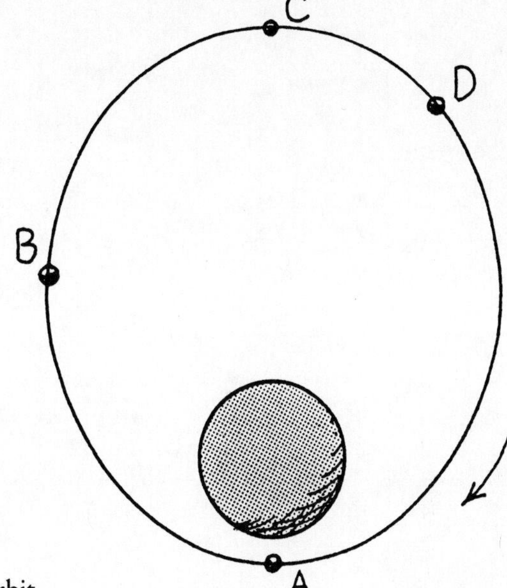

2. Answer the above questions for a satellite in circular orbit.

   a. _____ b. _____ c. _____ d. _____ e. _____ f. _____ g. _____ h. _____

3. In which position(s) is there momentarily no work done on the satellite by the force of gravity? Why?

   ____

4. Work changes energy. Let the equation for work, $W = Fd$, guide your thinking on these questions. Defend your answers in terms of $W = Fd$.

   a. In which position will a several-minutes thrust of rocket engines do the most work on the satellite and give it the greatest change in kinetic energy?

   ____

   b. In which position will a several-minutes thrust of rocket engines do the most work on the *exhaust gases* and give the *exhaust gases* the greatest change in kinetic energy?

   ____

   c. In which position will a several-minutes thrust of rocket engines give the satellite the least boost in kinetic energy?

   ____

Name ______________________________ Date ____________

Chapter 6: Heat

Temperature Mix

1. You apply heat to 1 L of water and raise its temperature by 10°C. If you add the same quantity of heat to 2 L of water, how much will the temperature rise? To 3 L of water?

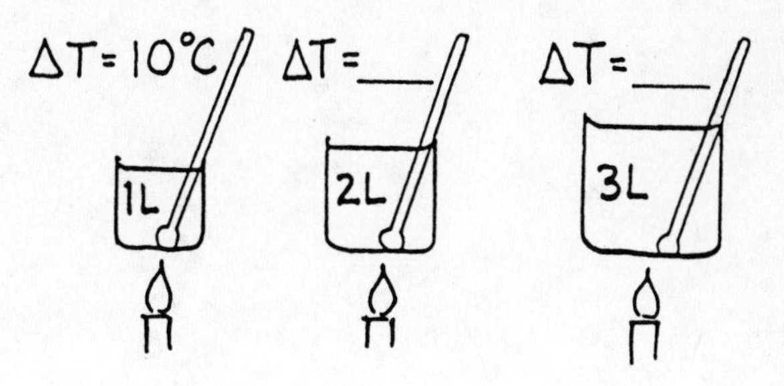

Record your answers on the blanks in the drawing at the right. (Hint: Heat transferred is directly proportional to its temperature change, $Q = mc\Delta T$.)

2. A large bucket contains 1 L of 20°C water.

   a. What will be the temperature of the mixture when 1 L of 20°C water is added?

   ____

   b. What will be the temperature of the mixture when 1 L of 40°C water is added?

   ____

   c. If 2 L of 40°C water were added, would the temperature of the mixture be greater or less than 30°C?

   ____

3. A red-hot iron kilogram mass is put into 1 L of cool water. Mark each of the following statements true (T) of false (F). (Ignore heat transfer to the container.)

   a. The increase in the water temperature is equal to the decrease in the iron's temperature.

   ____

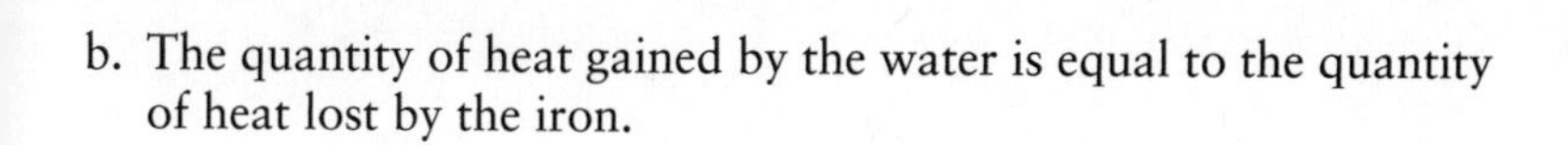

   b. The quantity of heat gained by the water is equal to the quantity of heat lost by the iron.

   ____

   c. The iron and the water will both reach the same temperature.

   ____

   d. The final temperature of the iron and water is about halfway between the initial temperatures of each.

   ____

4. *True or False:* When Queen Elizabeth throws the last sip of her tea over Queen Mary's rail, the ocean gets a little warmer. ____________

Name ______________________________ Date ____________

Chapter 6: Heat

Absolute Zero

A mass of air is contained so that the volume can change but the pressure remains constant. Table 1 shows air volumes at various temperatures when the air is heated slowly.

1. Plot the data in Table 1 on the graph and connect the points.

Table 1

TEMP. (°C)	VOLUME (mL)
0	50
25	55
50	60
75	65
100	70

2. The graph shows how the volume of air varies with temperature at constant pressure. The straightness of the line means that the air expands uniformly with temperature. From your graph, you can predict what will happen to the volume of air when it is cooled.

   Extrapolate (extend) the straight line of your graph to find the temperature at which the volume of the air would become zero. Mark this point on your graph. Estimate this temperature: ____________

3. Although air would liquify before cooling to this temperature, the procedure suggests that there is a lower limit to how cold something can be. This is the absolute zero of temperature.

   Careful experiments show that absolute zero is ____________°C.

4. Scientists measure temperature in *kelvins* instead of degrees Celsius, where the absolute zero of temperature is 0 kelvins. If you relabeled the temperature axis on the graph in Question 1 so that it shows temperature in kelvins, would your graph look like the one below? ____________

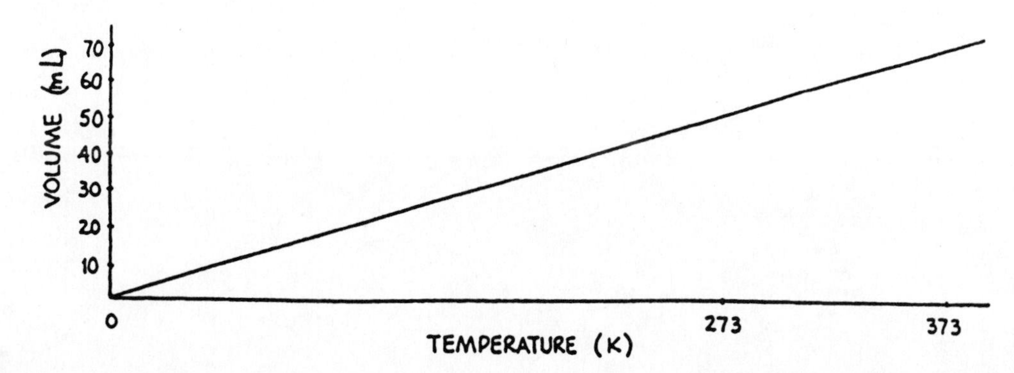

Name ____ Date ____________

Chapter 6: Heat

Thermal Expansion

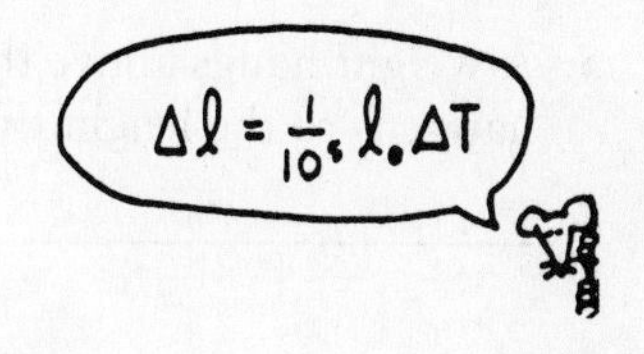

1. Steel expands by about 1 part in 100,000 for each 1°C increase in temperature.

   a. How much longer will a piece of steel 1000 mm long (1 meter) be when its temperature is increased by 10°C? __________

   b. How much longer will a piece of steel 1000 m long (1 kilometer) be when its temperature is increased by 10°C? __________

   c. You place yourself between a wall and the end of a 1-m steel rod when the opposite end is securely fastened as shown. No harm comes to you if the temperature of the rod is increased a few degrees. Discuss the consequences of doing this with a rod many meters long?

   ____

   ____

2. The Eiffel Tower in Paris is 298 meters high. On a cold winter night, it is shorter than on a hot summer day. What is its change in height for a 30°C temperature difference?

____

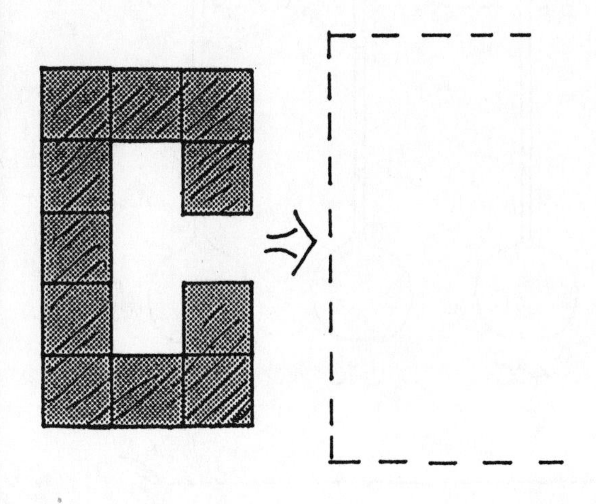

3. Consider a gap in a piece of metal. Does the gap become wider or narrower when the metal is heated? [Consider the piece of metal made up of 11 blocks—if the blocks are individually heated, each is slightly larger. Make a sketch of them, slightly enlarged, beside the sketch shown.]

____

4. The equatorial radius of Earth is about 6370 km. Consider a 40,000-km long steel pipe that forms a giant ring that fits snugly around the equator of the earth. Suppose people all along its length breathe on it so as to raise its temperature 1°C. The pipe gets longer. It is also no longer snug. How high does it stand above the ground? (Hint: Concentrate on the radial distance.)

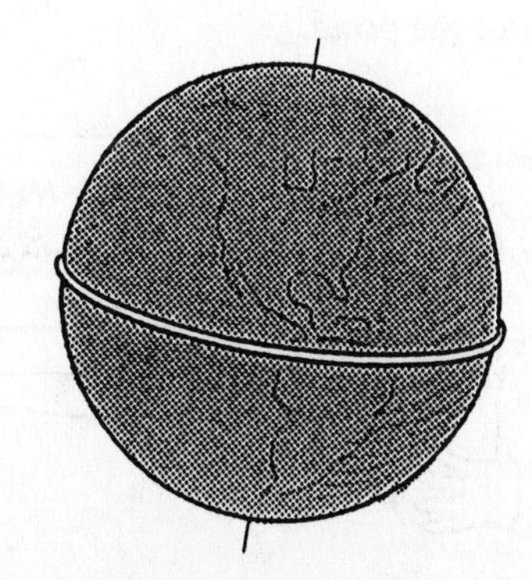

Name ______________________________ Date __________

Thermal Expansion—continued

5. A weight hangs above the floor from the copper wire. When a candle is moved along the wire and heats it, what happens to the height of the weight above the floor? Why?

____

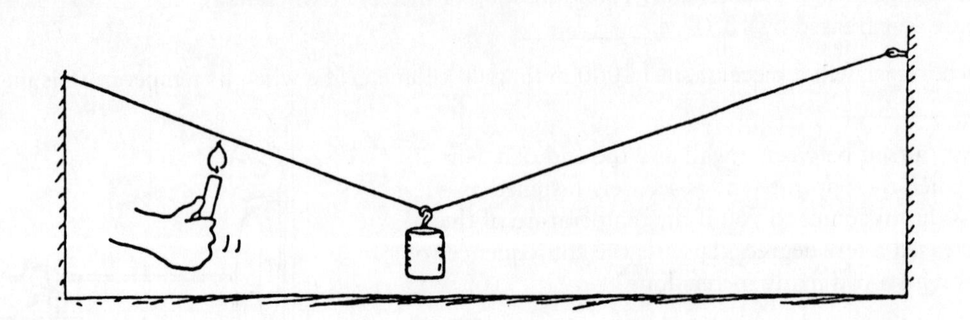

6. The levels of water at 0°C and 1°C are shown below in the first two flasks. At these temperatures there is microscopic slush in the water. There is slightly more slush at 0°C than at 1°C. As the water is heated, some of the slush collapses as it melts, and the level of the water falls in the tube. That's why the level of water is slightly lower in the 1°C tube. Make rough estimates and sketch in the appropriate levels of water at the other temperatures shown. What is important about the level when the water reaches 4°C?

____

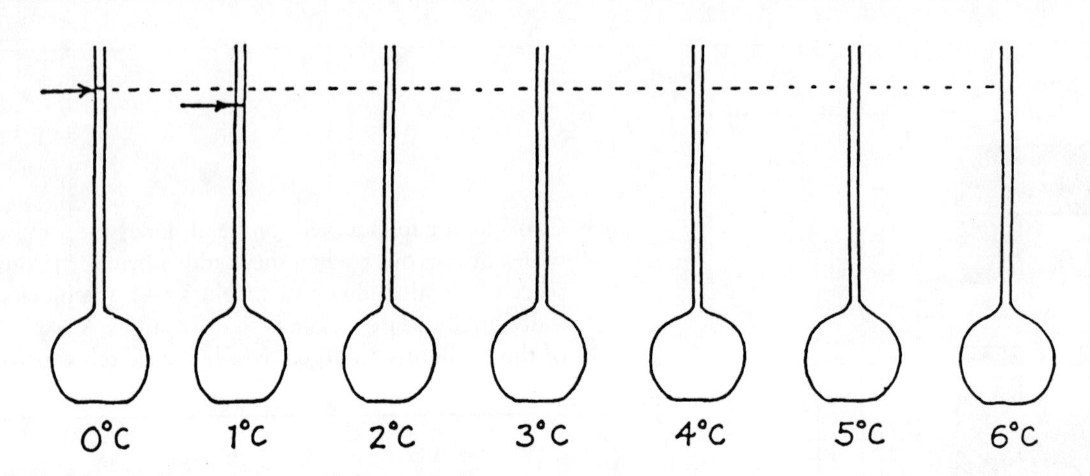
0°C
1°C
2°C
3°C
4°C
5°C
6°C

7. The diagram at right shows an ice-covered pond. Mark the probable temperatures of water at the top and bottom of the pond.

ICE

____°C

____°C

I CAN'T GET THIS METAL LID OFF THE JAR··· SHOULD I HEAT THE LID OR COOL IT? WHY? ____________
JAR

WHICH WILL WEIGH MORE, 1 LITER OF ICE OR 1 LITER OF WATER?

____

Name ______________________________ Date __________

Chapter 6: Heat

Transmission of Heat

1. The tips of both brass rods are held in the gas flame. *Mark the following true (T) or false (F).*

   a. Heat is conducted only along Rod A. _____

   b. Heat is conducted only along Rod B. _____

   c. Heat is conducted equally along both Rod A and Rod B. _____

   d. The idea that "heat rises" applies to heat transfer by *convection,* not by *conduction.* _____

2. Why does a bird fluff its feathers to keep warm on a cold day?

____

3. Why does a down-filled sleeping bag keep you warm on a cold night? Why is it useless if the down is wet?

____

4. What does *convection* have to do with the holes in the shade of the desk lamp?

____

5. When hot water rapidly evaporates, the result can be dramatic. Consider 4 g of boiling water spread over a large surface so that 1 g rapidly evaporates. Suppose further that the surface and surroundings are very cold so that all 540 calories for evaporation come from the remaining 3 g of water.

   a. How many calories are taken from each gram of water?

   ____

   b. How many calories are released when 1 g of 100°C water cools to 0°C?

   ____

   c. How many calories are released when 1 g of 0°C water changes to 0°C ice?

   ____

   d. What happens in this case to the remaining 3 g of boiling water when 1 g rapidly evaporates?

   ____

   ____

Name ______________________________ Date ____________

Chapter 7: Electricity and Magnetism

Electric Potential

Just as PE transforms to KE for a mass lifted against the gravitation field (left), the electric PE of an electric charge transforms to other forms of energy when it changes location in an electric field (right). In both cases, how does the KE acquired compare to the decrease in PE?

____

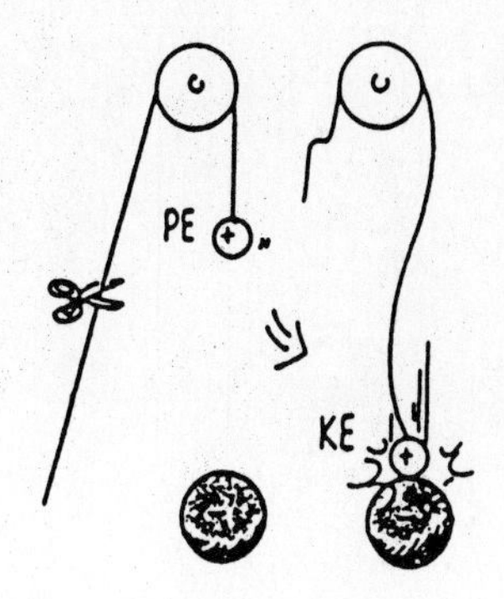

Complete the following statements:

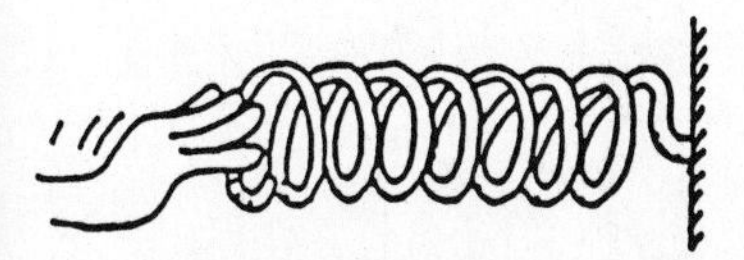

A force compresses the spring. The work done in compression is the product of the average force and the distance moved: $W = Fd$. This work increases the PE of the spring.

Similarly, a force pushes the charge (call it a *test charge*) closer to the charged sphere. The work done in moving the test charge is the product of the average ____________ and the ____________ moved. W = ____________. This work ____________ the PE of the test charge.

If the test charge is released, it will be repelled and fly past the starting point. Its gain in KE at this point is ____________ to its decrease in PE.

At any point, a greater amount of test charge means a greater amount of PE, but not a greater amount of PE *per amount* of charge. The quantities PE (measured in joules) and $\frac{\text{PE}}{\text{charge}}$ (measured in volts) are different concepts.

By definition: Electric Potential $= \frac{\text{PE}}{\text{charge}}$. 1 volt $= \frac{1 \text{ joule}}{1 \text{ coulomb}}$. So, 1 C of charge with a PE of 1 J has an electric potential of 1 V. 2 C of charge with a PE of 2 J has an electric potential of ____________V.

If a conductor connected to the terminal of a battery has an electric potential of 12 V, then each coulomb of charge on the conductor has a PE of ____________J.

You do very little work in rubbing a balloon on your hair to charge it. The PE of several thousand billion electrons (about one-millionth coulomb [10^{-6}C]) transferred may be a thousandth of a joule [10^{-3}J]. Impressively, however, the electric potential of the balloon is about ____________V!

Why is contact with a balloon charged to thousands of volts not as dangerous as contact with household 110 V?

____

____

Name ______________________________ Date ______________

Chapter 7: Electricity and Magnetism

Series Circuits

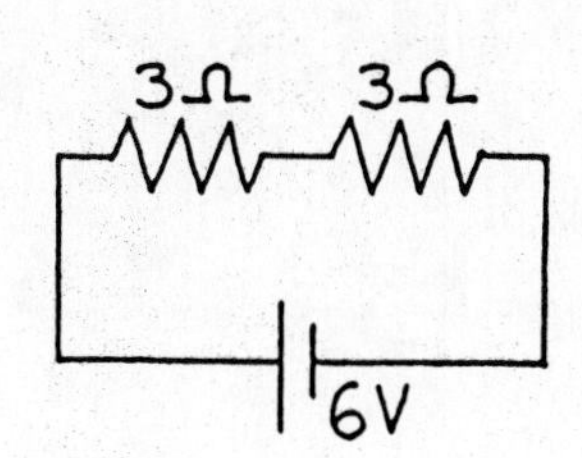

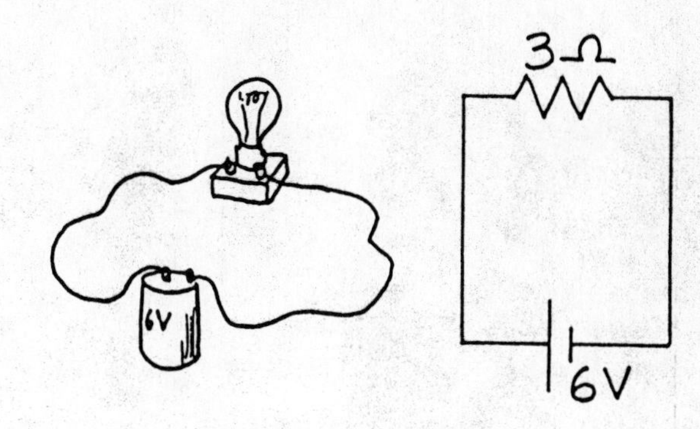

1. The simple circuit is a 6-V battery that pushes charge through a single lamp that has a resistance of 3 Ω. According to Ohm's law, the current in the lamp (and therefore the whole circuit) is _______A.

2. If a second identical lamp is added, the 6-V battery must push charge through a total resistance of _________Ω. The current in the circuit is then _______A.

3. If a third identical lamp is added in series, the total resistance of the circuit (neglecting any internal resistance in the battery) is _______Ω.

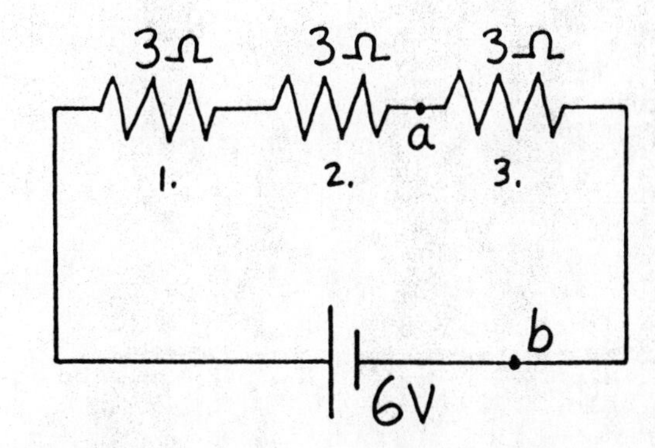

4. The current through all three lamps in series is _______A. The current through each individual lamp is _______A.

5. Does current in the lamps occur simultaneously, or does charge flow first through one lamp, then the other, and finally the last, in turn? ______________________________

6. Does current flow *through* a resistor, or *across* a resistor? ____________________ Is voltage established *through* a resistor, or *across* a resistor? ____________________

7. The voltage across all three lamps in the series is 6-V. The voltage (or commonly, *voltage drop*) across each individual lamp is _______V.

8. Suppose a wire connects points *a* and *b* in the circuit. The voltage drop across lamp 1 is now _______V, across lamp 2 is _______V, and across lamp 3 is _______V. So, the current through lamp 1 is now _______A, through lamp 2 is _______A, and through lamp 3 is _______A. The current in the battery (neglecting internal battery resistance) is _______A.

9. Which circuit dissipates more power, the 3-lamp circuit or the 2-lamp circuit? (Another way of asking this is which circuit would glow brightest and be best seen on a dark night from a great distance?) Defend your answer.

____

____

____

Name ____ Date __________

Chapter 7: Electricity and Magnetism

Parallel Circuits

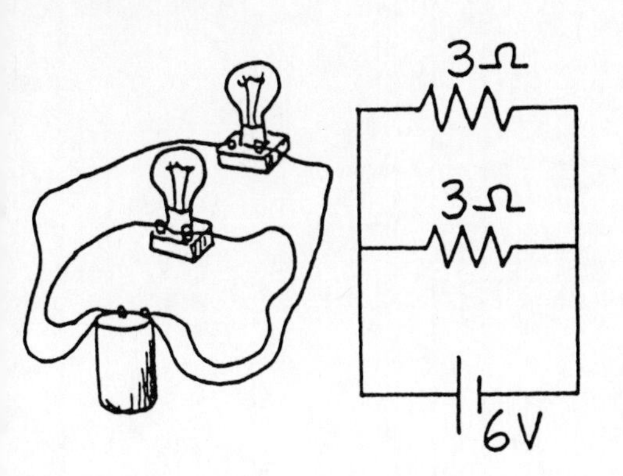

3Ω
3Ω
6V

1. In the circuit shown to the left there is a voltage drop of 6-V across each 3-Ω lamp. By Ohm's law, the current in each lamp is _______A. The current through the battery is the sum of the currents in the lamps, _______A.

THE SUM OF THE CURRENTS IN THE TWO BRANCH PATHS EQUALS THE CURRENT BOTH BEFORE AND AFTER IT DIVIDES!
WATER FLOW

2. Fill in the current in the eight blank spaces in the view of the same circuit shown again at the right.

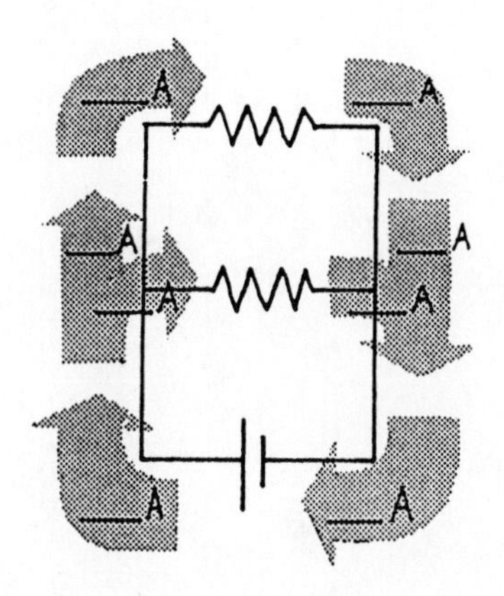

3. Suppose a third identical lamp is added in parallel to the circuit. Sketch a schematic diagram of the 3-lamp circuit in the space at the right.

4. For the three identical lamps in parallel, the voltage drop across each lamp is _______V. The current through each lamp is _______A. The current through the battery is now _______A. Is the circuit resistance now greater or less than before the third lamp was added? Explain.

____

5. Which circuit dissipates more power, the 3-lamp circuit or the 2-lamp circuit? (Another way of asking this is which circuit would glow brightest and be best seen on a dark night from a great distance?) Defend your answer and compare this to the similar case for 2-and 3-lamp series circuits.

____

____

Name ______________________________ Date ____________

Chapter 7: Electricity and Magnetism

Compound Circuits

The table beside circuit *a* below shows the current through each resistor, the voltage across each resistor, and the power dissipated as heat in each resistor. Find the similar correct values for circuits *b, c,* and *d,* and put your answers in the tables shown.

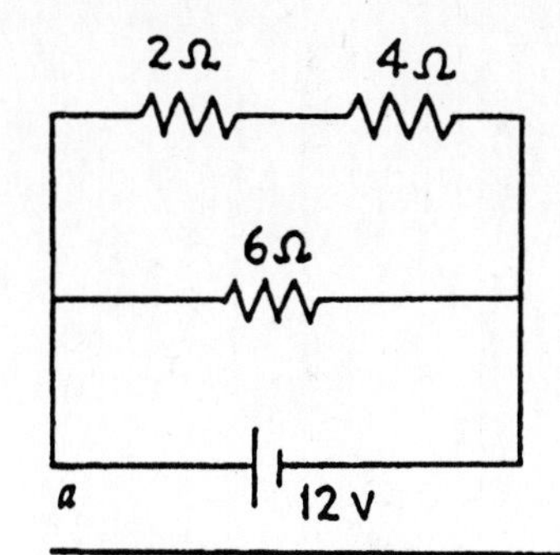

RESISTANCE	CURRENT ×	VOLTAGE =	POWER
2Ω	2 A	4 V	8 W
4Ω	2 A	8 V	16 W
6Ω	2 A	12 V	24 W

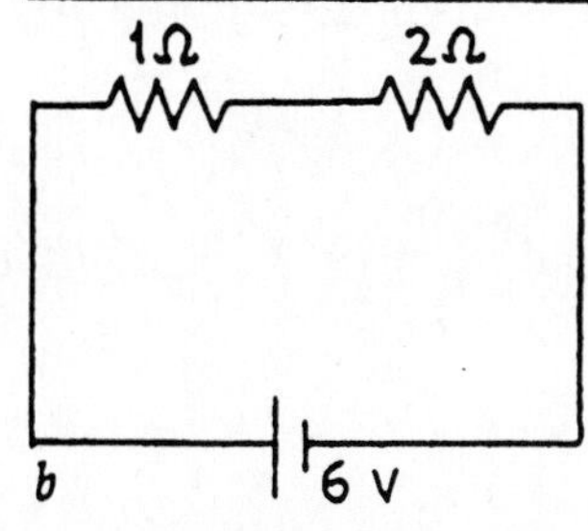

RESISTANCE	CURRENT ×	VOLTAGE =	POWER
1Ω			
2Ω			

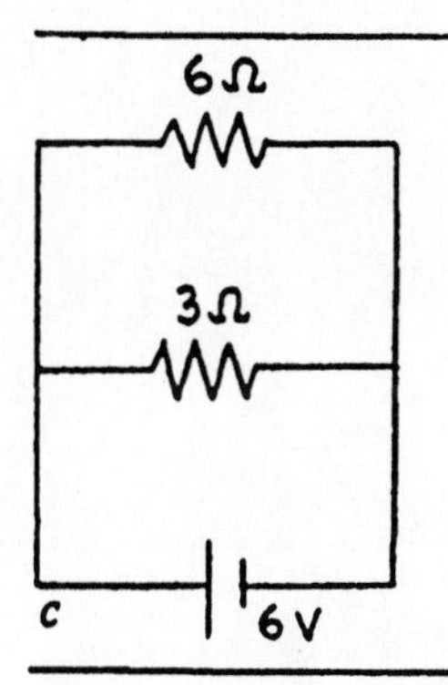

RESISTANCE	CURRENT ×	VOLTAGE =	POWER
6Ω			
3Ω			

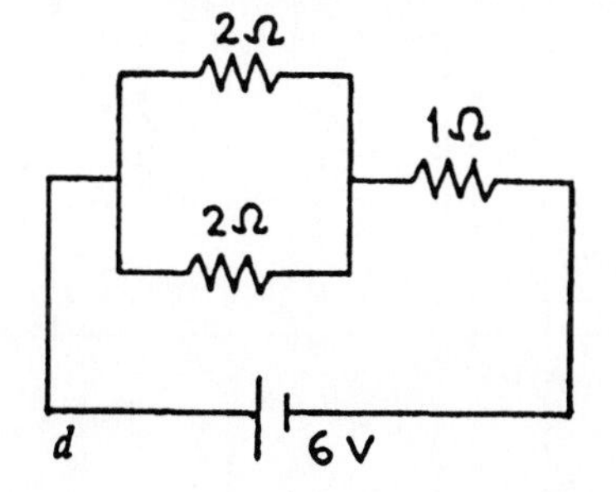

RESISTANCE	CURRENT ×	VOLTAGE =	POWER
2Ω			
2Ω			
1Ω			

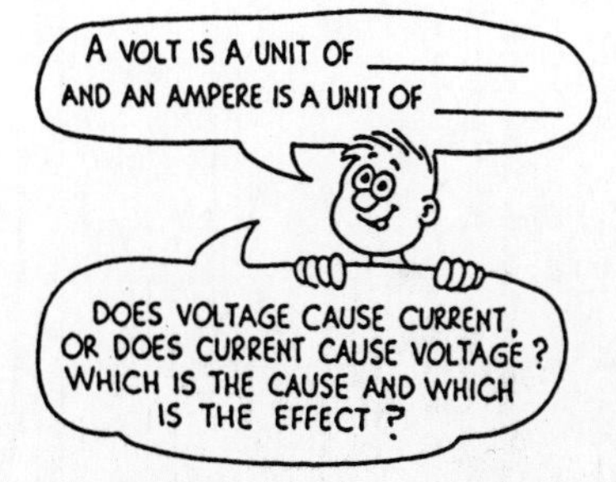

A VOLT IS A UNIT OF ______
AND AN AMPERE IS A UNIT OF ______
DOES VOLTAGE CAUSE CURRENT, OR DOES CURRENT CAUSE VOLTAGE? WHICH IS THE CAUSE AND WHICH IS THE EFFECT?

Name ____ Date ____________

Chapter 7: Electricity and Magnetism

Magnetism

Fill in each blank with the appropriate word:

1. Attraction or repulsion of charges depends on their *signs*, positives or negatives. Attraction or repulsion of magnets depends on their magnetic ____________: ____________ or ____________.

2. Opposite poles attract; like poles ____________.

3. A magnetic field is produced by the ____________ of electric charge.

4. Clusters of magnetically aligned atoms are magnetic ____________________.

5. A magnetic ____________ surrounds a current-carrying wire.

6. When a current-carrying wire is made to form a coil around a piece of iron, the result is an ____.

7. A charged particle moving in a magnetic field experiences a deflecting ____________________ that is maximum when the charge moves ____________________ to the field.

8. A current-carrying wire experiences a deflecting ____________ that is maximum when the wire and magnetic field are ____________ to one another.

9. A simple instrument designed to detect electric current is the ____________; when calibrated to measure current, it is an ____________; when calibrated to measure voltage, it is a ____________.

10. The largest size magnet in the world is the ____________ itself.

THEN TO REALLY MAKE THINGS "SIMPLE," THERE'S THE RIGHT-HAND RULE!
F
B
I

Name ______________________________ Date ____________

Chapter 7: Electricity and Magnetism

Field Patterns

1. The illustration below is similar to Figure 7.26 in your textbook. Iron filings trace out patterns of magnetic field lines about a bar magnet. In the field are some magnetic compasses. The compass needle in only one compass is shown. Draw in the needles with proper orientation in the other compasses.

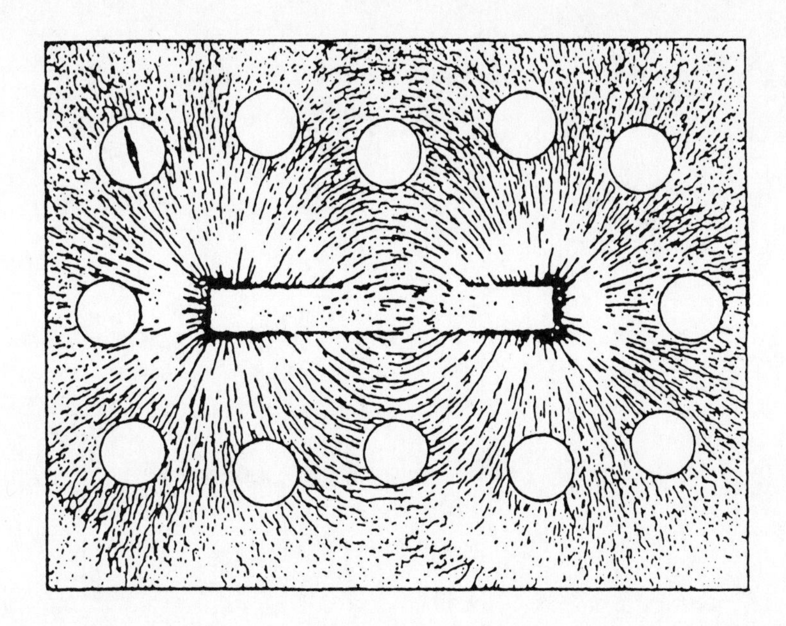

2. The illustration below is similar to Figure 7.33b in your textbook. Iron filings trace out the magnetic field pattern about the loop of current-carrying wire. Draw in the compass needle orientations for all the compasses.

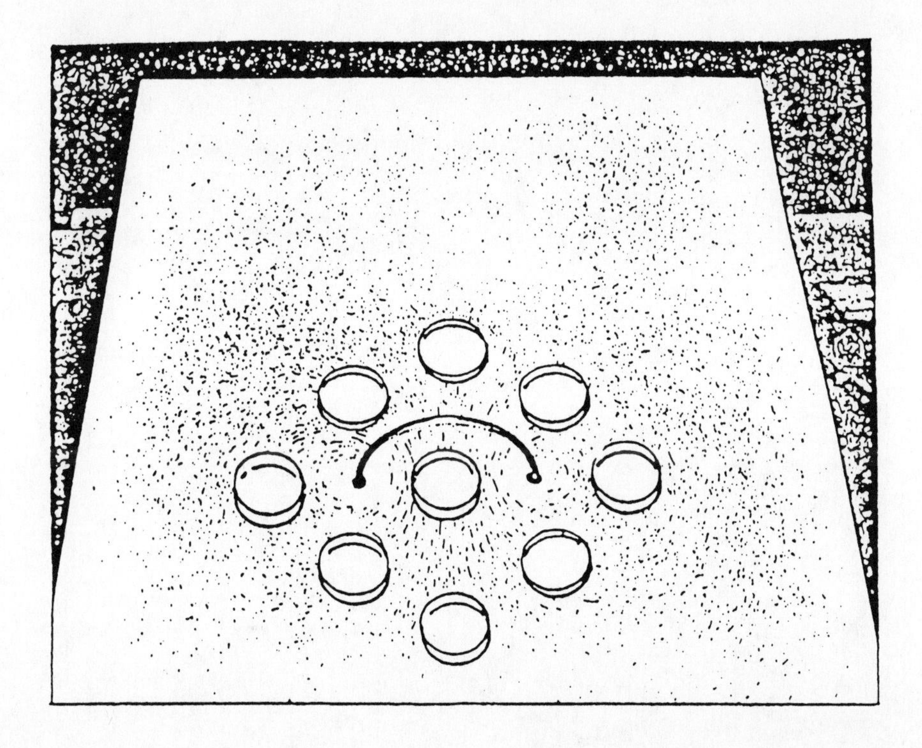

Name ______________________________ Date __________

Chapter 7: Electricity and Magnetism

Electromagnetism

1. Hans Christian Oersted discovered that magnetism and electricity are

   (related) (independent of each other).

   Magnetism is produced by

   (batteries) (the motion of electric charges).

Faraday and Henry discovered that electric current can be produced by

(batteries) (motion of a magnet).

More specifically, voltage is induced in a loop of wire if there is a change in the

(batteries) (magnetic field in the loop).

This phenomenon is called

(electromagnetism) (electromagnetic induction).

2. When a magnet is plunged in and out of a coil of wire, voltage is induced in the coil. If the rate of the in-and-out motion of the magnet is doubled, the induced voltage

   (doubles) (halves) (remains the same).

   If instead, the number of loops in the coil is doubled, the induced voltage

   (doubles) (halves) (remains the same).

3. A rapidly changing magnetic field in any region of space induces a rapidly changing

   (electric field) (magnetic field) (gravitational field),

   which in turn induces a rapidly changing

   (magnetic field) (electric field) (baseball field).

   This generation and regeneration of electric and magnetic fields makes up

   (electromagnetic waves) (sound waves) (both of these).

Name ______________________ Date __________

Chapter 8: Waves—Sound and Light

Vibration and Wave Fundamentals

1. A sine curve that represents a transverse wave is drawn below. With a ruler, measure the wavelength and amplitude of the wave.

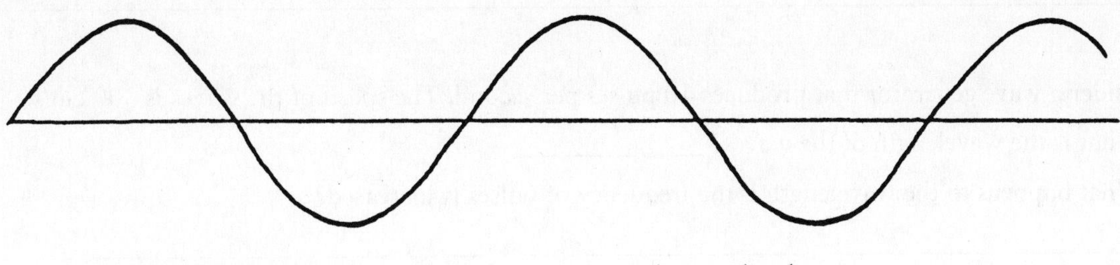

a. Wavelength = ____________ b. Amplitude = ____________

2. A girl on a playground swing makes a complete to-and-fro swing each 2 seconds. The frequency of swing is

(0.5 hertz) (1 hertz) (2 hertz)

and the period is

(0.5 second) (1 second) (2 seconds).

3. *Complete the following statements:*

THE PERIOD OF A 440-HERTZ SOUND WAVE IS ______ SECOND(S).

A MARINE WEATHER STATION REPORTS WAVES ALONG THE SHORE THAT ARE 8 SECONDS APART. THE FREQUENCY OF THE WAVES IS THEREFORE ______________ HERTZ.

4. The annoying sound from a mosquito occurs because it beats its wings at the average rate of 600 wingbeats per second.

a. What is the frequency of the soundwaves?

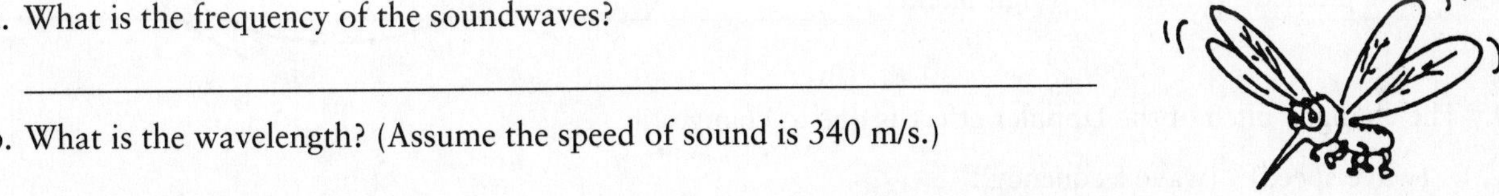

b. What is the wavelength? (Assume the speed of sound is 340 m/s.)

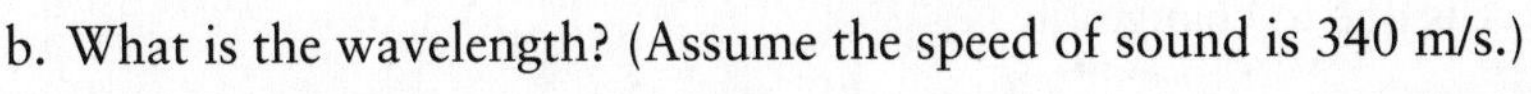

Name ______________________________ Date ____________

Vibration and Wave Fundamentals—continued

5. A machine gun fires 10 rounds per second. The speed of the bullets is 300 m/s.

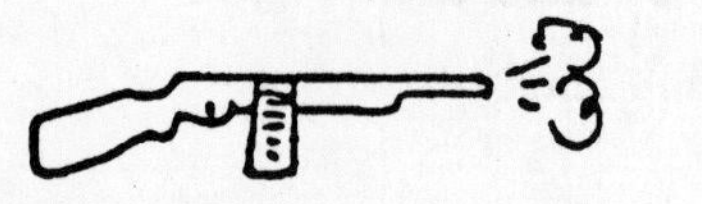

   a. What is the distance in the air between the flying bullets? ____________

   b. What happens to the distance between the bullets if the rate of fire is increased?

   ____

6. Consider a wave generator that produces 10 pulses per second. The speed of the waves is 300 cm/s.

   a. What is the wavelength of the waves? ____________

   b. What happens to the wavelength if the frequency of pulses is increased?

   ____

7. The bird at the right watches the waves. If the portion of a wave between 2 crests passes the pole each second, what is the speed of the wave?

   ____

   What is its period?

   ____

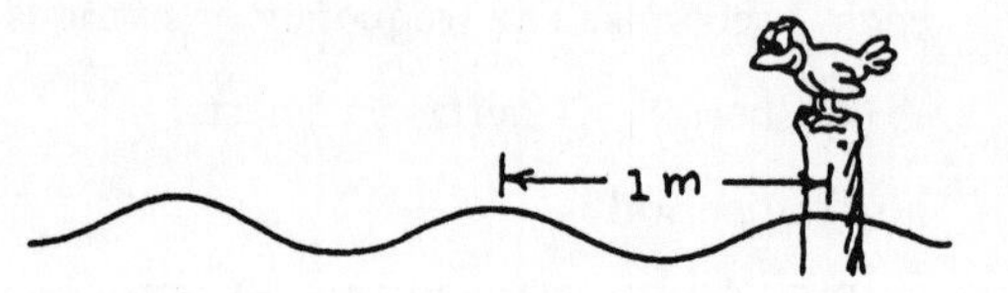

8. If the distance between crests in the above question were 1.5 meters apart, and 2 crests pass the pole each second, what would be the speed of the wave?

   ____

   What would be its period?

   ____

9. When an automobile moves toward a listener, the sound of its horn seems relatively

   (low pitched) (normal) (high pitched).

   When moving away from the listener, its horn seems

   (low pitched) (normal) (high pitched).

10. The changed pitch of the Doppler effect is due to changes in

    (wave speed) (wave frequency).

Name ______________________________ Date __________

Chapter 8: Waves—Sound and Light

Color

The sketch to the right shows the shadow of an instructor in front of a white screen in a dark room. The light source is red, so the screen looks red and the shadow looks black. Color the sketch, or label the colors with a pen or pencil.

A green lamp is added and makes a second shadow. The shadow cast by the red lamp is no longer black, but is illuminated by green light, so it is green. Color or mark it green. The shadow cast by the green lamp is not black because it is illuminated by the red lamp. Indicate its color. Do the same for the background, which receives a mixture of red and green light.

A blue lamp is added and three shadows appear. Indicate the appropriate colors of the shadows and the background.

The lamps are placed closer together so the shadows overlap. Indicate the colors of all screen areas.

Name ______________________________ Date __________

Color—continued

If you have colored pencils or markers, have a go at these.

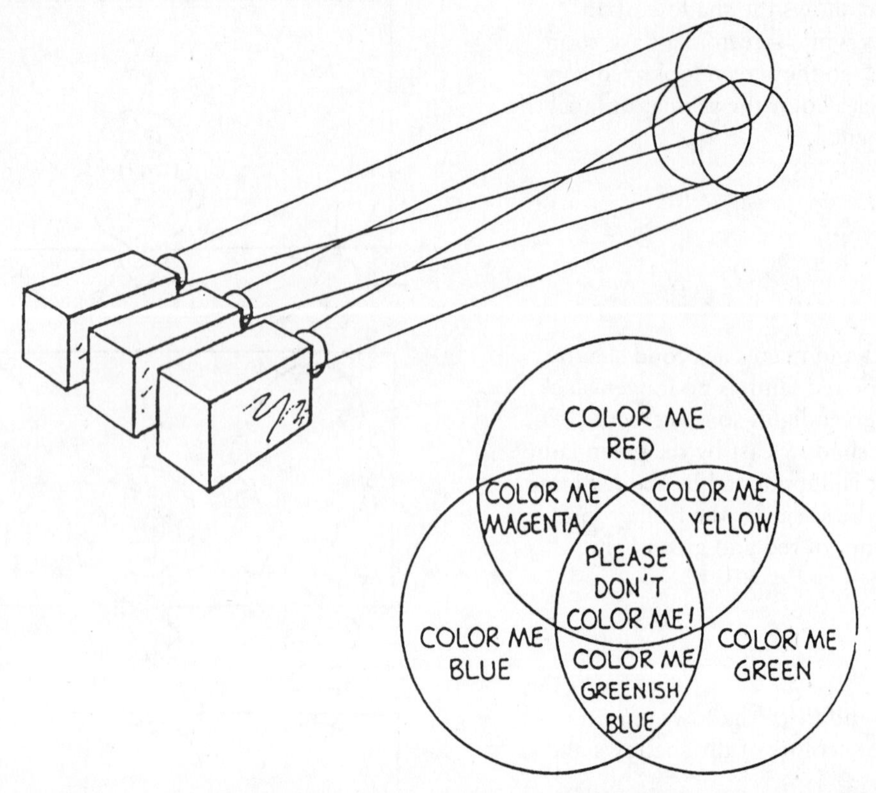

Name ______________________________ Date ____________

Chapter 8: Waves—Sound and Light

Diffraction and Interference

Shown below are concentric solid and dashed circles, each different in radius by 1 cm. Consider the circular pattern a top view of water waves, where the solid circles are crests and the dashed circles are troughs.

1. Draw another set of the same concentric circles with a compass. Choose any part of the paper for your center (except the present central point). Let the circles run off the edge of the paper.
2. Find where a dashed line crosses a solid line and draw a large dot at the intersection. Do this for ALL places where a solid and dashed line intersect.
3. With a wide felt marker, connect the dots with smooth lines. These *nodal lines* lie in regions where the waves have cancelled—where the crest of one wave overlaps the trough of another.

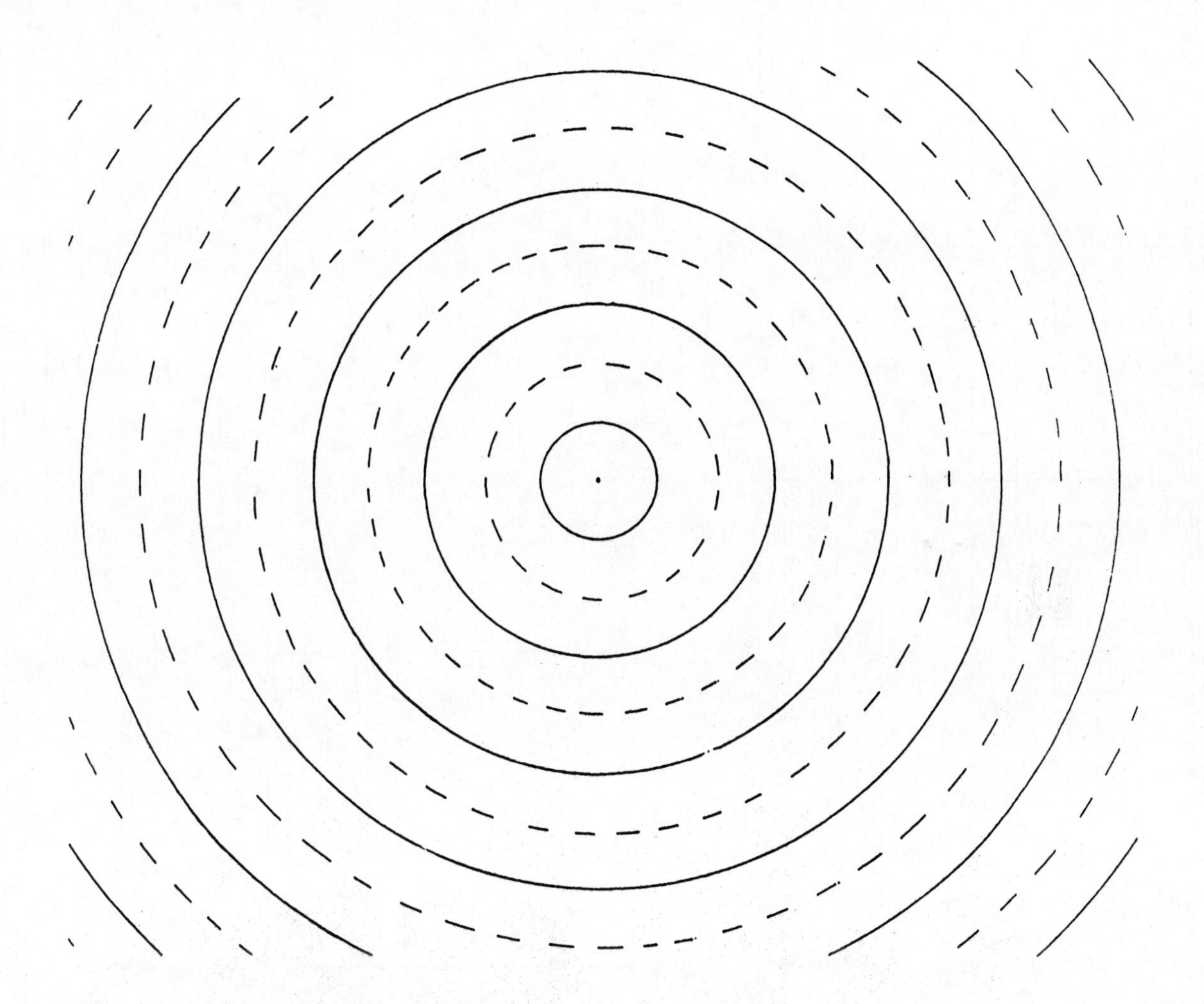

Name ______________________________ Date __________

Chapter 8: Waves—Sound and Light

Reflection

1. Light from a flashlight shines on a mirror and illuminates one of the cards. Draw the reflected beam to indicate the illuminated card.

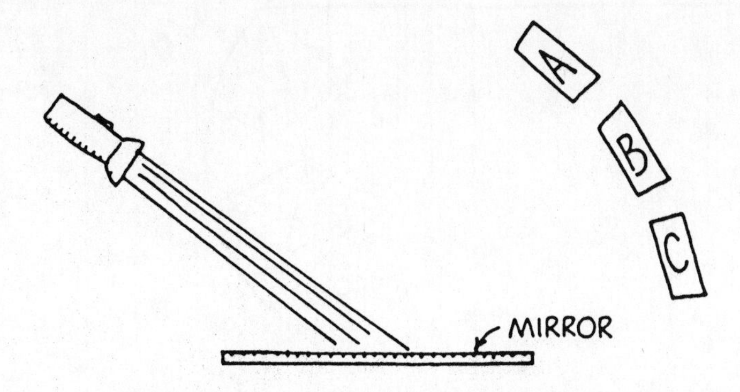

2. A periscope has a pair of mirrors in it. Draw the light path from the object "O" to the eye of the observer.

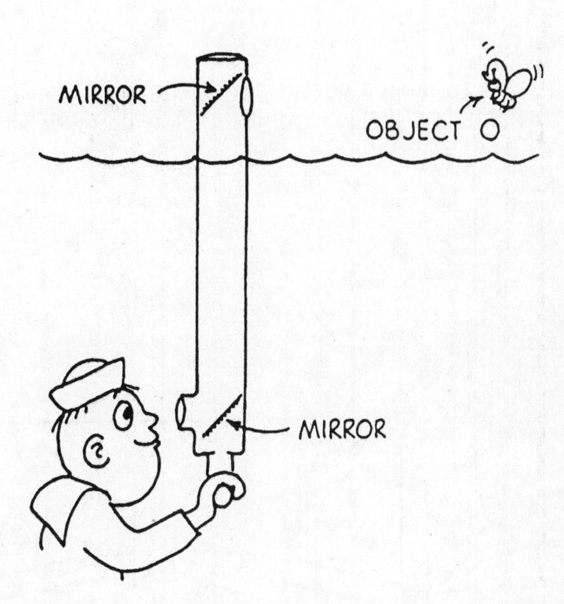

3. The ray diagram below shows the extension of one of the reflected rays from the plane mirror. Complete the diagram by (1) carefully drawing the three other reflected rays, and (2) extending them behind the mirror to locate the image of the flame. (Assume the candle and image are viewed by an observer on the left.)

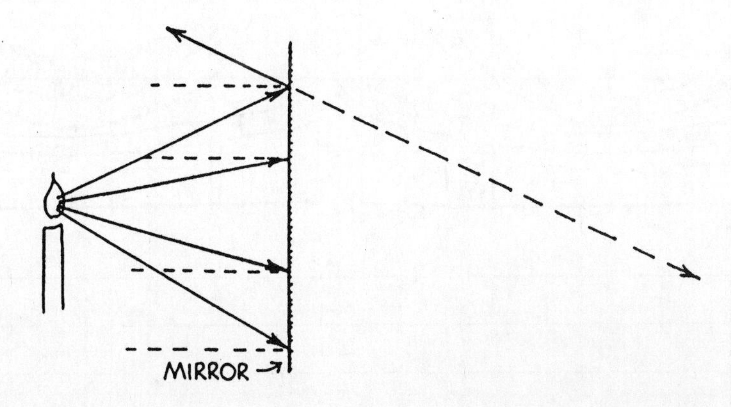

Name ______________________________ Date __________

Reflection—continued

4. The ray diagram below shows the reflection of one of the rays that strikes the parabolic mirror. Notice that the law of reflection is observed, and the angle of incidence (from the normal, the dashed line) equals the angle of reflection (from the normal). Complete the diagram by drawing the reflected rays of the other three rays that are shown. (Do you see why parabolic mirrors are used in automobile headlights?)

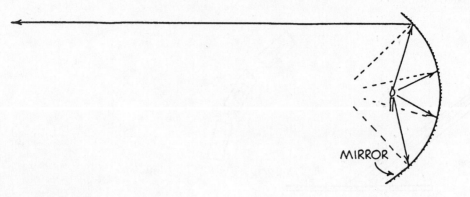

5. A girl takes a photograph of the bridge as shown. Which of the two sketches below correctly shows the reflected view of the bridge? Defend your answer.

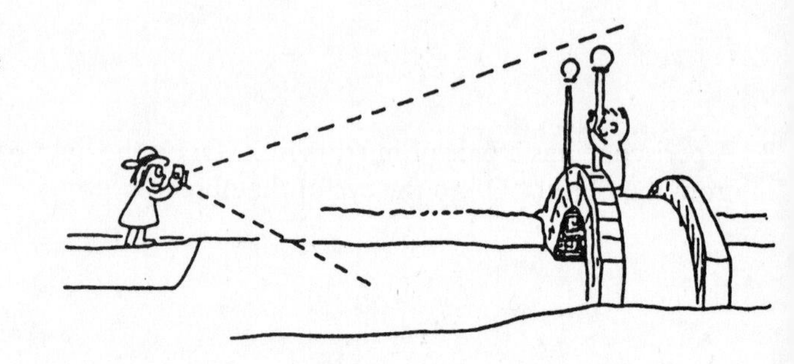

____

____

____

____

Name ______________________________ Date ____________

Chapter 8: Waves—Sound and Light

Refraction—Part 1

1. A pair of toy cart wheels are rolled obliquely from a smooth surface onto two plots of grass—a rectangular plot as shown at the left, and a triangular plot as shown at the right. The ground is on a slight incline so that after slowing down in the grass, the wheels speed up again when emerging on the smooth surface. Finish each sketch and show some positions of the wheels inside the plots and on the other side. Clearly indicate their paths and directions of travel.

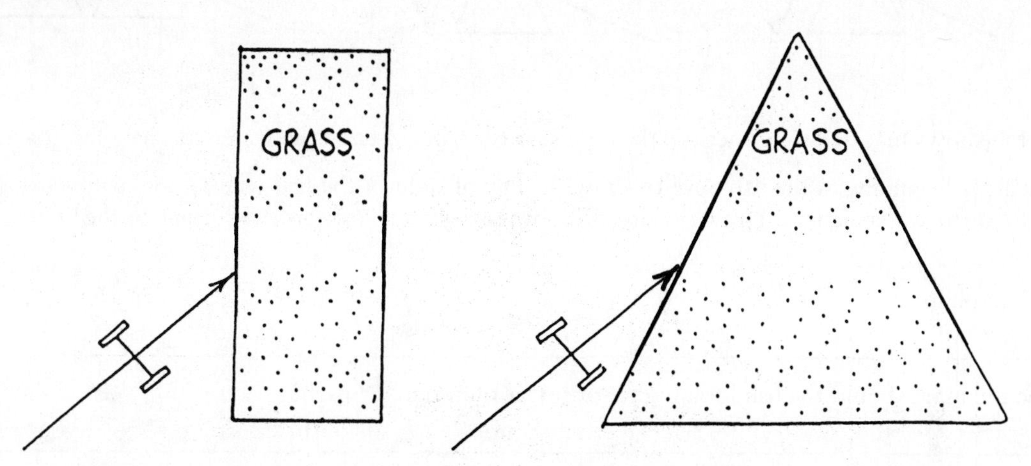

2. Red, green, and blue rays of light are incident upon a glass prism as shown. The average speed of red light in the glass is less than in air, so the red ray is refracted. When it emerges into the air it regains its original speed, and travels in the direction shown. Green light takes longer to get through the glass. Because of its slower speed, it is refracted as shown. Blue light travels even slower in glass. Complete the diagram by estimating the path of the blue ray.

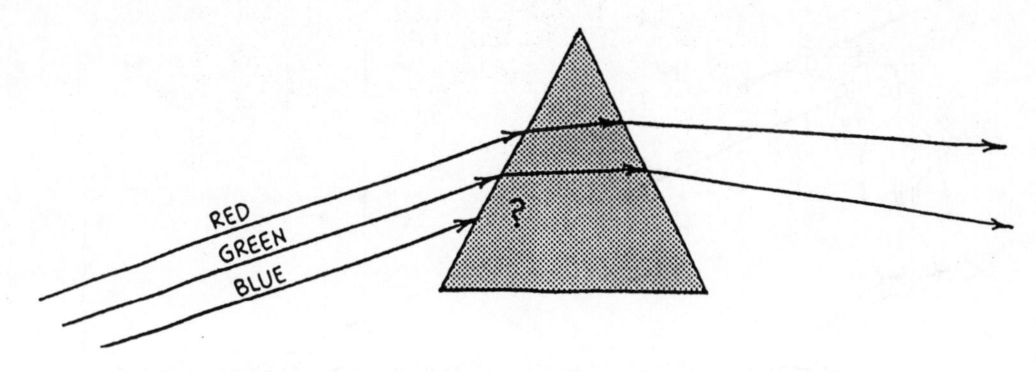

3. Below, we consider a prism-shaped hole in a piece of glass—that is, an "air prism." Complete the diagram; showing likely paths of the beams of red, green, and blue light as they pass through this "prism" and back to glass.

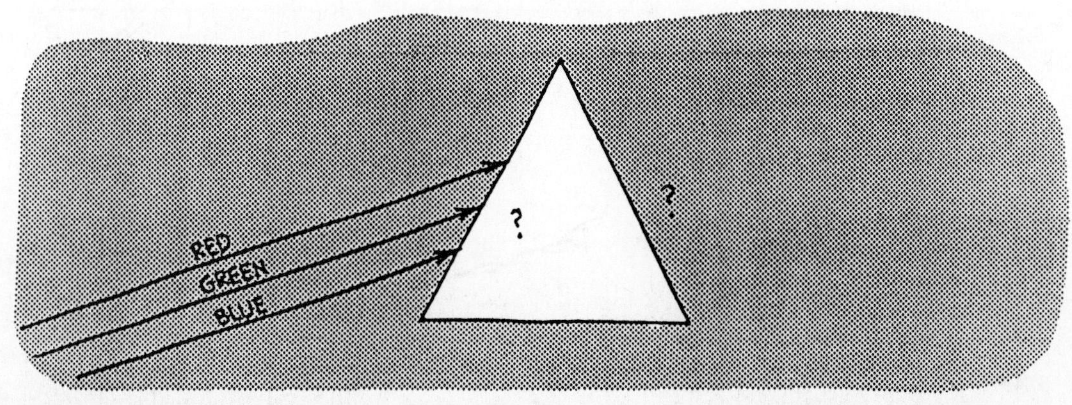

Name ____ Date __________

Refraction—Part 1—continued

4. Light of different colors diverges when emerging from a prism. Newton showed that with a second prism he could make the diverging beams become parallel again. Which placement of the second prism will do this?

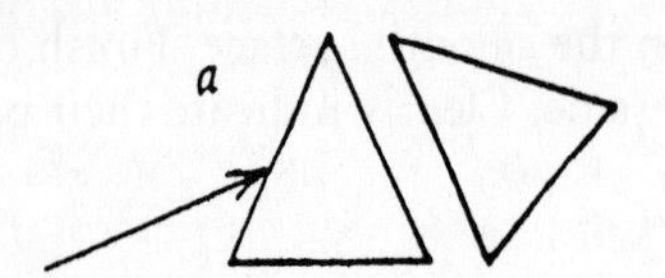

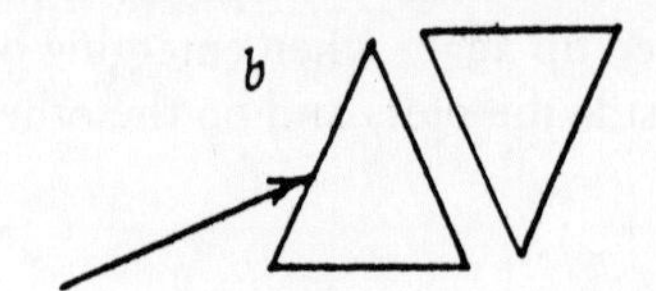

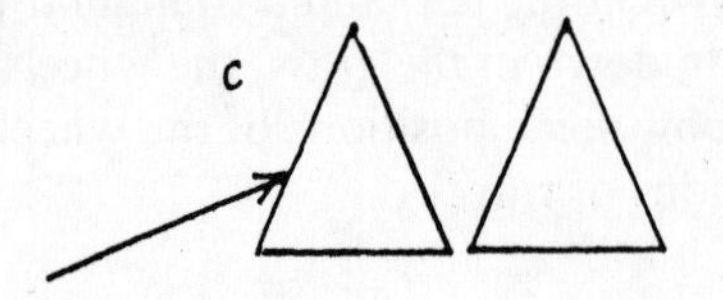

5. The sketch shows that due to refraction, the man sees the fish closer to the water surface than it actually is.

   a. Draw a ray beginning at the fish's eye to show the line of sight of the fish when it looks upward at 50° to the normal at the water surface. Draw the direction of the ray after it meets the surface of the water.

   b. At the 50° angle, does the fish see the man, or does it see the reflected view of the starfish at the bottom of the pond? Explain.

   ____

   c. To see the man, should the fish look higher or lower than the 50° path?

   ____

   d. If the fish's eye were barely above the water surface, it would see the world above in a 180° view, horizon to horizon. The fish's-eye view of the world above as seen beneath the water, however, is very different. Due to the 48° critical angle of water, the fish sees a normally 180° horizon-to-horizon view compressed within an angle of ______.

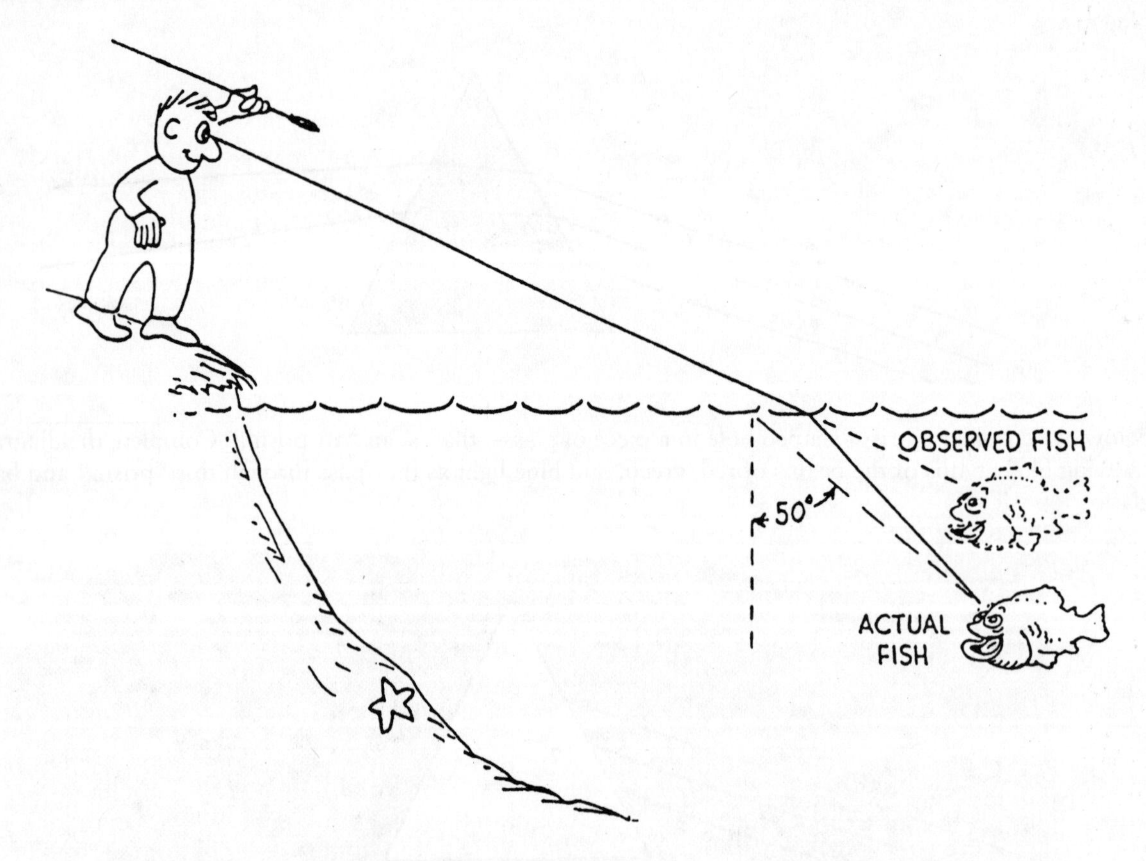

Name ______________________________ Date __________

Chapter 8: Waves—Sound and Light

Refraction—Part 2

1. The sketch to the right shows a light ray moving from air into water, at 45° to the normal. Which of the three rays indicated with capital letters is most likely the light ray that continues inside the water?

   ____

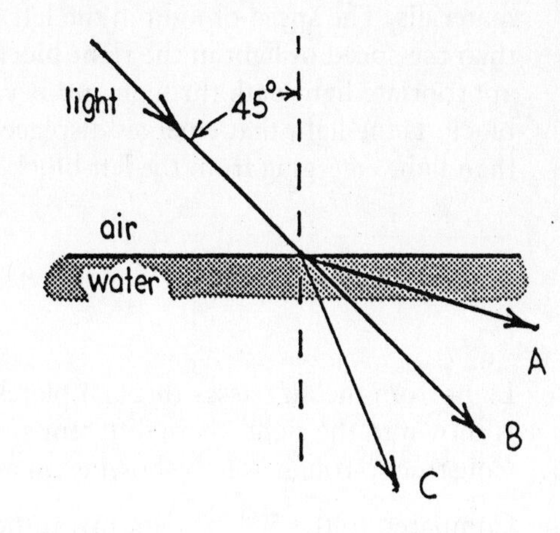

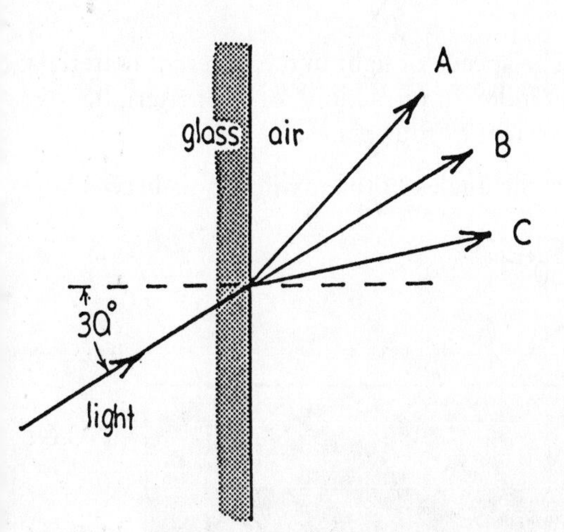

2. The sketch on the left shows a light ray moving from glass into air, at 30° to the normal. Which of the three is most likely the light ray that continues in the air?

   ____

3. To the right, a light ray is shown moving from air into a glass block, at 40° to the normal. Which of the three rays is most likely the light ray that travels in the air after emerging from the opposite side of the block?

   ____

   Sketch the path the light would take inside the glass.

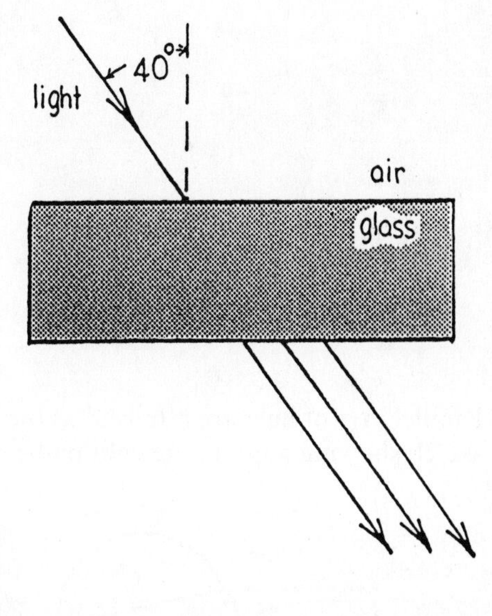

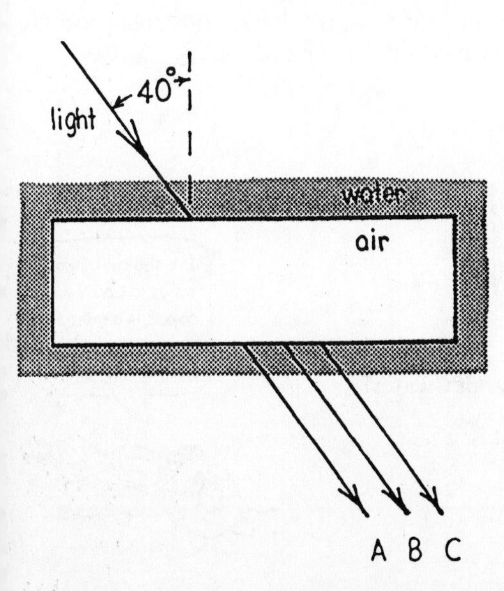

4. To the left, a light ray is shown moving from water into a rectangular block of air (inside a thin-walled plastic box), at 40° to the normal. Which of the three rays is most likely the light ray that continues into the water on the opposite side of the block?

   ____

   Sketch the path the light would take inside the air.

Name ______________________________ Date __________

Refraction—Part 2—continued

5. The two transparent blocks (right) are made of different materials. The speed of light in the left block is greater than the speed of light in the right block. Draw an appropriate light path through and beyond the right block. Is the light that emerges displaced more or less than light emerging from the left block?

____

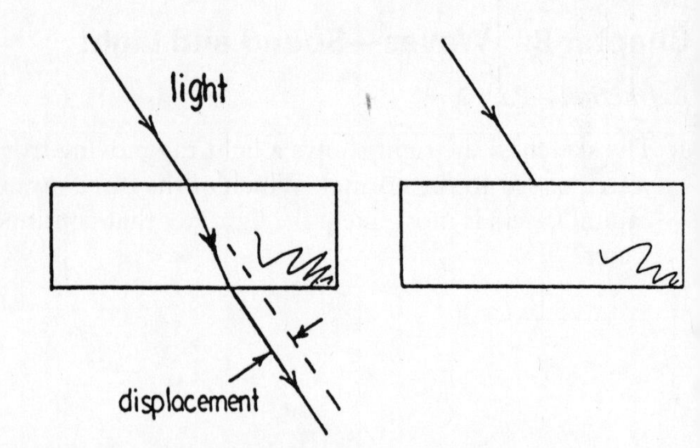

6. Light from the air passes through plates of glass and plastic below. The speeds of light in the different materials is shown to the right (these different speeds are often implied by the "index of refraction" of the material). Construct a rough sketch showing an appropriate path through the system of four plates.

   Compared to the 50° incident ray at the top, what can you say about the angles of the ray in the air between and below the block pairs?

   ____

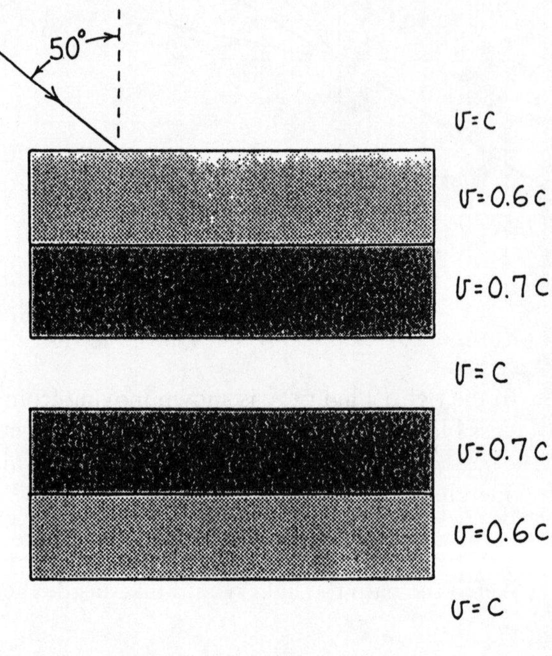

7. Parallel rays of light are refracted as they change speed in passing from air into the eye (left). Construct a rough sketch showing appropriate light paths when parallel light under water meets the same eye (right).

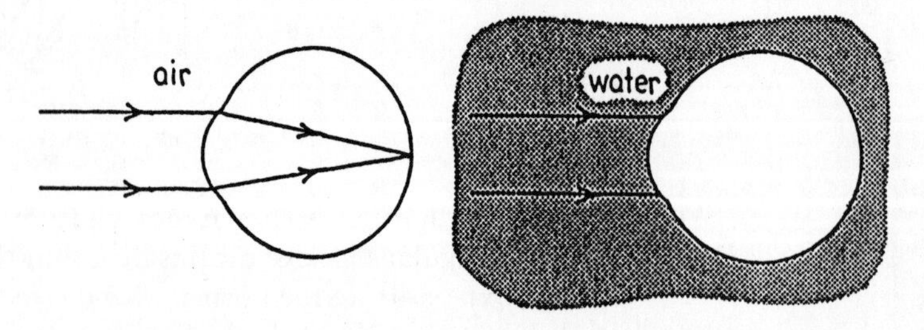

8. Why do we need to wear a face mask or goggles to see clearly when under water?

____

Name ______________________________ Date __________

Chapter 8: Waves—Sound and Light

Wave-Particle Duality

1. To say that light is quantized means that light is made up of

   (elemental units) (waves).

2. Compared to photons of low-frequency light, photons of higher-frequency light have more

   (energy) (speed) (quanta).

3. The photoelectric effect supports the

   (wave model of light) (particle model of light).

4. The photoelectric effect is evident when light shone on certain photosensitive materials ejects

   (photons) (electrons).

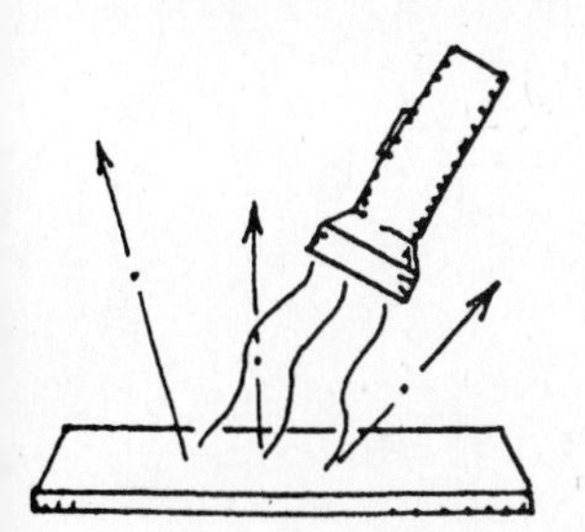

5. The photoelectric effect is more effective with violet light than with red light because the photons of violet light

   (resonate with the atoms in the material)

   (deliver more energy to the material)

   (are more numerous).

6. According to De Broglie's wave model of matter, a beam of light and a beam of electrons

   (are fundamentally different) (are similar).

7. According to De Broglie, the greater the speed of an electron beam, the

   (greater is its wavelength) (shorter is its wavelength).

8. The discreteness of the energy levels of electrons about the atomic nucleus is best understood by considering the electron to be a

   (wave) (particle).

9. Heavier atoms are not appreciably larger in size than lighter atoms. The main reason for this is the greater nuclear charge

   (pulls surrounding electrons into tighter orbits)

   (holds more electrons about the atomic nucleus)

   (produces a denser atomic structure).

10. Whereas in the everyday macroworld the study of motion is called *mechanics,* in the microworld the study of quanta is called

    (Newton mechanics) (quantum mechanics).

A QUANTUM MECHANIC!

Name ____________________ Date __________

Chapter 9: The Atom

Subatomic Particles

Three fundamental particles of the atom are the __________, __________, and __________. At the center of each atom lies the atomic __________, which consists of __________ and __________. The **atomic number** refers to the number of __________ in the nucleus. All atoms of the same element have the same number of __________, hence, the same atomic number.

Isotopes are atoms that have the same number of __________, but a different number of __________. An isotope is identified by its **atomic mass number,** which is the total number of __________ and __________ in the nucleus. A carbon isotope that has 6 __________ and 6 __________ is identified as carbon-12, where 12 is the atomic mass number. A carbon isotope having 6 __________ and 8 __________, on the other hand, is carbon-14.

1. Complete the following table:

	Number of...		
Isotope	Electrons	Protons	Neutrons
Hydrogen-1	1		
Chlorine-36		17	
Nitrogen-14			7
Potassium-40	19		
Arsenic-75		33	
Gold-197			118

2. Which results in a more valuable product—*adding* or *subtracting* protons from gold nuclei?

3. Which has more mass, a helium atom or a neon atom?

4. Which has a greater number of atoms, a gram of helium or a gram of neon?

Name ______________________________ Date ____________

Chapter 10: Nuclear Physics

Radioactivity

1. Complete the following statements.

   a. A lone neutron spontaneously decays into a proton plus an ____________________.

   b. Alpha and beta rays are made of streams of particles, whereas gamma rays are streams of ____________________.

   c. An electrically charged atom is called an ____________________.

   d. Different ____________________ of an element are chemically identical but differ in the number of neutrons in the nucleus.

   e. Transuranic elements are those beyond atomic number ____________________.

   f. If the amount of a certain radioactive sample decreases by half in four weeks, in four more weeks the amount remaining should be ____________________ the original amount.

   g. Water from a natural hot spring is warmed by ____________________ inside Earth.

2. The gas in the little girl's balloon is made up of former alpha and beta particles produced by radioactive decay.

   a. If the mixture is electrically neutral, how many more beta particles than alpha particles are in the balloon?

   ____

   ____

   b. Why is your answer not "same"?

   ____

   ____

   c. Why are the alpha and beta particles no longer harmful to the child?

   ____

   ____

   d. What element does this mixture make?

   ____

Name ______________________________ Date __________

Radioactivity—continued

Draw in a decay-scheme diagram below, similar to Figure 10.15 in your text. In this case, you begin at the upper right with U-235 and end up with a different isotope of lead. Use the table at the left and identify each element in the series by its chemical symbol.

Step	Particle Emitted
1	Alpha
2	Beta
3	Alpha
4	Alpha
5	Beta
6	Alpha
7	Alpha
8	Alpha
9	Beta
10	Alpha
11	Beta
12	Stable

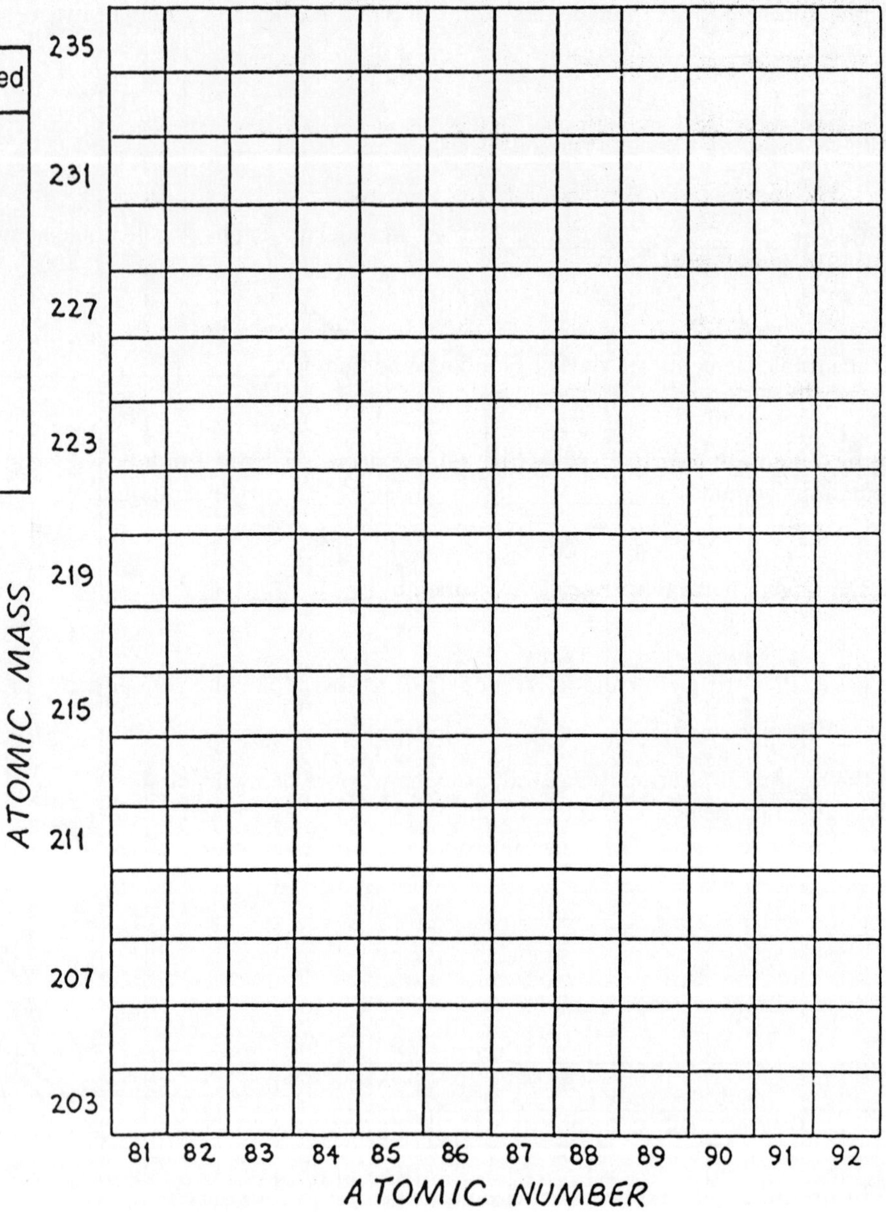

ATOMIC MASS: 235, 231, 227, 223, 219, 215, 211, 207, 203

ATOMIC NUMBER: 81, 82, 83, 84, 85, 86, 87, 88, 89, 90, 91, 92

What isotope is the final product? ______________

Name ____ Date ____________

Chapter 10: Nuclear Physics

Radioactive Half-Life

You and your classmates will now play the "half-life game." Each of you should have a coin to shake inside cupped hands. After it has been shaken for a few seconds, the coin is tossed on the table or on the floor. Students with tails up fall out of the game. Only those who consistently show heads remain in the game. Finally, everybody has tossed a tail and the game is over.

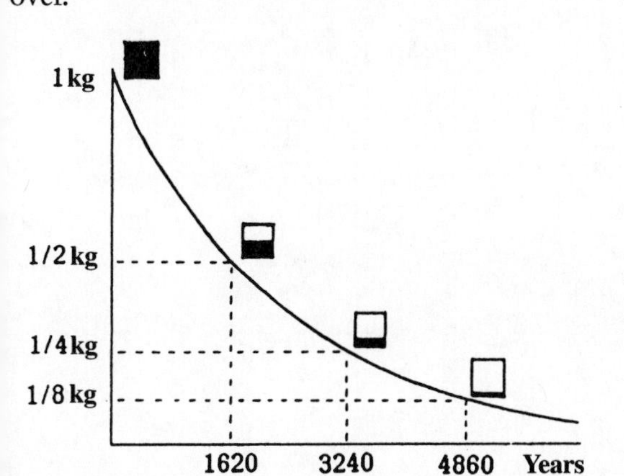

1. The graph to the left shows the decay of Radium-226 with time. Note that each 1620 years, half remains (the rest changes to other elements). In the grid below, plot the number of students left in the game after each toss. Draw a smooth curve that passes close to the points on your plot. What is the similarity of your curve with that of the curve of Radium-226?

____

____

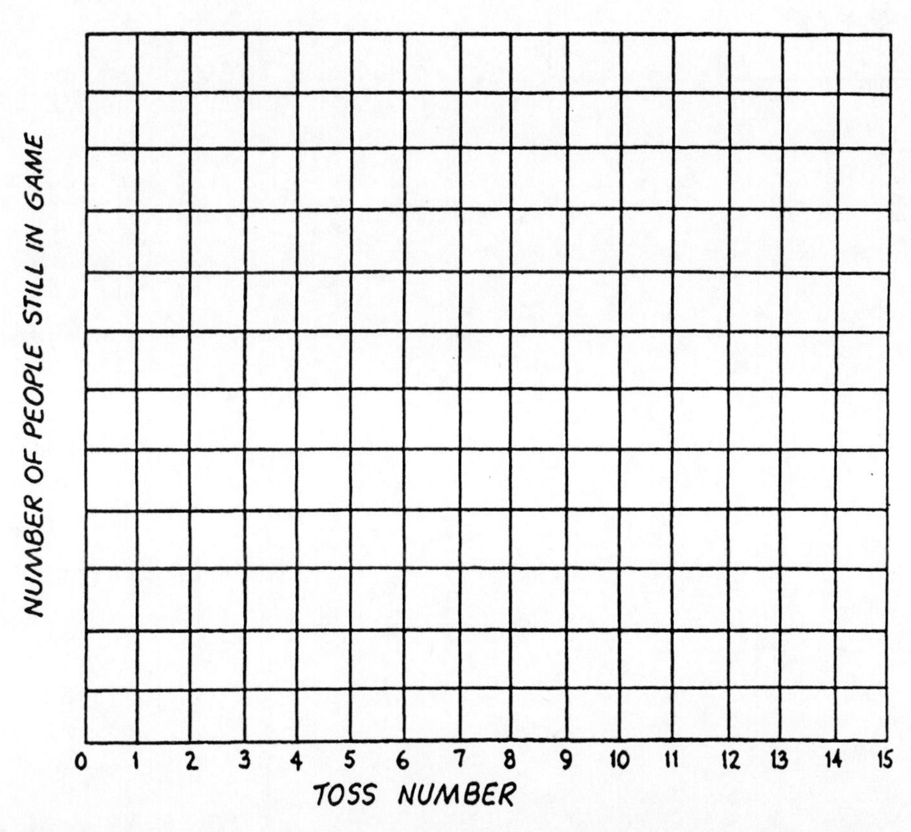

2. Was the person to last longest in the game *lucky*, with some sort of special powers to guide the long survival? What test could you make to decide the answer to this question?

____

Chapter 10: Nuclear Physics

Radioactive Half-Life

Name ______________________________ Date ____________

Chapter 10: Nuclear Physics

Nuclear Fission and Fusion

1. Complete the table for a chain reaction in which two neutrons from each step individually cause a new reaction.

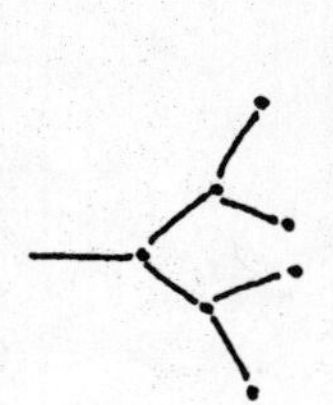

EVENT	1	2	3	4	5	6	7
NO. OF REACTIONS	1	2	4				

2. Complete the table for a chain reaction where three neutrons from each reaction cause a new reaction.

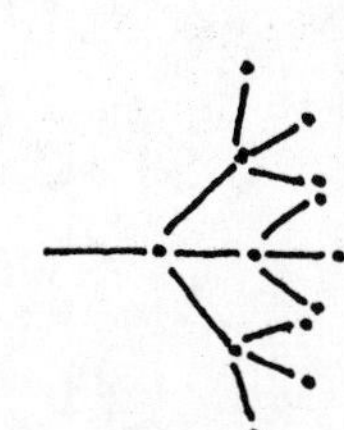

EVENT	1	2	3	4	5	6	7
NO. OF REACTIONS	1	3	9				

3. Complete these beta reactions, which occur in a fission breeder reactor.

$$^{239}_{92}U \rightarrow \_\_\_\_ + ^{0}_{-1}e$$

$$^{239}_{93}Np \rightarrow \_\_\_\_ + ^{0}_{-1}e$$

4. Complete the following fission reactions.

$$^{1}_{0}n + ^{235}_{92}U \rightarrow ^{143}_{54}Xe + ^{90}_{38}Sr + \_\_\_(^{1}_{0}n)$$

$$^{1}_{0}n + ^{235}_{92}U \rightarrow ^{152}_{60}Nd + \_\_\_\_ + 4(^{1}_{0}n)$$

$$^{1}_{0}n + ^{239}_{94}Pu \rightarrow \_\_\_\_ + ^{97}_{40}Zr + 2(^{1}_{0}n)$$

5. Complete the following fusion reactions.

$$^{2}_{1}H + ^{2}_{1}H \rightarrow ^{3}_{2}He + \_\_\_\_$$

$$^{2}_{1}H + ^{3}_{1}H \rightarrow ^{4}_{2}He + \_\_\_\_$$

Name ______________________________ Date ____________

Chapter 10: Nuclear Physics

Nuclear Reactions

Complete these nuclear reactions:

1. $^{230}_{90}Th \longrightarrow ^{226}_{88}Ra + \_\_\_\_$

2. $^{218}_{85}At \longrightarrow \_\_\_\_ + ^{4}_{2}He$

3. $^{14}_{6}C \longrightarrow \_\_\_\_ + ^{14}_{7}N$

4. $^{80}_{35}Br \longrightarrow ^{80}_{36}Kr + \_\_\_\_$

5. $^{214}_{83}Bi \longrightarrow ^{4}_{2}He + \_\_\_\_$

6. $^{212}_{83}Bi \longrightarrow ^{0}_{-1}e + \_\_\_\_$

7. $^{80}_{35}Br \longrightarrow ^{0}_{-1}e + \_\_\_\_$

8. $^{80}_{35}Br \longrightarrow ^{0}_{+1}e + \_\_\_\_$

9. $^{1}_{1}H + ^{7}_{3}Li \longrightarrow ^{4}_{2}He + \_\_\_\_$

10. $^{2}_{1}H + ^{3}_{1}H \longrightarrow ^{4}_{2}He + \_\_\_\_$

NUCLEAR PHYSICS··· IT'S THE SAME TO ME WITH THE FIRST TWO LETTERS INTERCHANGED!

Name ____ Date ______________

Chapter 11: Investigating Matter

Melting Points of the Elements

There is a remarkable degree of organization in the periodic table. As discussed in your textbook, elements within the same atomic group (vertical column) share similar properties. Also, the chemical reactivity of an element can be deduced from its position in the periodic table. Two additional examples of the periodic table's organization are the melting points and densities of the elements.

The periodic table below shows the melting points of nearly all the elements. Note the melting points are not randomly oriented, but, with only a few exceptions, either gradually increase or decrease as you move in any particular direction. This can be clearly illustrated by color coding each element according to its melting point.

Use colored pencils to color in each element according to its melting point. Use the suggested color legend. Color lightly so that symbols and numbers are still visible.

Color	Temperature Range, °C	Color	Temperature Range, °C
Violet	-273 — -50	Yellow	1400 — 1900
Blue	-50 — 300	Orange	1900 — 2900
Cyan	300 — 700	Red	2900 — 3500
Green	700 — 1400		

Melting Points of the Elements (°C)

1	2	3	4	5	6	7	8	9	10	11	12	13	14	15	16	17	18
H -259																	He -272
Li 180	Be 1278											B 2079	C 3550	N -210	O -218	F -219	Ne -248
Na 97	Mg 648											Al 660	Si 1410	P 44	S 113	Cl -100	Ar -189
K 63	Ca 839	Sc 1541	Ti 1660	V 1890	Cr 1857	Mn 1244	Fe 1535	Co 1495	Ni 1453	Cu 1083	Zn 419	Ga 30	Ge 937	As 817	Se 217	Br -7	Kr -156
Rb 39	Sr 769	Y 1522	Zr 1852	Nb 2468	Mo 2617	Tc 2172	Ru 2310	Rh 1966	Pd 1554	Ag 961	Cd 320	In 156	Sn 231	Sb 630	Te 449	I 113	Xe -111
Cs 28	Ba 725	La 921	Hf 2227	Ta 2996	W 3410	Re 3180	Os 3045	Ir 2410	Pt 1772	Au 1064	Hg -38	Tl 303	Pb 327.	Bi 271	Po 254	At 302	Rn -71
Fr 27	Ra 700	Ac 1050	--	--	--	--	--	--									

Lanthanides:	Ce 799	Pr 931	Nd 1021	Pm 1168	Sm 1077	Eu 822	Gd 1313	Tb 1356	Dy 1412	Ho 1474	Er 1159	Tm 1545	Yb 819	Lu 1663
Actinides:	Th 1750	Pa 1600	U 1132	Np 640	Pu 641	Am 994	Cm 1340	Bk --	Cf --	Es --	Fm --	Md --	No --	Lr --

1. Which elements have the highest melting points?

____

2. Which elements have the lowest melting points?

____

3. Which atomic groups tend to go from higher to lower melting points reading from top to bottom? (Identify each group by its group number.)

____

4. Which atomic groups tend to go from lower to higher melting points reading from top to bottom?

____

Name ______________________________ Date ____________

Chapter 11: Investigating Matter

Densities of the Elements

The periodic table below shows the densities of nearly all the elements. As with the melting points, the densities of the elements either gradually increase or decrease as you move in any particular direction. Use colored pencils to color in each element according to its density. Shown below is a suggested color legend. Color lightly so that symbols and numbers are still visible. (Note: All gaseous elements are marked with an asterisk and should be the same color. Their densities, which are given in units of g/L, are much less than the densities nongaseous elements, which are given in units of g/mL.)

Color	Density (g/mL)	Color	Density (g/mL)
Violet	gaseous elements	Yellow	16 — 12
Blue	5 — 0	Orange	20 — 16
Cyan	8 — 5	Red.	23 — 20
Green	12 — 8		

Densities of the Elements

(g/mL)

1	2	3	4	5	6	7	8	9	10	11	12	13	14	15	16	17	18
H * 0.09																	He * 0.18
Li 0.5	Be 1.8											B 2.3	C 2.0	N * 1.25	O * 1.43	F * 1.70	Ne * 0.90
Na 1.0	Mg 1.7											Al 2.7	Si 2.3	P 1.8	S 2.1	Cl * 3.21	Ar * 1.78
K 0.9	Ca 1.6	Sc 3.0	Ti 4.5	V 6.1	Cr 7.2	Mn 7.3	Fe 7.8	Co 8.9	Ni 8.9	Cu 9.0	Zn 7.1	Ga 6.1	Ge 5.3	As 5.7	Se 4.8	Br * 7.59	Kr * 3.73
Rb 1.5	Sr 2.5	Y 4.5	Zr 6.5	Nb 8.5	Mo 6.8	Tc 11.5	Ru 12.4	Rh 12.4	Pd 12.0	Ag 10.5	Cd 8.7	In 7.3	Sn 5.7	Sb 6.7	Te 6.2	I 4.9	Xe * 5.89
Cs 1.9	Ba 3.5	La 6.2	Hf 13.3	Ta 16.6	W 19.3	Re 21.0	Os 22.6	Ir 22.4	Pt 21.5	Au 18.9	Hg 13.5	Tl 11.9	Pb 11.4	Bi 9.7	Po 9.3	At --	Rn * 9.73
Fr --	Ra 5.0	Ac 10.1	Unq --	Unp --	Unh --	Uns --	Uno --	Une --									

* density of gaseous phase in g/L

Lanthanides:	Ce 6.7	Pr 6.7	Nd 6.8	Pm 7.2	Sm 7.5	Eu 5.2	Gd 7.9	Tb 8.2	Dy 8.6	Ho 8.8	Er 9.1	Tm 9.3	Yb 6.9	Lu 9.8
Actinides:	Th 11.7	Pa 15.4	U 19.0	Np 20.1	Pu 19.8	Am 13.7	Cm 13.5	Bk 14	Cf --	Es --	Fm --	Md --	No --	Lr --

1. Which elements are the most dense?

____

2. How variable are the densities of the lanthanides compared to the densities of the actinides?

____

3. Which atomic groups tend to go from higher to lower densities reading from top to bottom? (Identify each group by its group number).

____

4. Which atomic groups tend to go from lower to higher densities reading from top to bottom?

____

Name ______________________________ Date ____________

Chapter 11: Investigating Matter

The Submicroscopic

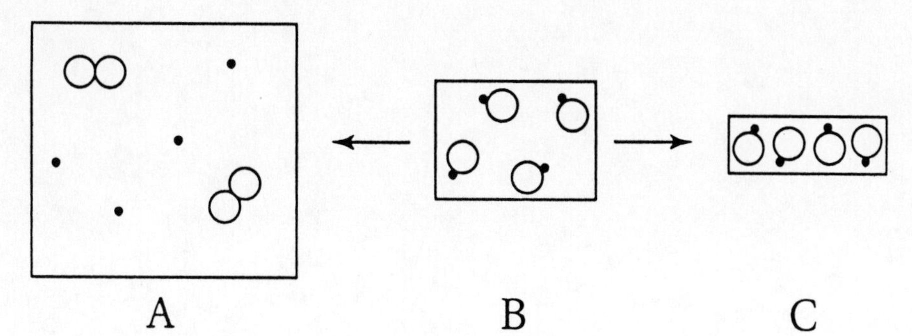

1. How many molecules are shown in A _____ B _____ C _____
2. How many atoms are shown in A _____ B _____ C _____
3. Which represents a physical change? B ⟶ A B ⟶ C (*circle one*)
4. Which represents a chemical change? B ⟶ A B ⟶ C (*circle one*)
5. Which box(es) represent(s) a mixture? A _____ B _____ C _____
6. Which box contains the most mass? A _____ B _____ C _____
7. Which box is coldest? A _____ B _____ C _____
8. Which box contains the most air between molecules? A _____ B _____ C _____

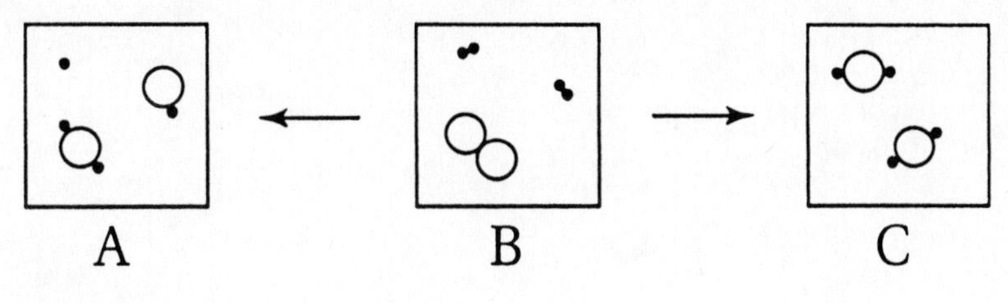

9. How many molecules are shown in A _____ B _____ C _____
10. How many atoms are shown in A _____ B _____ C _____
11. Which represents a physical change? B ⟶ A B ⟶ C (*circle one*)
12. Which represents a chemical change? B ⟶ A B ⟶ C (*circle one*)
13. Which box(es) represent(s) a mixture? A _____ B _____ C _____
14. Which box contains the most mass? A _____ B _____ C _____
15. Which should take longer? B ⟶ A B ⟶ C (*circle one*)
16. Which box most likely contains ions? A _____ B _____ C _____

Name ______________________________ Date ____________

Chapter 11: Investigating Matter

Physical and Chemical Changes

1. What distinguishes a chemical change from a physical change?

2. Based upon observations alone, why is distinguishing a chemical change from a physical change not always so straight-forward?

Try your hand at categorizing the following processes as either chemical or physical changes. Some of these examples are debatable! Be sure to discuss your reasoning with fellow classmates or your instructor.

	(*circle one*)	
3. A cloud grows dark. ______________	chemical	physical
4. Leaves produce oxygen. ______________	chemical	physical
5. Food coloring is added to water. ______________	chemical	physical
6. Tropical coral reef dies. ______________	chemical	physical
7. Dead coral reef is pounded by waves into beach sand. ______________	chemical	physical
8. Oil and vinegar separate. ______________	chemical	physical
9. Soda drink goes flat. ______________	chemical	physical
10. Sick person develops a fever. ______________	chemical	physical
11. Compost pit turns into mulch. ______________	chemical	physical
12. A computer is turned on. ______________	chemical	physical
13. An electrical short melts a computer's integrated circuits. ______________	chemical	physical
14. A car battery runs down. ______________	chemical	physical
15. A pencil is sharpened. ______________	chemical	physical
16. Mascara is applied to eyelashes. ______________	chemical	physical
17. Sunbather gets tan lying in the sun. ______________	chemical	physical
18. Invisible ink turns visible upon heating. ______________	chemical	physical
19. A light bulb burns out. ______________	chemical	physical
20. Car engine consumes a tank of gasoline. ______________	chemical	physical
21. B vitamins turn urine yellow. ______________	chemical	physical

Name ______________________________ Date ___________

Chapter 12: The Nature of Chemical Bonds

Losing Valence Electrons

The shell model described in Section 12.1 can be used to explain a wide variety of properties of atoms. Using the shell model, for example, we can explain how atoms within the same group tend to lose (or gain) the same number of electrons. Let's consider the case of three group 1 elements: lithium, sodium, and potassium. Look to a periodic table and find the nuclear charge of each of these atoms:

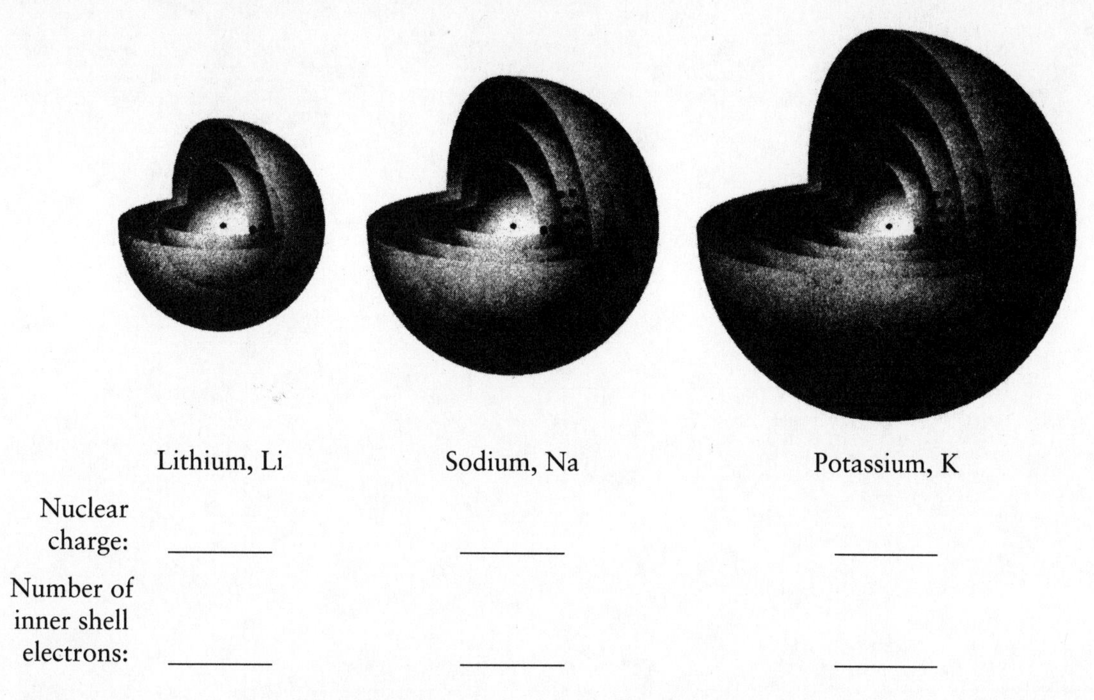

	Lithium, Li	Sodium, Na	Potassium, K
Nuclear charge:	______	______	______
Number of inner shell electrons:	______	______	______

How strongly the valence electron is held to the nucleus depends on the strength of the nuclear charge—the stronger the charge, the stronger the valence electron is held. There's more to it, however, because inner-shell electrons weaken the attraction outer-shell electrons have for the nucleus. The valence shell in lithium, for example, doesn't experience the full effect of three protons. Instead, it experiences a diminished nuclear charge of about +1. We get this by subtracting the number of inner-shell electrons from the actual nuclear charge. What do the valence electrons for sodium and potassium experience?

Diminished nuclear charge:	______	______	______

Question: Potassium has a nuclear charge many times greater than that of lithium. Why is it actually *easier* for a potassium atom to lose its valence electron than it is for a lithium atom to lose its valence electron?

Hint: Remember from Chapter 7 what happens to the electric force as distance is increased!

Name ______________________________ Date __________

Chapter 12: The Nature of Chemical Bonds

Drawing Shells

Atomic shells can be represented by a series of concentric circles as shown in your textbook. With a little effort, however, it's possible to show these shells in three dimensions. Grab a pencil and blank sheet of paper and follow the steps shown below. Practice makes perfect.

1. Lightly draw a diagonal guideline. Then, draw a series of seven semicircles. Note how the ends of the semicircles are not perpendicular to the guideline. Instead, they are parallel to the length of the page, as shown in Figure 1.

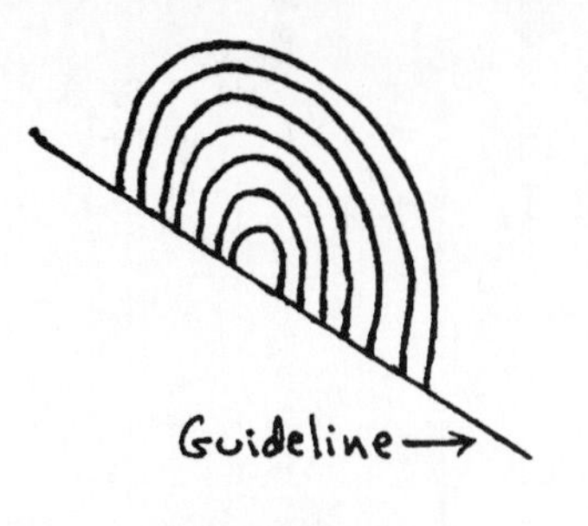

Figure 1

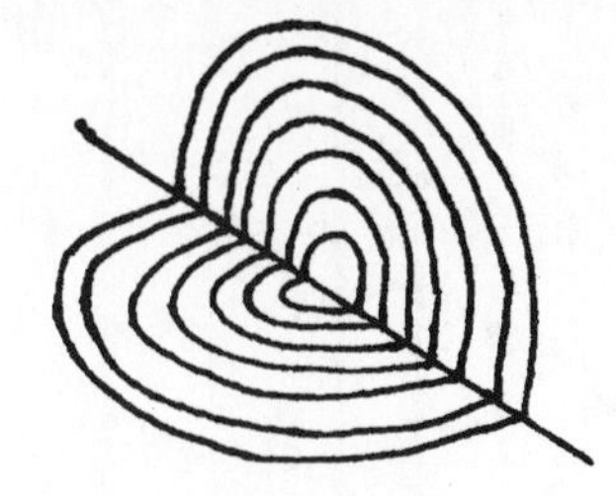

Figure 2

2. Connect the ends of each semicircle with another semicircle such that a series of concentric hearts is drawn. The ends of these new semicircles should be drawn perpendicular to the ends of the previously drawn semicircles, as shown in Figure 2.
3. Now the hard part. Draw a portion of a circle that connects the apex of the largest vertical and horizontal semicircles, as in Figure 3.

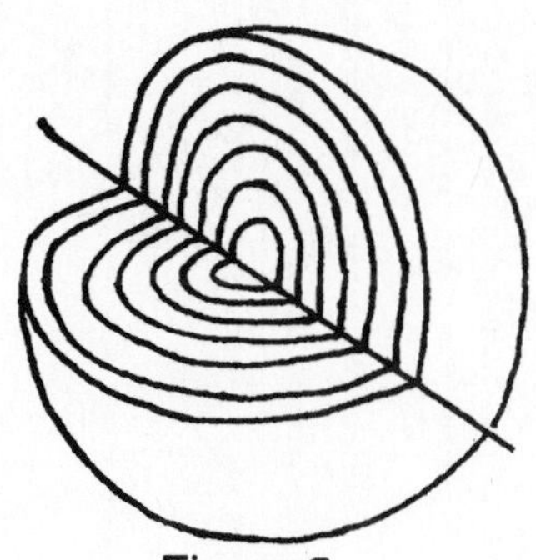

Figure 3

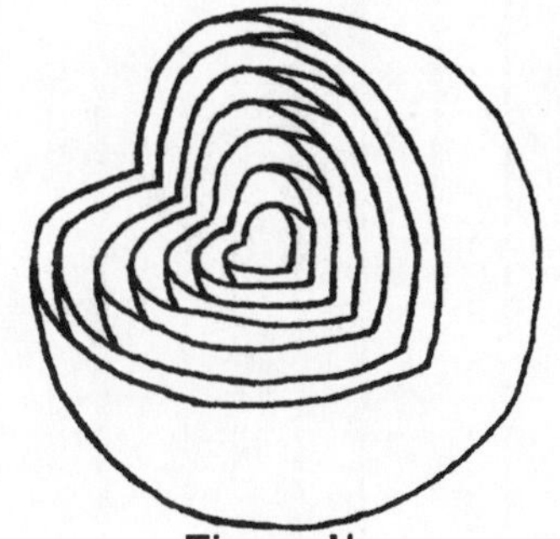

Figure 4

4. Now the fun part. Erase the pencil guideline drawn in Step 1, then add the internal lines, as shown in Figure 4, that create a series of concentric shells.

You need not draw all the shells for each atom. Oxygen, for example, is nicely represented drawing only the first two inner shells, which are the only ones that contain electrons. Remember that these shells are not to be taken literally. Rather, they are a highly simplified view of how electrons tend to organize themselves with an atom. You should know that each shell represents a set of atomic orbitals of similar energy levels as shown in your textbook.

Name ____ Date ____________

Chapter 12: The Nature of Chemical Bonds

Atomic Size

1. Complete the shells for the following atoms using arrows to represent electrons.

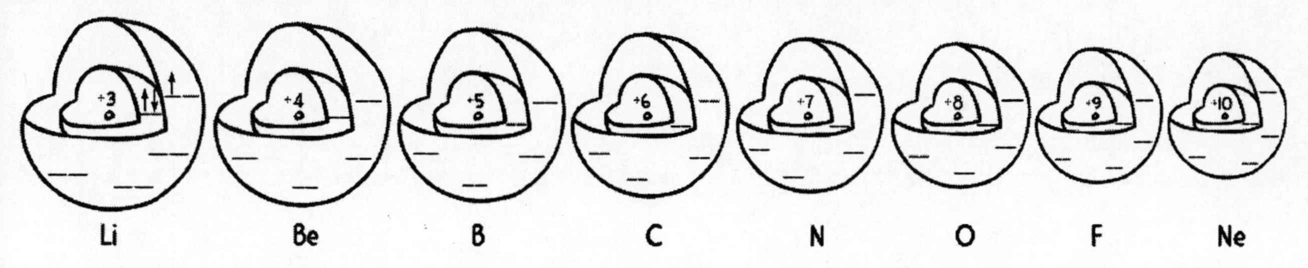

2. Neon, Ne, has many more electrons than lithium, Li, yet it is a much smaller atom. Why?

3. Draw the shell model for a sodium atom, Na (atomic number 11), adjacent to the neon atom in the box shown below. Use a pencil because you may need to erase.

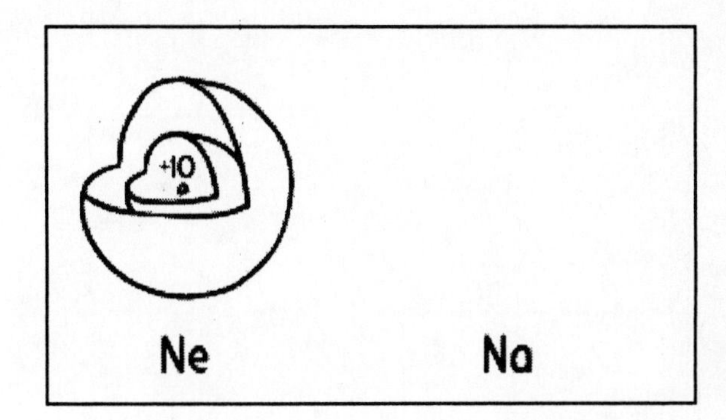

   a. Which should be larger: neon's first shell or sodium's first shell. Why? Did you represent this accurately within your drawing?

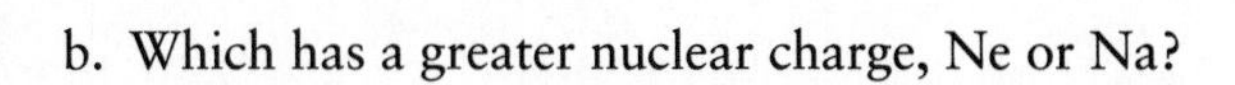

   b. Which has a greater nuclear charge, Ne or Na?

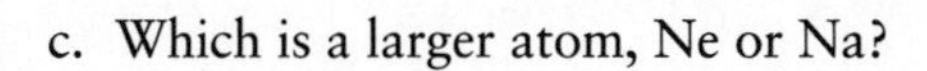

   c. Which is a larger atom, Ne or Na?

4. Moving from left to right across the periodic table, what happens to the nuclear charge within atoms? What happens to atomic size?

5. Moving from top to bottom down the periodic table, what happens to the number of occupied shells? What happens to atomic size?

6. Where in the periodic table are the smallest atoms found? Where are the largest atoms found?

Name ______________________________ Date ____________

Chapter 12: The Nature of Chemical Bonds

Effective Nuclear Charge

The magnitude of the nuclear charge sensed by an orbiting electron depends upon several factors, including the number of positively–charged protons in the nucleus, the number of inner shell electrons shielding it from the nucleus, and its distance from the nucleus.

1. Place the proper number of electrons in each shell for carbon and silicon (use arrows to represent electrons).

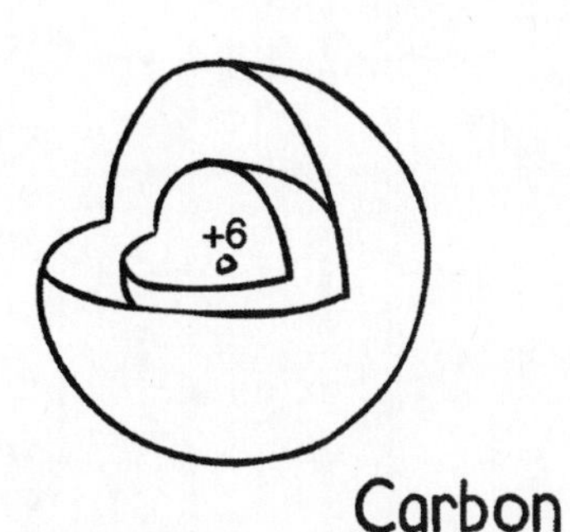

Carbon

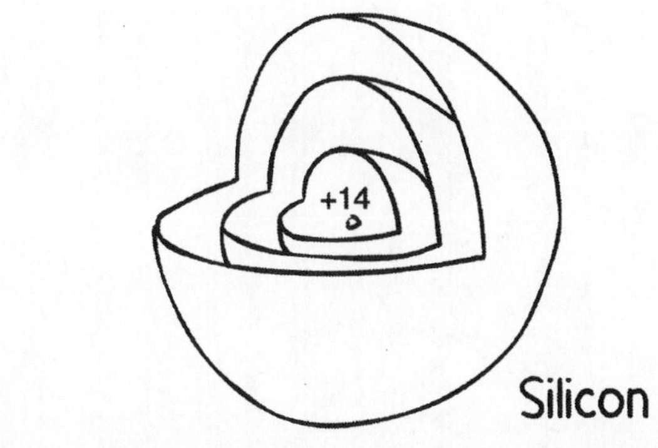

Silicon

2. According to the shell model, which should experience the greater effective nuclear charge: an electron in
   a. carbon's 1st shell or silicon's 1st shell? (circle one)
   b. carbon's 2nd shell or silicon's 2nd shell? (circle one)
   c. carbon's 2nd shell or silicon's 3rd shell? (circle one)

3. List the shells of carbon and silicon in order of decreasing effective nuclear charge.

   ________ > ________ > ________ > ________ > ________

4. Which should have the greater ionization energy, the carbon atom or the silicon atom?
   Defend your answer.

5. How many additional electrons are able to fit in the outermost shell of carbon? _______ silicon? _______

6. Which should be stronger, a C-H bond or an Si-H bond? Defend your answer.

7. Which should be larger in size, the ion C^{4+} or the ion Si^{4+}? Why?

Name ______________________________ Date __________

Chapter 12: The Nature of Chemical Bonds

Solutions

1. Use these terms to complete the following sentences. Some terms may be used more than once.

solution	**solvent**	**solute**
dissolve	**concentrated**	**dilute**
saturated	**concentration**	**mole**
molarity	**solubility**	**soluble**
insoluble	**precipitate**	

Sugar is __________ in water for the two can be mixed homogeneously to form a __________. The __________ of sugar in water is so great that __________ homogeneous mixtures are easily prepared. Sugar, however, is not infinitely __________ in water for when too much of this __________ is added to water, which behaves as the __________, the solution becomes __________. At this point any additional sugar is __________ for it will not __________. If the temperature of a saturated sugar solution is lowered, the __________ of the sugar in water is also lowered. If some of the sugar comes out of solution, it is said to form a __________. If, however, the sugar remains in solution despite the decrease in solubility, then the solution is said to be supersaturated. Adding only a small amount of sugar to water results in a __________ solution. The __________ of this solution or any solution can be measured in terms of __________, which tells us the number of solute molecules per liter of solution. If there are 6.022×10^{23} molecules in 1 liter of solution, then the __________ of the solution is 1 __________ per liter.

2. Temperature has a variety of effects on the solubilities of various solutes. With some solutes, such as sugar, solubility increases with increasing temperature. With other solutes, such as sodium chloride (table salt), changing temperature has no significant effect. With some solutes, such as lithium sulfate, Li_2SO_4, the solubility actually decreases with increasing temperature.

   a. Describe how you would prepare a supersaturated solution of lithium sulfate.

   ____

   b. How might you cause a saturated solution of lithium sulfate to form a precipitate?

   ____

Name ______________________________ Date ____________

Chapter 12: The Nature of Chemical Bonds

Pure Mathematics

Using a scientist's definition of *pure,* identify whether each of the following is 100% pure:

	100% pure?	
Freshly squeezed orange juice....	Yes	No
Country air	Yes	No
Ocean water................	Yes	No
Fresh drinking water	Yes	No
Skim milk..................	Yes	No
Stainless steel	Yes	No
A single water molecule	Yes	No

A glass of water contains on the order of a trillion trillion (1×10^{24}) molecules. If the water in this were 99.9999% pure, you could calculate the percent of impurities by subtracting from 100.0000%.

  100.0000% water + impurity molecules
− 99.9999% water molecules

0.0001% impurity molecules

Pull out your calculator and calculate the number of impurity molecules in the glass of water. Do this by finding 0.0001% of 1×10^{24}, which is the same as multiplying 1×10^{24} by 0.000001.

$$(1 \times 10^{24})(0.000001) = \_\_\_\_\_\_\_\_\_\_$$

1. How many impurity molecules are there in a glass of water that's 99.9999% pure?
   a. 1000 (one thousand: 10^3)
   b. 1,000,000 (one million: 10^6)
   c. 1,000,000,000 (one billion: 10^9)
   d. 1,000,000,000,000,000,000 (one million trillion: 10^{18}).
2. How does your answer make you feel about drinking water that is 99.9999 percent free of some poison, such as pesticide?

   ____
3. For every one impurity molecule, how many water molecules are there? (Divide the number of water molecules by the number of impurity molecules.)

   ____
4. Would you describe these impurity molecules within water that's 99.9999% pure as "rare" or "common"?

   ____
5. A friend argues that he or she doesn't drink tap water because it contains thousands of molecules of some impurity in each glass. How would you respond in defense of the water's purity, if it indeed does contain thousands of molecules of some impurity per glass?

   ____

Name ______________________________ Date ____________

Chapter 12: The Nature of Chemical Bonds

Chemical Bonds

1. Based upon their positions in the periodic table, predict whether each pair of elements will form an ionic bond, covalent bond, or neither (atomic number in parenthesis).
   a. Gold (79) and platinum (78) _____
   b. Rubidium (37) and iodine (53) _____
   c. Sulfur (16) and chlorine (17) _____
   d. Sulfur (16) and magnesium (12) _____
   e. Calcium (20) and chlorine (17) _____
   f. Germanium (32) and arsenic (33) _____
   g. Iron (26) and chromium (24) _____
   h. Chlorine (17) and iodine (53) _____
   i. Carbon (6) and bromine (35) _____
   j. Barium (56) and astatine (85) _____

2. The most common ions of lithium, magnesium, aluminum, chlorine, oxygen, and nitrogen and their respective charges are as follows:

Positively Charged Ions	Negatively Charged Ions
Lithium ion: Li^{1+}	Chloride ion: Cl^{1-}
Barium ion: Ba^{2+}	Oxide ion: O^{2-}
Aluminum ion: Al^{3+}	Nitride ion: N^{3-}

Use this information to predict the chemical formulas for the following ionic compounds:
   a. Lithium chloride:_____
   b. Barium chloride:_____
   c. Aluminum chloride:_____
   d. Lithium oxide:_____
   e. Barium oxide:_____
   f. Aluminum oxide:_____
   g. Lithium nitride:_____
   h. Barium nitride:_____
   i. Aluminum nitride:_____
   j. How are elements that form positive ions grouped in the periodic table relative to elements that form negative ions?______________________________

3. Specify whether the following chemical structures are polar or nonpolar:

H–O–O–H
____

H–O–C(=O)–O–H
____

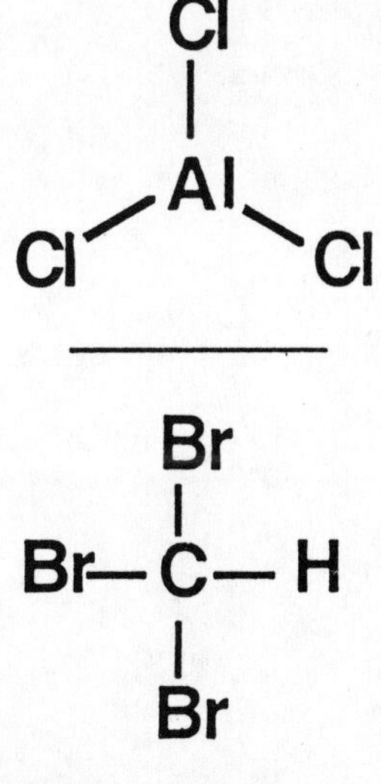

Br
Br—C—Br
Br
____

H—C≡C—H
____

F— F
____

Name ______________________________ Date ____________

Chapter 12: The Nature of Chemical Bonds

Shells and the Covalent Bond

When atoms bond covalently, their atomic shells overlap so that shared electrons can occupy both shells at the same time.

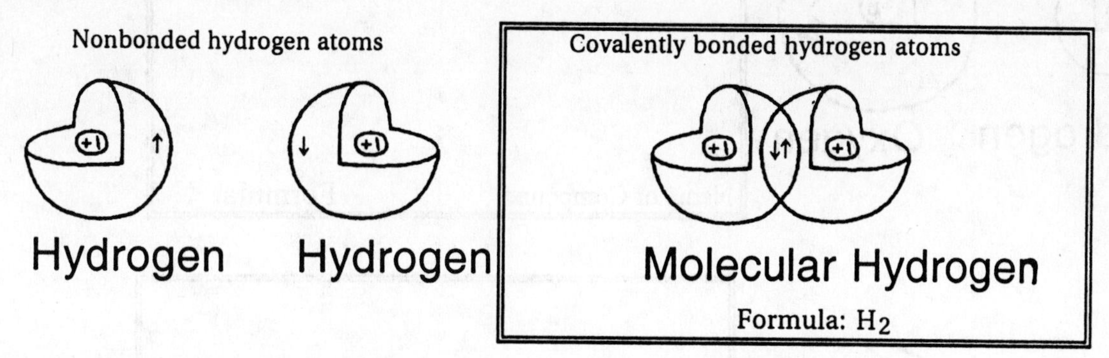

Fill each shell model shown below with enough electrons to make each atom electrically neutral. Use arrows to represent electrons. Within the box draw a sketch showing how the two atoms bond covalently. Draw hydrogen shells more than once when necessary so that no electrons remain unpaired. Write the name and chemical formula for each compound.

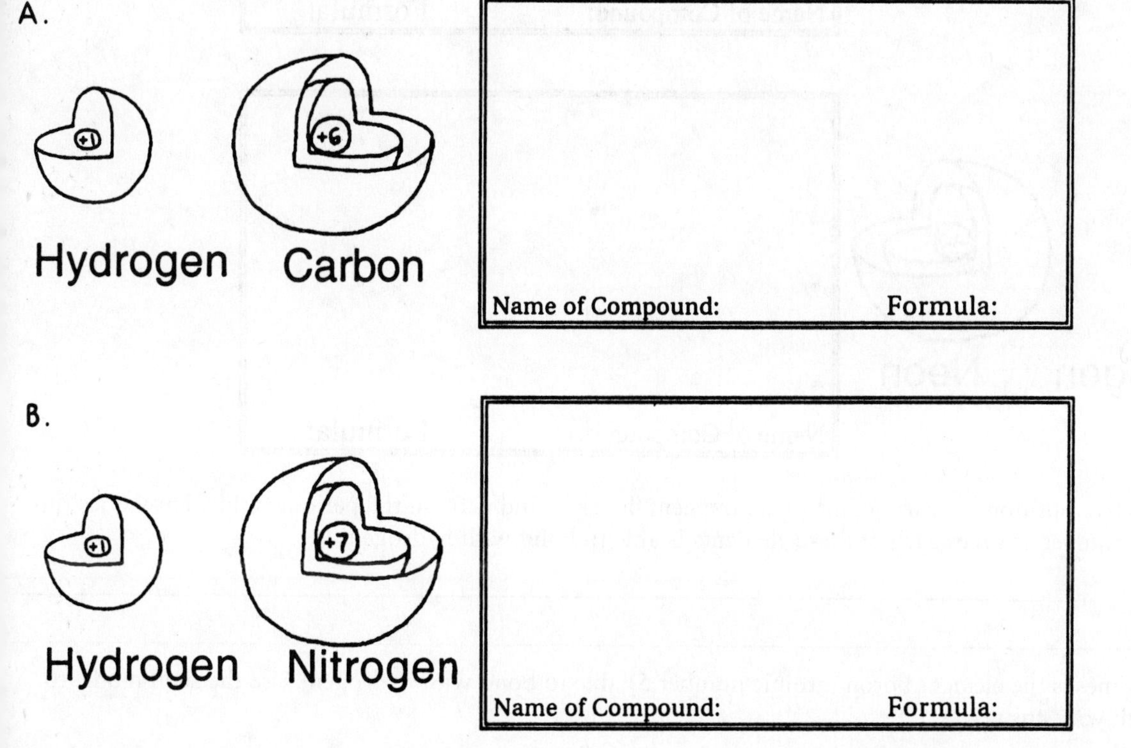

Name ______________________________ Date __________

Shells and the Covalent Bond—continued

C.

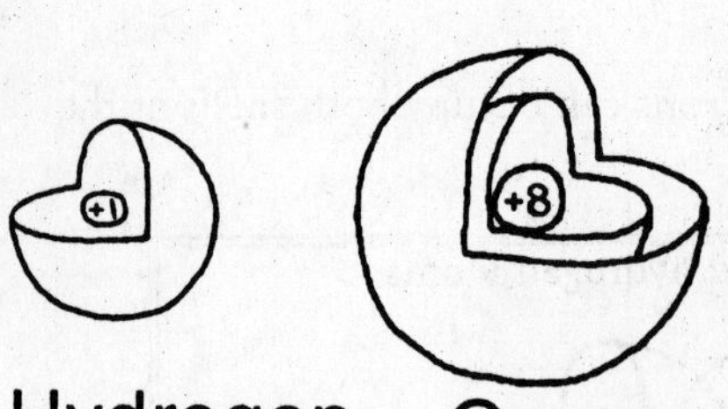

Name of Compound: Formula:

D.

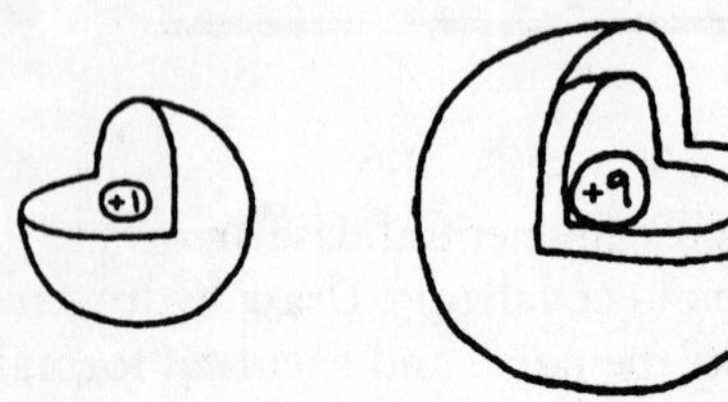

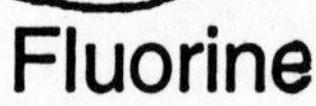

Name of Compound: Formula:

E.

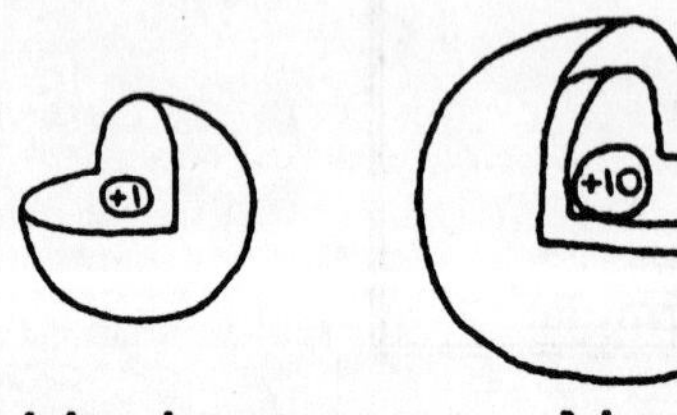

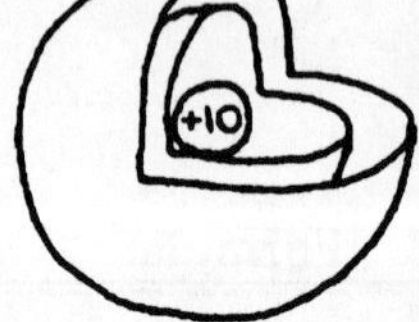

Name of Compound: Formula:

1. Note the relative positions of carbon, nitrogen, oxygen, fluorine, and neon in the periodic table. How does this relate to the number of times each of these elements is able to bond with hydrogen?

____

____

2. How many times is the element boron (atomic number 5) able to bond with hydrogen? Use the shell model to help you with your answer.

____

____

Name ______________________________ Date __________

Chapter 12: The Nature of Chemical Bonds

Bond Polarity

Pretend you are one of two electrons being shared by a hydrogen atom and a fluorine atom. Say, for the moment, you are centrally located between the two nuclei. You find that both nuclei are attracted to you. Hence, because of your presence, the two nuclei are held together.

You are here.

H : F

1. Why are the nuclei of these atoms attracted to you?______________________

2. What type of chemical bonding is this?______________________

You are held within hydrogen's 1st shell and at the same time within fluorine's 2nd shell. Draw a sketch using the shell models below to show how this is possible. Represent yourself and all other electrons using arrows. Note your particular location with a circle.

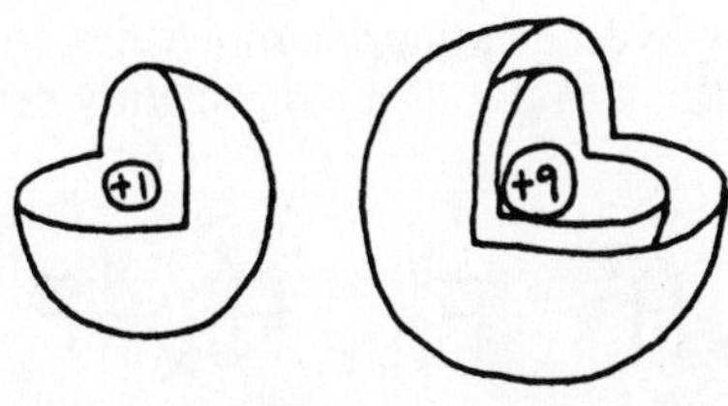

Hydrogen Fluorine

According to the laws of physics, if the nuclei are both attracted to you, then you are attracted to both of the nuclei.

3. You are pulled toward the hydrogen nucleus, which has a positive charge. How strong is this charge from your point of view—what is its *electronegativity?*______

4. You are also attracted to the fluorine nucleus. What is its electronegativity?______

You are being shared by the hydrogen and fluorine nuclei. But as a moving electron you have some choice as to your location.

5. Consider the electronegativities you experience from both nuclei. Which nucleus would you tend to be closest to? ______________

Name ______________________________ Date __________

Bond Polarity—continued

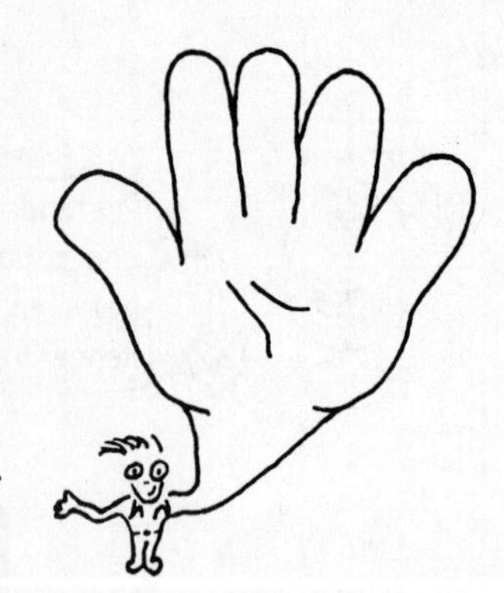

Stop pretending you are an electron and observe the hydrogen-fluorine bond from outside the hydrogen fluoride molecule. Bonding electrons tend to congregate to one side because of the differences in effective nuclear charges. This makes one side slightly negative in character and the opposite side slightly positive. Indicate this on the following structure for hydrogen fluoride using the symbols δ- and δ+

H : F

By convention, bonding electrons are not shown. Instead, a line is simply drawn connecting the two bonded atoms. Indicate the slightly negative and positive ends.

H—F

6. Would you describe hydrogen fluoride as a polar or nonpolar molecule?______
7. If two hydrogen fluoride molecules were thrown together, would they stick or repel? (Hint: What happens when you throw two small magnets together?)______
8. Place bonds between the hydrogen and fluorine atoms to show many hydrogen fluoride molecules grouped together. Each element should be bonded only once. Circle each molecule and indicate the slightly negative and slightly positive ends.

H F H F H F H F H F

F H F H F H F H F H

H F H F H F H F H F

F H F H F H F H F H

H F H F H F H F H F

F H F H F H F H F H

Name ______________________________ Date ______________

Chapter 12: The Nature of Chemical Bonds

Atoms to Molecules

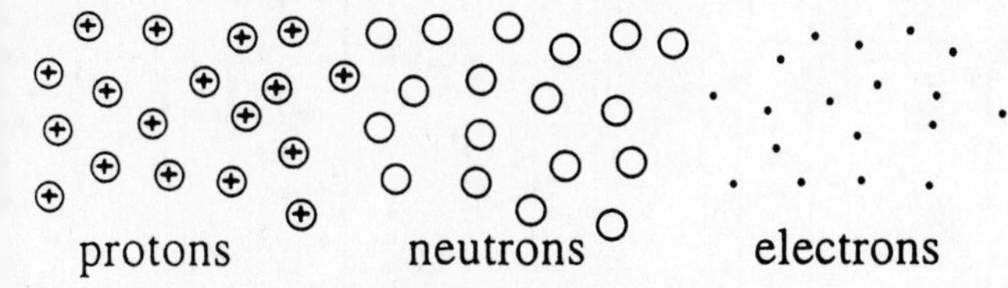

SUBATOMIC PARTICLES

Subatomic particles are the fundamental building blocks of all ____________.

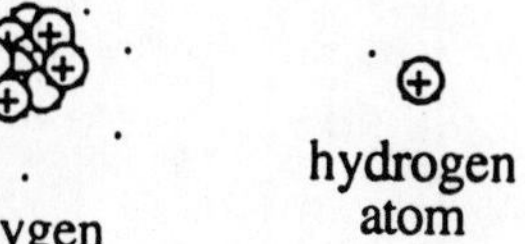

ATOMS

An atom is a group of ____________ held tightly together. An oxygen atom is a group of 8 ____________, 8 ____________, and 8 ____________. A hydrogen atom is a group of only 1 ____________ and 1 ____________.

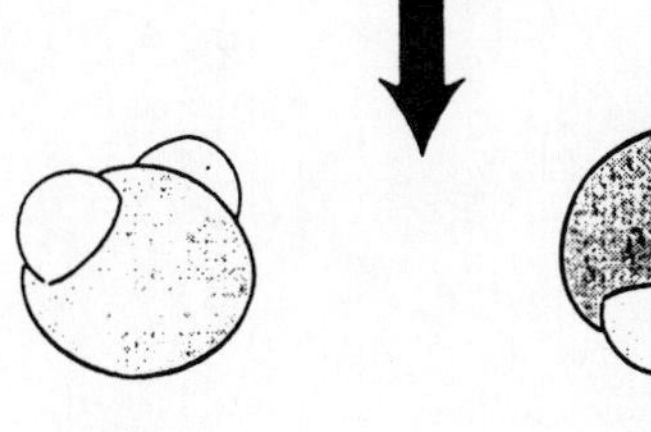

water molecule

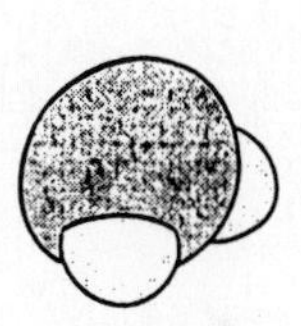

water molecule

MOLECULES

A ____________ is a group of atoms held tightly together. A water ____________ consists of 2 ____________ atoms and 1 ____________ atom.

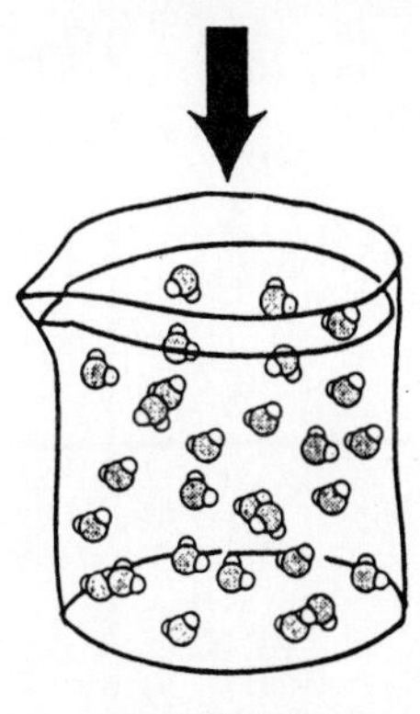

WATER

Water is a material made up of billions upon billions of water ____________. The physical properties of water are based upon how these water ____________ interact with one another. The electronic attractions between ____________ is one of the major topics of Chapter 12.

Name ______________________________ Date __________

Chapter 13: Chemical Reactions

Balancing Chemical Equations

In a balanced chemical equation the number of times each element appears as a reactant is equal to the number of times it appears as a product. For example,

$$2\ H_2 + O_2 \text{ ---> } 2\ H_2O$$

Recall that *coefficients* (the integer appearing before the chemical formula) indicate the number of times each chemical formula is to be counted and *subscripts* indicate when a particular element occurs more than once within the formula.

Check whether or not the following chemical equations are balanced.

$3\ NO$ ---> $N_2O + NO_2$ ☐ balanced ☐ unbalanced

$SiO_2 + 4\ HF$ ---> $SiF_4 + 2\ H_2O$ ☐ balanced ☐ unbalanced

$4\ NH_3 + 5\ O_2$ ---> $4\ NO + 6\ H_2O$ ☐ balanced ☐ unbalanced

Unbalanced equations are balanced by changing the coefficients. Subscripts, however, should never be changed because this changes the chemical's identity—H_2O is water, but H_2O_2 is hydrogen peroxide! The following steps may help guide you:

1. Focus on balancing only one element at a time. Start with the left-most element and modify the coefficients such that this element appears on both sides of the arrow the same number of times.
2. Move to the next element and modify the coefficients so as to balance this element. Do not worry if you incidentally unbalance the previous element. You will come back to it in subsequent steps.
3. Continue from left to right balancing each element individually.
4. Repeat steps 1–3 until all elements are balanced.

Use the above methodology to balance the following chemical equations.

____N_2O + ____N_2 ---> ____O_2

____$NaClO_3$ ---> ____$NaCl$ + ____O_2

____$MnCl_2$ + ____Al ---> ____Mn + ____$AlCl_3$

____K + ____H_2O ---> ____H_2 + ____KOH

____Al_2O_3 + ____C ---> ____Al + ____CO_2

____NH_3 + ____F_2 ---> ____NH_4F + ____NF_3

This is just one of the many methods that chemists have developed to balance chemical equations.

Knowing how to balance a chemical equation is a useful technique, but understanding why a chemical equation needs to be balanced in the first place is far more important.

Name ____ Date ____________

Chapter 13: Chemical Reactions

Exothermic and Endothermic Reactions

During a chemical reaction atoms are neither created nor destroyed. Instead, atoms rearrange—they change partners. This rearrangement of atoms necessarily involves the input and output of energy. First, energy must be supplied to break chemical bonds that hold atoms together. Separated atoms then form new chemical bonds, which involves the release of energy. In an **exothermic** reaction more energy is released than is consumed. Conversely, in an **endothermic** reaction more energy is consumed than is released.

Table 1 Bond Energies

Bond	Bond Energy*	Bond	Bond Energy*
H—H	436	Cl—Cl	243
H—C	414	N—N	159
H—N	389	O=O	498
H—O	464	O=C	803
H—Cl	431	N≡N	946

*In kJ/mol

Table 1 shows bond energies—the amount of energy required to break a chemical bond, and also the amount of energy released when a bond is formed. Use these bond energies to determine whether the following chemical reactions are exothermic or endothermic.

H—H + Cl—Cl ⟶ H—Cl + H—Cl

Hydrogen   Chlorine   Hydrogen chloride

Total Amount of Energy Required to Break Bonds ______ kJ/mol

Total Amount of Energy Released Upon Bond Formation ______ kJ/mol

Net Energy Change of Reaction: ____________ kJ/mole (absorbed) (released)
circle one

1. Is this reaction exothermic or endothermic?

2. Write the balanced equation for this reaction using chemical formulas and coefficients. If it is exothermic, write "Energy" as a product. If it is endothermic, write "Energy" as a reactant.

Name ____ Date ____________

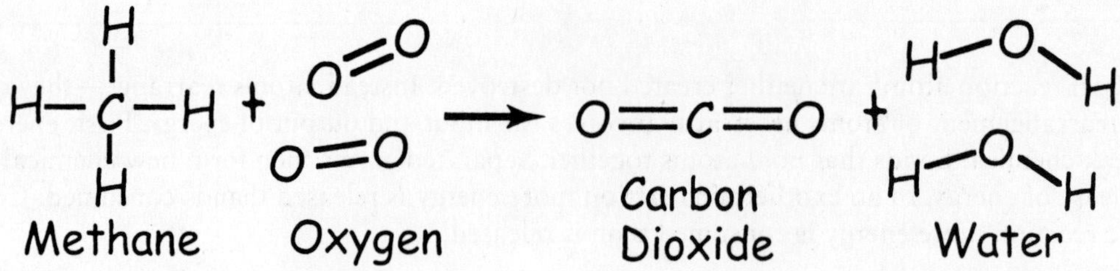

H–C(H)(H)–H + O=O + O=O → O=C=O + H–O–H + H–O–H

Methane Oxygen Carbon Dioxide Water

Total Amount of Energy Required to Break Bonds ______kJ/mol

Total Amount of Energy Released Upon Bond Formation ______kJ/mol

Net Energy Change of Reaction: ____________kJ/mole (absorbed/released) *circle one*

3. Is this reaction exothermic or endothermic?

4. Write the balanced equation for this reaction using chemical formulas and coefficients. If it is exothermic write "Energy" as a product. If it is endothermic write "Energy" as a reactant.

N≡N + H–H + H–H → $H_2N–NH_2$

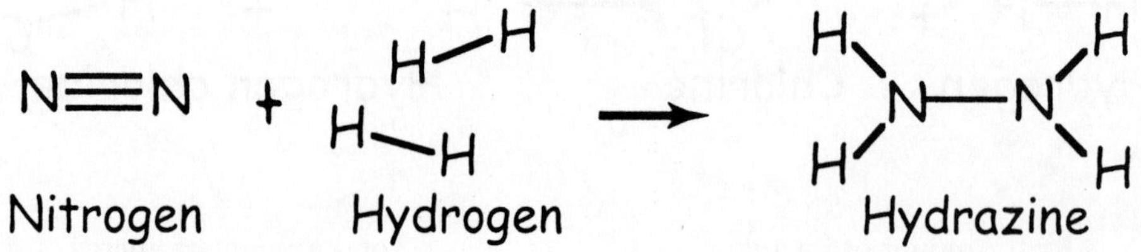

Nitrogen Hydrogen Hydrazine

Total Amount of Energy Required to Break Bonds _____ kJ/mol

Total Amount of Energy Released Upon Bond Formation _____ kJ/mol

Net Energy Change of Reaction: ____________kJ/mole (absorbed/released) *circle one*

5. Is this reaction exothermic or endothermic?

6. Write the balanced equation for this reaction using chemical formulas and coefficients. If it is exothermic write "Energy" as a product. If it is endothermic write "Energy" as a reactant.

Name ______________________________ Date ____________

Chapter 13: Chemical Reactions

Donating and Accepting Hydrogen Ions

A chemical reaction that involves the transfer of a hydrogen ion from one molecule to another is classified as an acid-base reaction. The molecule that donates the hydrogen ion behaves as an acid. The molecule that accepts the hydrogen ion behaves as a base.

On paper, the acid-base process can be depicted through a series of frames:

Frame 1

ammonium ion   hydroxide ion

Ammonium and hydroxide ions in close proximity.

Frame 2

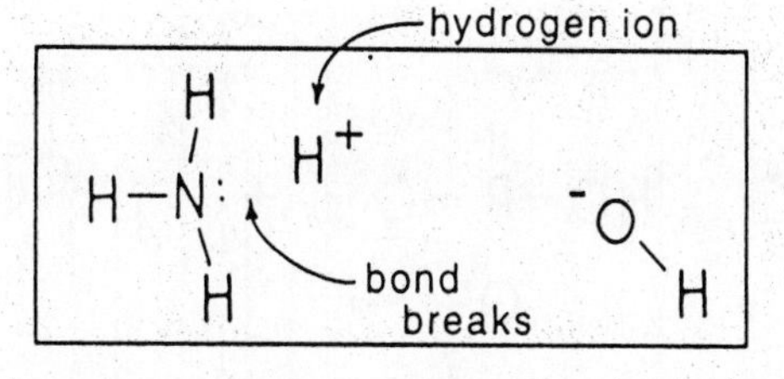

Bond is broken between the nitrogen and a hydrogen of the ammonium ion. The two electrons of the broken bond stay with the nitrogen leaving the hydrogen with a positive charge.

Frame 3

The hydrogen ion migrates to the hydroxide ion.

Frame 4

The hydrogen ion bonds with the hydroxide ion to form a water molecule.

In equation form we abbreviate this process by only showing the before and after:

frame 1   frame 4

Name ______________________________ Date __________

Donating and Accepting Hydrogen Ions—continued

We see from the previous reaction that because the ammonium ion donated a hydrogen ion, it behaved as an acid. Conversely, the hydroxide ion by accepting a hydrogen ion behaved as a base. How do the ammonia and water molecules behave during the reverse process?

$$NH_4^+ + {}^-OH \rightleftharpoons NH_3 + H_2O$$

acid ______ base ______ ______ ammonia ______ water

Identify the following molecules as behaving as an acid or a base:

$$H_2PO_4^- + H_2O \rightleftharpoons HPO_4^{2-} + H_3O^+$$

______ ______ ______ ______

$$NH_3 + NaH \rightleftharpoons Na^+\,{}^-NH_2 + H\text{-}H$$

______ ______ ______ ______

$$H\text{-}H + {}^-H \rightleftharpoons H^- + H\text{-}H$$

______ ______ ______ ______

$$HNO_3 + NH_3 \rightleftharpoons {}^-NO_3 + {}^+NH_4$$

______ ______ ______ ______

Name ______________________________ Date ____________

Chapter 13: Chemical Reactions

Loss and Gain of Electrons

A chemical reaction that involves the transfer of an electron is classified as an oxidation–reduction reaction. Oxidation is the process of losing electrons, while reduction is the process of gaining them. Any chemical that causes another chemical to lose electrons (become oxidized) is called an *oxidizing agent.* Conversely, any chemical that causes another chemical to gain electrons is called a *reducing agent.*

1. What is the relationship between an atom's ability to behave as an oxidizing agent and its electron affinity?

2. Relative to the periodic table, which elements tend to behave as strong oxidizing agents?

3. Why don't the noble gases behave as oxidizing agents?

4. How is it that an oxidizing agent is itself reduced?

5. Specify whether each reactant is about to be oxidized or reduced.

$2\ K + H_2O \longrightarrow 2\ K^+ + {}^-OH$

_____ _____

$2\ Mg + O_2 \longrightarrow 2\ Mg^{2+}O^{2-}$

_____ _____

$2\ Na + Cl_2 \longrightarrow 2\ Na^+Cl^-$

_____ _____

$CH_4 + 2\ O_2 \longrightarrow O{=}C{=}O + H{-}O{-}H$

_____ _____

6. Which oxygen atom enjoys a greater negative charge?

this one → O=O   or   that one → H—O—H   (*circle one*)

7. Relate your answer to Question 6 to how it is that O_2 is reduced upon reacting with CH_4 to form carbon dioxide and water.

Name ______________________________ Date ____________

Chapter 14: Organic Chemistry

Structures of Organic Compounds

1. What are the chemical formulas for the following structures?

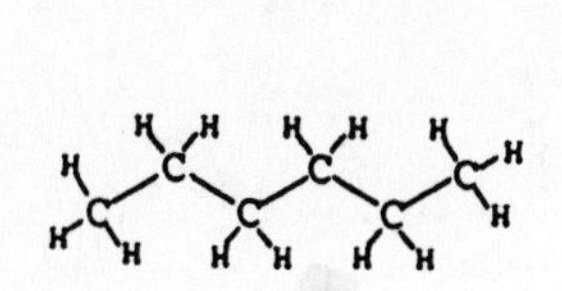

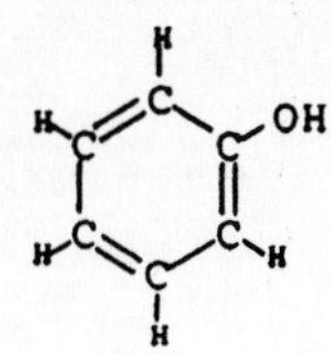

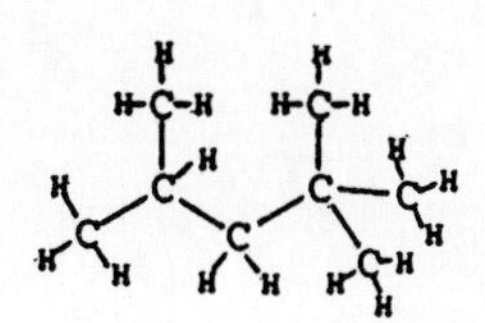

Formula: ______ ______ ______ ______

2. How many covalent bonds is carbon able to form? ______

3. What is wrong with the structure shown in the box at right?

____

4. a. Draw a hydrocarbon that contains 4 carbon atoms.

b. Redraw your structure and transform it into an amine.

c. Transform your amine into an amide. You may need to relocate the nitrogen.

d. Redraw your amide, transforming it into a carboxylic acid.

e. Redraw your carboxylic acid, transforming it into an alcohol.

f. Rearrange the carbons of your alcohol to make an ether.

Name ____ Date __________

Chapter 14: Organic Chemistry

Polymers

1. Circle the monomers that may be useful for forming an addition polymer and draw a box around the ones that may be useful for forming a condensation polymer.

$H_2C{=}CH_2$

$HO{-}C({=}O){-}CH_2CH_2CH_2CH_3$

H_2N NH_2

$H_2N{-}CH(CH_3){-}C({=}O){-}OH$

$HO{-}C({=}O){-}CH_2{-}\underset{H}{C}{=}CH_3$

$HO{-}C({=}O){-}CH_2CH_2CH_2CH_2{-}C({=}O){-}OH$

2. Which type of polymer always weighs less than the sum of its parts? Why?

3. Would a material with the following arrangement of polymer molecules have a relatively high or low melting point? Why?

Name ____ Date ___________

Chapter 15: The Basic Unit of Life—The Cell

Features of Prokaryotic and Eukaryotic Cells

1. Are the following associated with prokaryotic cells, eukaryotic cells, or both?

   a. nucleic acids

   b. cell membrane

   c. nucleus

   d. organelles

   e. mitochondria

   f. chloroplasts

   g. bacteria

   h. circular chromosome

   i. cytoplasm

   j. human cells

2. Match the following organelles with their functions:

   ribosome

   rough endoplasmic reticulum

   smooth endoplasmic reticulum

   Golgi apparatus

   lysosome

   mitochondrion

   chloroplast

   cytoskeleton

   a. assembles proteins destined to go either to the cell membrane or to leave the cell

   b. obtains energy for the cell to use

   c. in plant cells, captures energy from sunlight to build organic molecules

   d. receives products from the endoplasmic reticulum and packages them for transport

   e. helps cell hold its shape

   f. assembles membranes and performs other specialized functions in certain cells

   g. assembles proteins for the cell

   h. breaks down organic materials

Name ______________________________ Date __________

Features of Prokaryotic and Eukaryotic Cells—continued

3. What are the three components of a cell membrane? Draw a portion of a cell membrane showing each of these components.

4. What are some functions carried out by membrane proteins?

5. What are the functions of short carbohydrates?

Name ______________________________ Date __________

Chapter 15: The Basic Unit of Life—The Cell

Transport In and Out of Cells

1. Assume that the square-shaped molecules shown below can pass freely across the cell membrane. What is the name of the process by which they move across the cell membrane?

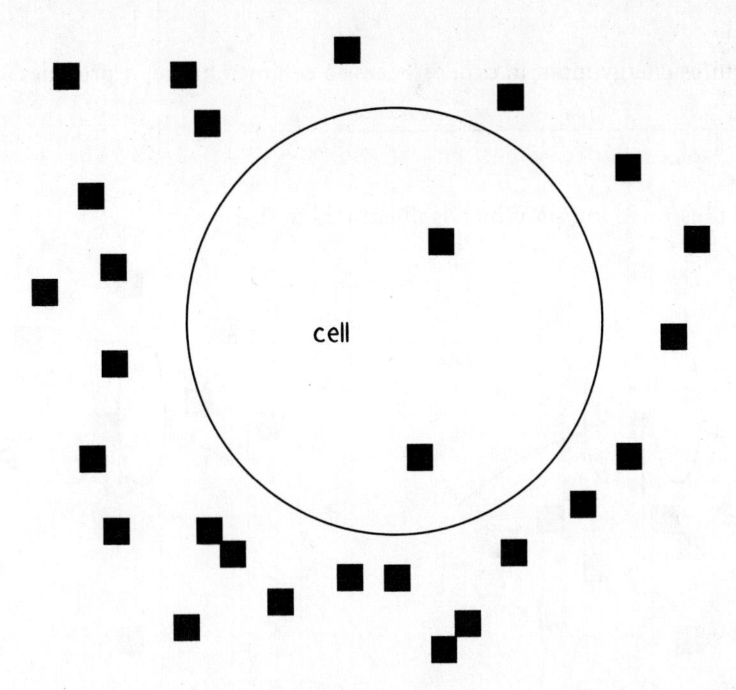

Will the square-shaped molecules tend to move out of the cell or into the cell? Why?

2. The diffusion of water has a special name. It is called ______________________.

   In the figure below, a membrane allows water to move freely between two compartments. The dark circles represent solute molecules, which are not free to move between the two compartments. Which way will water tend to flow? Why?

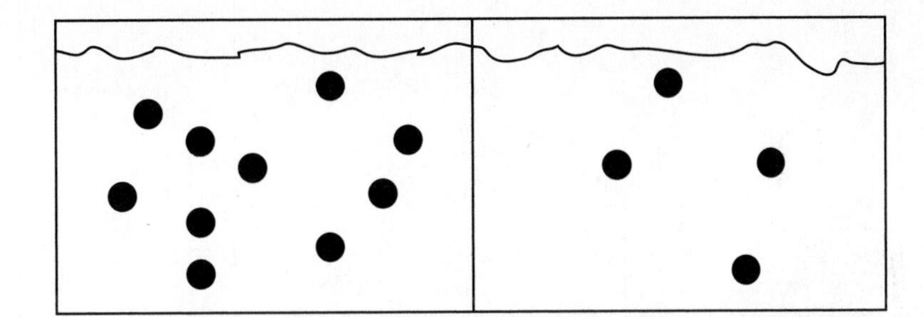

Name ______________________________ Date __________

Transport In and Out of Cells—continued

3. a. If a molecule needs a carrier protein, but no energy input in order to cross a cell membrane, it provides an example of ______________________.

   b. If a molecule requires energy input in order to cross a cell membrane, it provides an example of ______________________.

   c. Which process is illustrated in (a)? Which is illustrated in (b)?

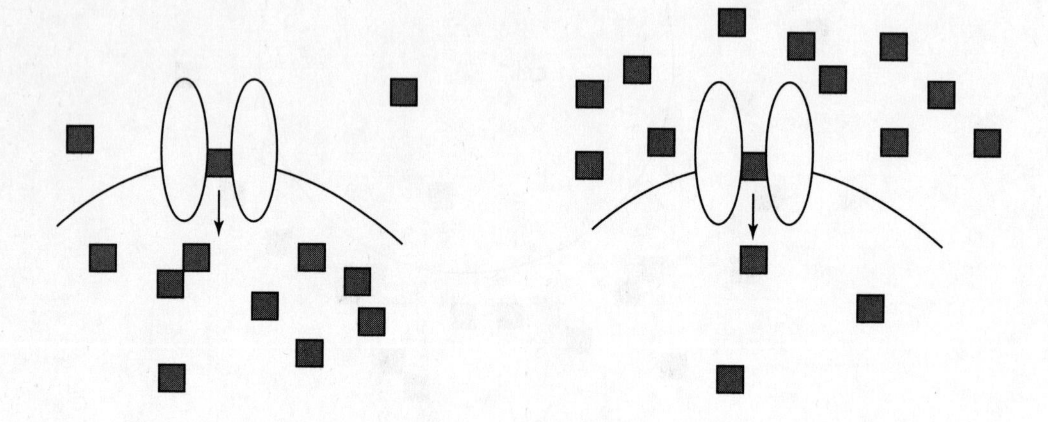

4. The cell below is going to take in the triangular molecule below via endocytosis. Draw what happens.

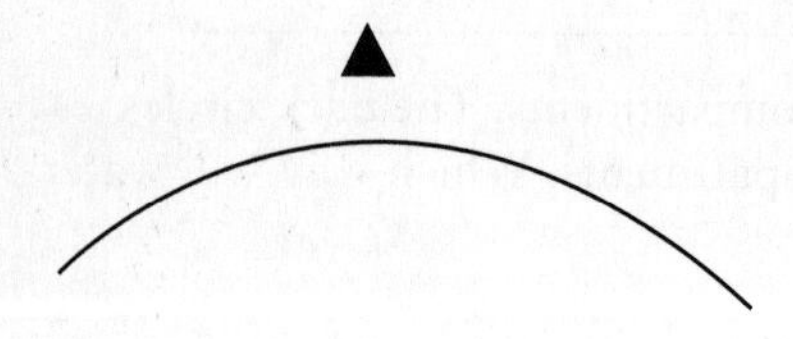

Name ____ Date ____________

Chapter 15: The Basic Unit of Life—The Cell

Photosynthesis and Cellular Respiration

1. The chemical reaction for photosynthesis is (use words or chemical formulae):

   _____ + ______ + ______ → _____ + ______

2. a. Where in plant cells does photosynthesis take place?

   b. Where do plants get carbon dioxide from?

   c. Where do plants get water from?

   d. What parts of plants capture sunlight?

3. For each of the following events that occur during photosynthesis, indicate whether it occurs during the light-dependent or light-independent reactions:

   a. sunlight strikes a chlorophyll molecule

   b. Calvin cycle

   c. energy is generated in the form of molecules of ATP and NADPH

   d. free oxygen is produced

   e. carbon is fixed

   f. glucose is produced

Photosynthesis and Cellular Respiration—continued

4. The chemical reaction for cellular respiration is

_____ + ______ + ______ → _____ + ______ + _______

5. Cellular respiration allows cells to produce ATP. How do cells later obtain energy from ATP?

6. Which of the following processes requires oxygen?

   glycolysis

   Krebs cycle and electron transport

   alcoholic fermentation

   lactic acid fermentation

Name ______________________________ Date __________

Chapter 16: Genetics

DNA Replication, Transcription, and Translation

1. Let's start with the following strand of DNA:

AT
GC
CG
TA
TA
AT
CG
CG
GC
TA
AT
CG
GC

The strand is unwound, so that the DNA can be replicated. Fill in the nucleotides on the new strands.

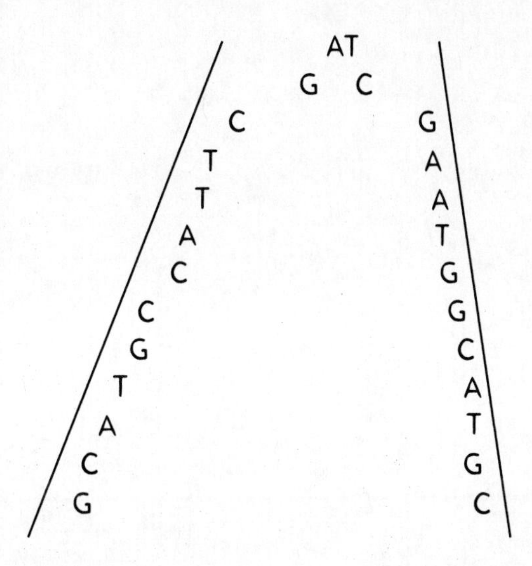

Name ______________________________ Date ____________

DNA Replication, Transcription, and Translation—continued

2. Transcription takes place in the ______________________.
   During transcription, DNA is used to make a molecule of ______________________________.
   If the following length of DNA is being transcribed, what are the bases found on the transcript?

   ATGGTCATACGTACAATG

Name ____ Date __________

DNA Replication, Transcription, and Translation—continued

3. Translation takes place in the _____ of cells in organelles called ______.

   During translation, ____ is used to build a ____.

   Divide the transcript from Problem 2 into codons and indicate the sequence of amino acids that is assembled in the ribosome. How many amino acids are coded for by this sequence?

   For reference, the genetic code table is shown below:

First base	Second base: U	Second base: C	Second base: A	Second base: G	Third base
U	UUU Phenylalanine (Phe)	UCU Serine (Ser)	UAU Tyrosine (Tyr)	UGU Cysteine (Cys)	U
U	UUC Phenylalanine (Phe)	UCC Serine (Ser)	UAC Tyrosine (Tyr)	UGC Cysteine (Cys)	C
U	UUA Leucine (Leu)	UCA Serine (Ser)	UAA Stop	UGA Stop	A
U	UUG Leucine (Leu)	UCG Serine (Ser)	UAG Stop	UGG Tryptophan (Trp)	G
C	CUU Leucine (Leu)	CCU Proline (Pro)	CAU Histidine (His)	CGU Arginine (Arg)	U
C	CUC Leucine (Leu)	CCC Proline (Pro)	CAC Histidine (His)	CGC Arginine (Arg)	C
C	CUA Leucine (Leu)	CCA Proline (Pro)	CAA Glutamine (Gln)	CGA Arginine (Arg)	A
C	CUG Leucine (Leu)	CCG Proline (Pro)	CAG Glutamine (Gln)	CGG Arginine (Arg)	G
A	AUU Isoleucine (Ile)	ACU Threonine (Thr)	AAU Asparagine (Asn)	AGU Serine (Ser)	U
A	AUC Isoleucine (Ile)	ACC Threonine (Thr)	AAC Asparagine (Asn)	AGC Serine (Ser)	C
A	AUA Isoleucine (Ile)	ACA Threonine (Thr)	AAA Lysine (Lys)	AGA Arginine (Arg)	A
A	AUG Met or start	ACG Threonine (Thr)	AAG Lysine (Lys)	AGG Arginine (Arg)	G
G	GUU Valine (Val)	GCU Alanine (Ala)	GAU Aspartic acid(Asp)	GGU Glycine (Gly)	U
G	GUC Valine (Val)	GCC Alanine (Ala)	GAC Aspartic acid(Asp)	GGC Glycine (Gly)	C
G	GUA Valine (Val)	GCA Alanine (Ala)	GAA Glutamic acid (Glu)	GGA Glycine (Gly)	A
G	GUG Valine (Val)	GCG Alanine (Ala)	GAG Glutamic acid (Glu)	GGG Glycine (Gly)	G

Name ____ Date ____________

Chapter 16: Genetics

Meiosis

Consider the following diploid cell. The long, dot-filled chromosomes are homologous, but are shaded differently to distinguish them from each other. This is true for the shorter checkerboard chromosomes as well.

1. a. How many chromosomes are there in this diploid cell?

   b. If this cell were to undergo meiosis, how many chromosomes would there be in the resulting haploid cells?

2. Draw the cell during the following phases of meiosis:

   a. What does the cell look like when it has duplicated its genetic material in preparation for meiosis?

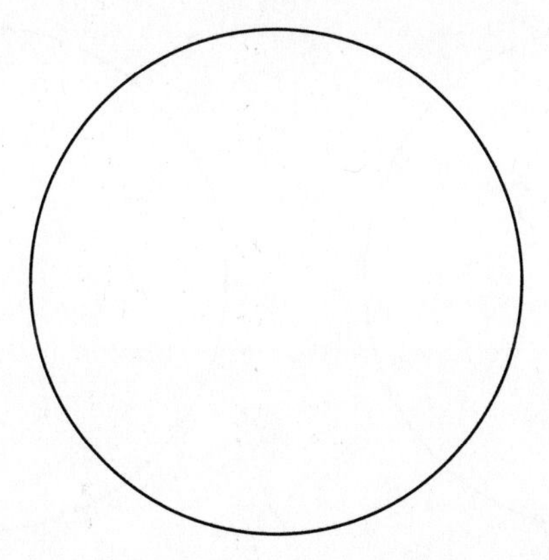

Name ______________________________ Date __________

Meiosis—continued

b. What does the cell look like during metaphase I, before crossing over and recombination have occurred?

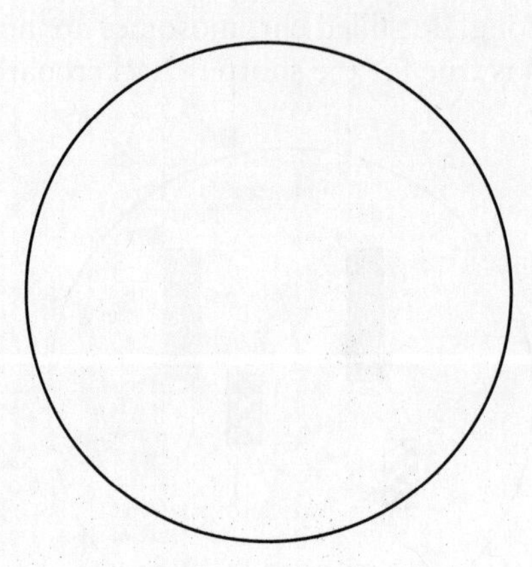

c. Suppose each chromosome experiences a single crossing over event with its homologue. Draw the cell during metaphase I, after crossing over has occurred.

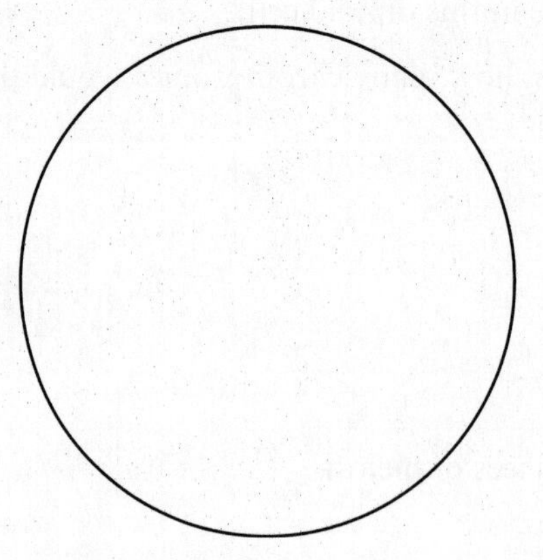

d. Draw the two daughter cells at the end of meiosis I.

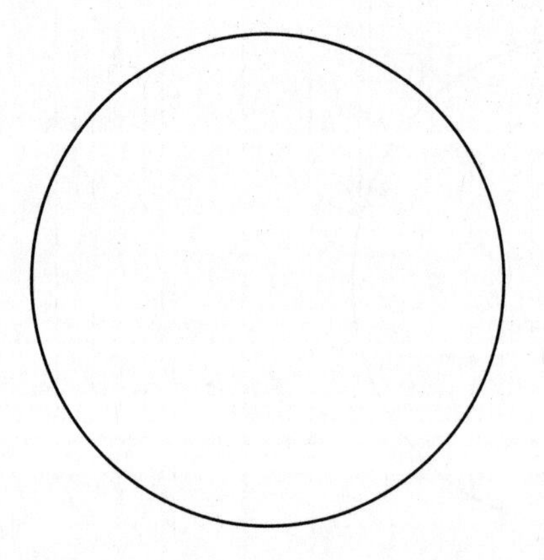

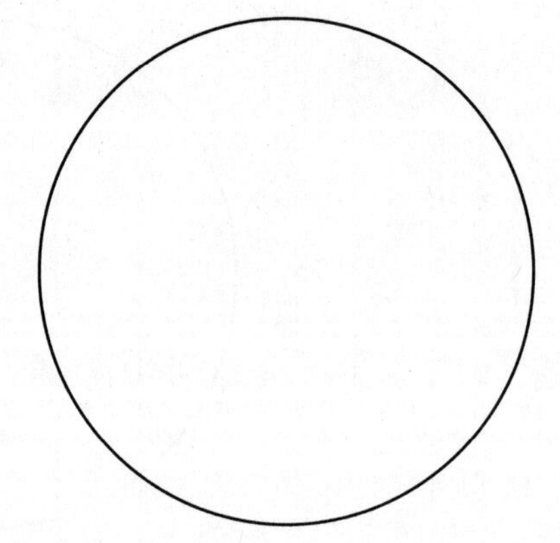

Name ______________________________ Date ____________

Meiosis—continued

e. Draw the four daughter cells at the end of meiosis II.

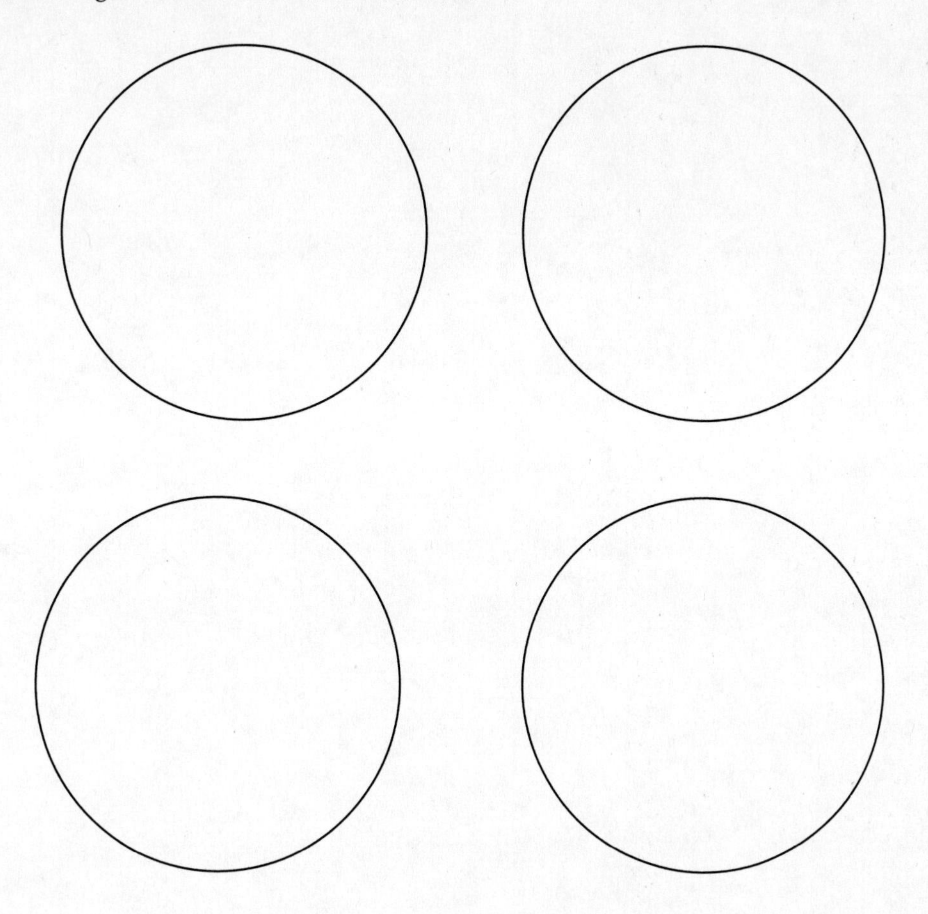

Name ______________________________ Date __________

Chapter 16: Genetics

Inheritance

1. Suppose there exists a species of small woodland creature in which fur is either spotted or striped. It turns out that fur pattern is determined by a single gene, and that the striped phenotype is dominant to the spotted phenotype. Which of the following must be a homozygote?

2. Suppose, in fact, both the woodland creatures above are homozygotes. Their genotypes are aa and AA. Fill in the blanks below:

Genotype ________ ________

Phenotype ________ ________

3. What phenotype would an Aa heterozygote have? Draw it below:

Name ____ Date __________

Inheritance—continued

4. Now, you breed together an aa individual with an AA individual. Draw the cross below:

Genotype ________ ________

Phenotype ________ ________

What are the progeny like?

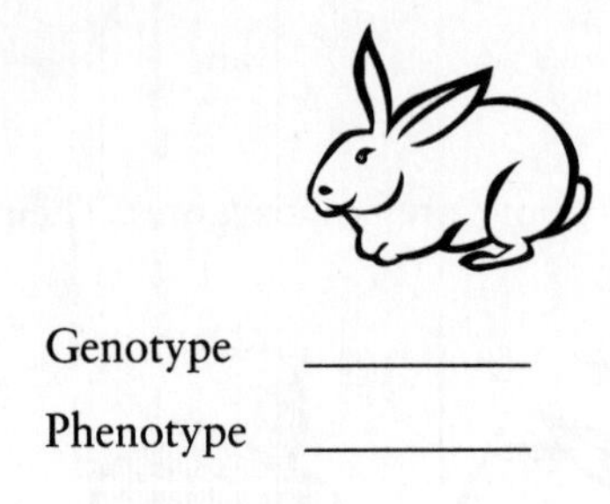

Genotype ________

Phenotype ________

5. Cross together two of the progeny from Problem 4 above. Fill in the boxes below.

Allele received from mother	Allele received from father: A	Allele received from father: a
A	Genotype: Phenotype:	Genotype: Phenotype:
a	Genotype: Phenotype:	Genotype: Phenotype:

So, the progeny are spotted/striped/both, and found in a ratio ______:_______.

Name ______________________________ Date __________

Chapter 17: The Evolution of Life

Natural Selection

1. a. One of the best-documented instances of natural selection in the wild is the evolution of bird beak sizes during a severe drought on the Galápagos Islands in 1977. Before the drought, there was natural variation in beak size in a species of finch found on the islands. Draw a series of finches, showing this variation in beak size.

   b. What happened was this: The drought made seeds scarce. Small seeds were quickly eaten up, leaving only larger, tougher seeds. Birds with larger, stronger beaks were better at cracking these larger seeds. Many birds died during the drought. Which were most likely to survive? Mark X's through some of the individuals in your drawing likely to have died, and circle individuals likely to have survived.

Name ______________________________ Date __________

Natural Selection—continued

c. Beak size is a trait that is partly genetically determined, so parents with larger, stronger beaks tend to have offspring with larger, stronger beaks. Draw the offspring of the individuals you circled as surviving the drought.

d. How does the population you drew in part (c) compare with the population you drew in part (a)?

Name ______________________________ Date __________

Natural Selection—continued

2. Of the human traits listed below, put a V by traits that are variable and put an H by traits that are heritable. Which traits have the potential of evolving via natural selection?

   a. age

   b. eye color

   c. number of toes

   d. curliness or straightness of hair

   e. presence or absence of dimples

   f. upright posture

   g. owning versus not owning a dog

   h. height

Name ______________________________ Date ____________

Chapter 17: The Evolution of Life

Adaptation

1. The imaginary mammal below occupies temperate forests in the Eastern United States.

If a population of these mammals moved to and successfully colonized an Arctic habitat, how might you predict that it would evolve? Draw the Arctic form below.

If a population of these mammals moved to and successfully colonized a desert habitat, how might you predict that it would evolve? Draw the desert form below.

Name ______________________________ Date __________

Adaptation—continued

Explain your drawings.

____

____

____

____

____

____

____

____

____

____

____

____

____

____

2. You are studying peppered moth populations in various locales. In order to determine whether light moths or dark moths survive better in different habitats, you mark 500 light moths and 500 dark moths and release them in different places.

   If light moths survive better, you expect to recapture more ______________________.

   Similarly, if dark moths survive better, you expect to recapture more ______________________.

   Do you expect light moths or dark moths to survive better in the following habitats?

   a. polluted areas
   b. unpolluted areas
   c. industrial centers before pollution laws were passed
   d. industrial centers some time after pollution laws were passed
   e. the countryside

Name ____________________ Date ________

Chapter 17: The Evolution of Life

Speciation

1. What's the difference between a prezygotic reproductive barrier and a postzygotic reproductive barrier?

2. Which of the following are prezygotic reproductive barriers and which are postzygotic reproductive barriers?
   a. different courtship rituals in different bird species
   b. incompatible anatomical structures that prevent copulation in insects
   c. different-sounding calls by males in different frog species
   d. sterility in offspring produced when members of two different species mate
   e. two species that mate at different times of year
   f. two species that use different mating sites
   g. offspring that are unable to survive when members of two species mate

3. What's the difference between allopatric speciation and sympatric speciation?

4. Are the following examples of allopatric or sympatric speciation?
   a. speciation after developing glaciers divide a population
   b. speciation by hybridization
   c. speciation after a river cuts through a population's habitat
   d. speciation after plate tectonics causes a continent to split
   e. speciation by polyploidy

Name ______________________________ Date __________

Chapter 18: Biological Diversity

Classification

1. Linnaean classification groups species together based on _____.

2. Fill in the levels of Linnaean classification from the largest group to the smallest group below.

   Domain

   _____

   _____

   _____

   _____

   _____

   _____

   Species

3. A species' scientific name consists of its _____ name and its _____ name.

4. Cladistic classification groups species together based on their ____.

5. The following cladogram shows evolutionary relationships between the rufous hummingbird, the honey mushroom, and the bristlecone pine.

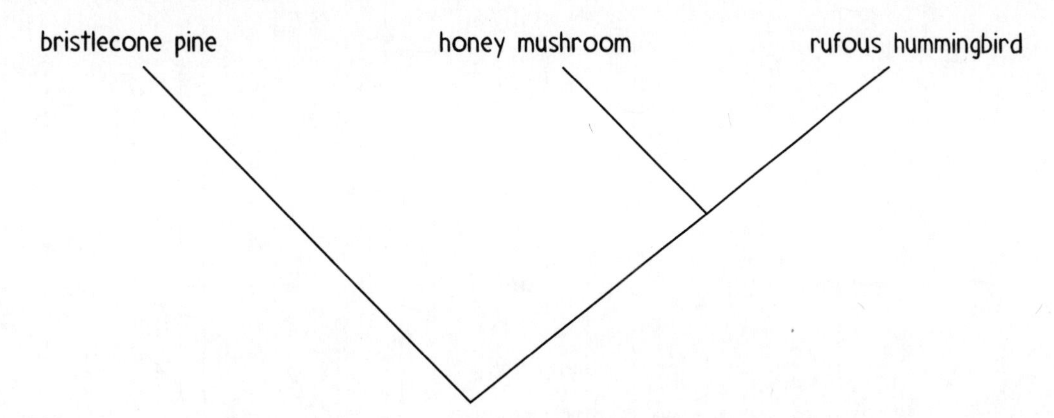

   This cladogram suggests that _____ and _____ should be classified together to the exclusion of ________.

Name ______________________________ Date __________

Chapter 18: Biological Diversity

Biological Diversity I: Bacteria, Archaea, Protists

1. What are the three domains of life?

   ____

   Of the three domains, __________ and __________ consist of prokaryotes and __________ consists of eukaryotes.

2. The four kingdoms that make up Eukarya are:

   ____

3. Are there any bacteria that can photosynthesize? Are there any heterotrophic bacteria?

   ____

4. How do bacteria typically reproduce?

   ____

5. Can bacteria exchange genetic material? If so, how?

   ____

   ____

   ____

6. Do the many bacteria that live in and on our bodies benefit us in any way?

   ____

7. Are archaea more closely related to bacteria or to eukaryotes? What evidence supports this?

   ____

   ____

   ____

Name ______________________________ Date ____________

8. What's an extremophile? Are all archaea extremophiles?

9. What is a chemoautotroph?

10. Are each of the following groups of protists autotrophs or heterotrophs?
    a. diatoms
    b. amoebas
    c. kelp
    d. dinoflagellates
    e. *Plasmodium* (the protist that causes malaria)

Name ____ Date __________

Chapter 18: Biological Diversity

Biological Diversity II: Plants, Fungi, Animals

1. Match the following plant structures with their function:

stomata	a. move water and nutrients up from the roots
roots	b. take in carbon dioxide
shoots	c. conduct photosynthesis
xylem	d. move sugars produced during photosynthesis
phloem	e. transport resources to different parts of plant
vascular system	f. absorb water and nutrients from soil

2. The life history of plants involves a(n) ______ in which plants move between a haploid ______ stage and a diploid ______ stage.

3. Mosses are unique among plants in that the ______ is much larger than the ______. When you see a moss in the forest, you are looking at a ______. The sperm are released by the male ______ directly into the environment, where they use their flagella to swim through a film of water to eggs in the female gametophyte. Sperm and egg then fuse and grow into a tiny (haploid/diploid) ______ that is completely dependent on the female gametophyte for nutrients and water. Eventually, cells in the sporophyte undergo meiosis to produce (haploid/diploid) spores that scatter and grow into new ______ (moss plants).

4. Why are ferns less tied to a moist environment than mosses? Why are they more tied to a moist environment than seed plants?

____

____

____

____

5. What is pollen? What is a seed? What is a fruit? In what groups of plants are each of these structures found?

____

____

____

____

6. Fungi are autotrophs/heterotrophs/both. Fungi are unicellular/multicellular/both. Fungi reproduce sexually/asexually/both.

____

Name ______________________________ Date ____________

7. Match the following animal groups with the list of features. Some groups may have more than one feature.

sponges
cnidarians
flatworms
roundworms
arthropods
mollusks
annelids
echinoderms
chordates
catilaginous fishes
ray-finned fishes
amphibians
reptiles
turtles
snakes
birds
mammals

a. muscular foot responsible for locomotion
b. swim bladder
c. the only animals that lack tissues
d. adaptations for subduing large prey and swallowing them whole
e. muscles all run longitudinally—from head to tail—down the body, resulting in a flailing whiplike motion
f. polyp stage and medusa stage alternate
g. leeches
h. segmented worms
i. segmented bodies and jointed legs
j. terrestrial vertebrates restricted to moist environments because their skins are composed of living cells that are vulnerable to drying out, and lay eggs without shells
k. maintain a constant flow of water in through numerous pores, into the central cavity, and out the top, whose purpose is for food capture
l. birds and crocodiles
m. tube feet
n. tentacles armed with barbed stinging cells
o. hollow bones, air sacs in the body; and a four-chambered heart
p. includes the most diverse group of living things on Earth, the insects
q. a notochord, gill slits, and a tail that extends beyond the anus
r. starfish
s. have a skeleton made of cartilage
t. have hair and feed their young milk
u. tapeworms
v. platypus
w. squeezes its entire body inside its ribcage
x. flying endotherms
y. frogs

Name ______________________________ Date __________

Chapter 19: Human Biology I—Control and Development

The Nervous System

1. Match parts of the brain with the body functions they are responsible for. Some parts may correspond with more than one function:

brainstem	a. processing visual information
cerebellum	b. balance, posture, coordination
frontal lobes of cerebrum	c. sensory areas for temperature, touch, and pain
parietal lobes of cerebrum	d. involuntary activities such as heartbeat, respiration, and digestion
occipital lobes of cerebrum	e. comprehending language
temporal lobes of cerebrum	f. control of voluntary movements
thalamus	g. sorting and filtering of incoming information, which it then passes to the cerebral cortex
hypothalamus	h. emotions, such as pleasure and rage, and bodily drives, such as hunger and thirst
	i. speech
	j. body's internal clock

2. The two divisions of the nervous system are the __________ and __________. The central nervous system consists of the __________ and __________.

__________ carry messages from the senses to the central nervous system. __________ connect neurons to other neurons. __________ carry messages from the central nervous system to muscle cells or to other responsive organs.

Motor neurons are further divided into two groups, the __________, which controls voluntary actions and stimulates our voluntary muscles, and the __________, which controls involuntary actions and stimulates involuntary muscles and other internal organs.

The autonomic nervous system includes a __________ that promotes a "fight or flight" response and a __________ that operates in times of relaxation.

Name ______________________________ Date __________

The Nervous System—continued

3.

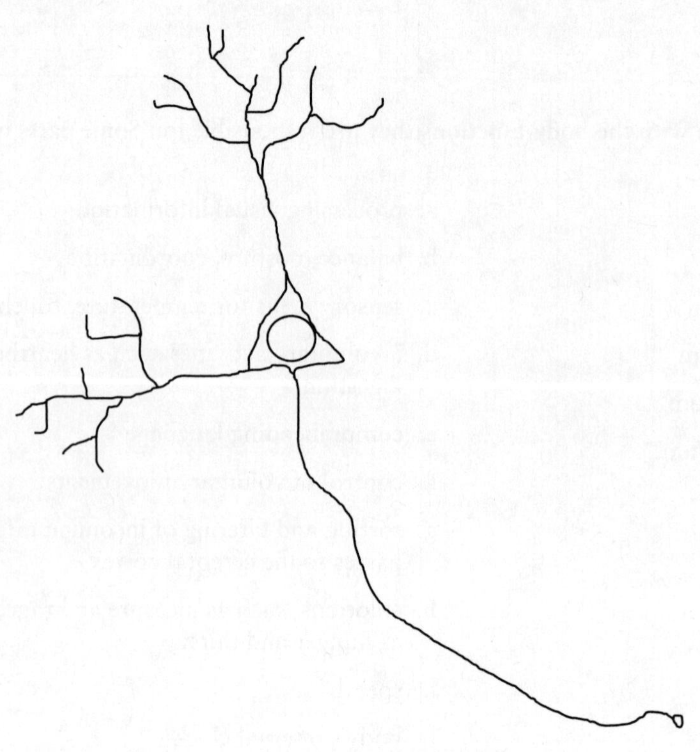

Identify the dendrites, cell body, and axon in the neuron above. What is the function of each of these parts of a neuron?

Name ______________________________ Date ____________

The Nervous System—continued

4. Place the following events that occur during an action potential in the correct order:

   membrane potential decreases

   membrane potential increases

   potassium channels open

   sodium channels open

   membrane potential spikes

   membrane potential reaches threshold

   membrane potential returns to resting potential

   sodium ions flow into neuron

   potassium ions flow out of neuron

5. What is the difference between an electric synapse and a chemical synapse? How does each allow a neuron to signal a target cell?

   ____

   ____

   ____

   ____

   ____

   ____

Name ______________________________ Date __________

Chapter 19: Human Biology I—Control and Development

Senses

1. The light-sensitive cells in the eye are found in the _________. There are two types of light-sensitive cells, _________ and _________.

2. Does each of the following describe rods or cones?

   responsible for vision at night or in dim light

   detect color

   3 types

   cannot discriminate colors

   very sensitive to light, responding to even a single photon

   not very good at making out fine details

   nonfunctional form of these causes colorblindness

   cone-shaped

3. How does sound move through the ear? Place the following items in the correct order:

   middle ear bones

   pinna

   cochlea

   eardrum

4. What are the five basic tastes?

   ____

5. What is the role of prostaglandins in sensing pain?

   ____

   ____

   ____

Name ____ Date ____________

Chapter 19: Human Biology I—Control and Development

Skeleton and Muscles

1. Label the three layers of bone below:

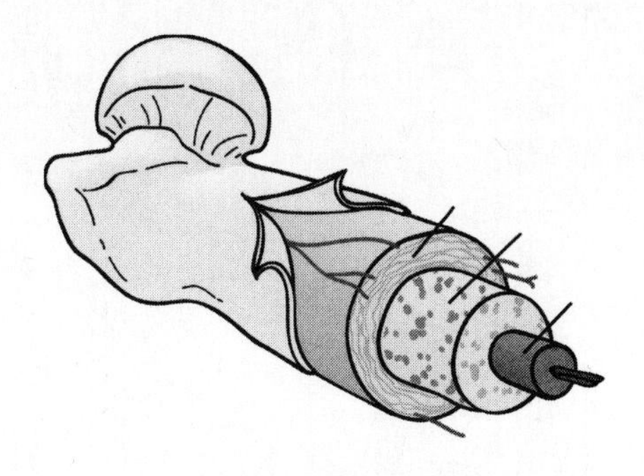

What are the two types of bone marrow?

____

What is the function of each type?

____

2. Muscles are made up of a series of contractile units called _________. These contain carefully arranged fibers of two proteins, thin filaments called _________ and thick filaments called _________. When an action potential arrives at a muscle cell, _________ are released from the cell's_______. These ions allow a series of pivoting heads on the ______ fibers to attach to_________. The heads attach and pivot, _________ the length of the sarcomere a tiny bit—about 10 nanometers, to be exact—and, consequently, the length of the muscle as a whole. After pulling, the heads release, recock, reattach, and pull again. This cycle repeats until ________________________________. Muscle contractions require energy, of course. ATP is required for__________________, an essential step in the contraction cycle.

Name ____ Date __________

Chapter 20: Human Biology II—Care and Maintenance

Circulatory System

1. Each heartbeat begins in a part of the right atrium called the _________, or pacemaker. The pacemaker initiates an action potential that sweeps quickly through the_____________, which contract simultaneously. The signal also passes to the___________, and from there to the two_________, which also contract simultaneously.

2. Why does the heart make a "lub-dubb" sound as it beats? What is the "lub"? What is the "dubb"?

____

____

____

3. How does blood flow around the body? Place the following in the correct order, beginning with the right atrium:

   Arteries to lungs

   Right atrium

   Left atrium

   Veins from body tissues

   Veins from lungs

   Capillaries that supply body tissues

   Arteries to body tissues

   Arterioles

   Right ventricle

   Venules

   Left ventricle

   Capillaries around alveoli

4. The three types of cells found in blood are_________, _________, and _________. _________ carry oxygen. _________ are part of the immune system and help our bodies defend against disease. _________ are involved in blood clotting.

5. The molecule in red blood cells that carries oxygen is _________. It can carry up to _________ oxygen molecules, each bound to a(n) _________ atom.

Name ______________________________ Date __________

Chapter 20: Human Biology II—Care and Maintenance

Respiration, Digestion

1. Match each part of the respiratory system with its description:

nasal passages	a. site of gas exchange
larynx	b. another word for trachea
trachea	c. allows us to speak
alveoli	d. raises ribcage during breathing
diaphragm	e. air is warmed and moistened
muscles between ribs	f. dome-shaped muscle involved in breathing
bronchi	g. tubes leading to right and left lungs
windpipe	h. stiffened by cartilaginous rings

2. Where does each of the following events important in digestion occur?

   a. bile is produced here

   b. a highly acidic mix of hydrochloric acid and digestive enzymes is added

   c. absorption of nutrients

   d. food is chewed, breaking it into smaller pieces

   e. reabsorption of water

   f. food is churned by muscular action

   g. saliva begins digestion of starches in food

   h. bile is added to food here

   i. pancreatic enzymes are added to food

   j. home to large numbers of *Escherichia coli* and other bacteria

   k. bile is stored here

   l. absorption of many minerals

   m. synthesis of vitamin K and some of the B vitamins by bacteria

Name ____ Date ____________

Chapter 20: Human Biology II—Care and Maintenance

Excretory System, Immune System

1. We excrete nitrogen-containing wastes in the form of ____.

2. The functional unit of a kidney is the _________. Each of these units is associated with a cluster of capillaries called the _________. Blood pressure in the _________ pushes fluid out of the capillaries and into _________. This fluid is called the_________ and is pretty similar to_________. From Bowman's capsule, the filtrate flows into the _____________. After "good" molecules are removed from the filtrate and "bad" molecules are added to it, the filtrate moves to the _________, whose primary function is to _________. Then the filtrate moves into the _________, where additional wastes are transported into it. Finally, the filtrate moves down the _________, where _________ may be absorbed if _________ is present. Urine drips into the _________ and flows down the _________ to the _________, where it is temporarily stored. Finally, urine flows down the _________ and out the body.

3. Is each of the following associated with innate or acquired immunity?

   skin

   T cells

   B cells

   acidic secretions from hair follicles in skin

   antibodies

   enzymes in tears and milk

   on the order of 10 million different receptors

   mucus

   large Y-shaped proteins

   memory cells

   immediate response

   antigen

   inflammatory response

   clones

   histamine

   vaccines

   maximum response delayed

Name ______________________________ Date ____________

Chapter 21: Ecosystems and Environment

Species Interactions

Define the following terms.

1. A population is ______________________________.

   A community is ______________________________.

   An ecosystem is ______________________________.

2. Look at the food chain below:

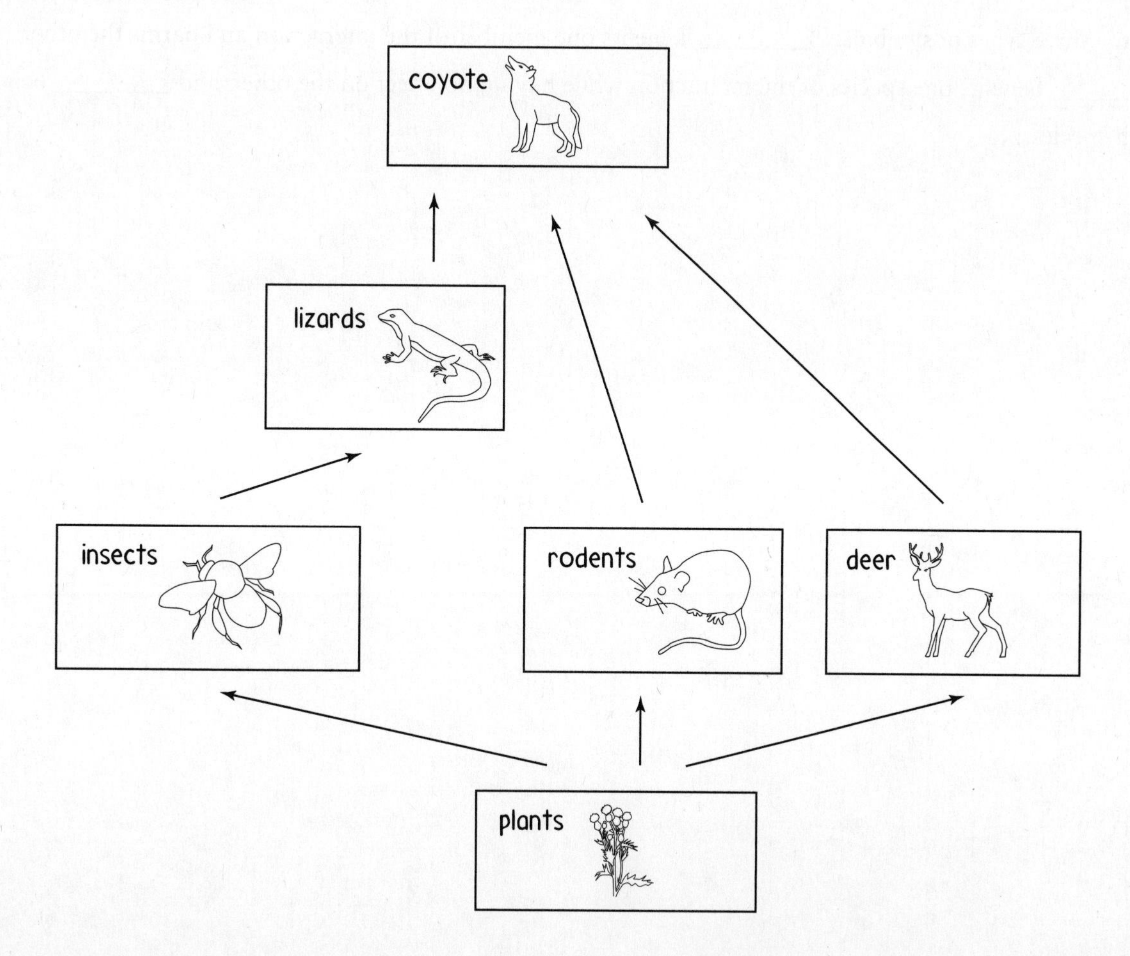

Name ______________________________ Date __________

Species Interactions—continued

a. Who are the producers in this community?

____

b. Who are the primary consumers?

____

c. Who are the secondary consumers?

____

d. Who are the tertiary consumers?

____

e. Who is the top predator?

____

3. What is a species's niche?

____

4. What is symbiosis?

____

Of the three types of symbiosis, ________ benefits one member of the interaction and harms the other, ________ benefits one species of the interaction while having no effect on the other, and ________ benefits both species.

Name ______________________________ Date __________

Chapter 21: Ecosystems and Environment

Ecosystems

1. Match each of the following features with the appropriate biome or aquatic life zone:

tropical forest	a. tropical grassland
temperate forest	b. habitats that receive very little precipitation, may be cold or hot
coniferous forest	c. permafrost
tundra	d. plants that are adapted to surviving in changing salinity conditions
savanna	e. in the water column
temperate grassland	f. evergreen trees with needlelike leaves
desert	g. inhabitants have adaptations that allow them to deal with exposure, temperature fluctuations, and the action of waves
littoral zone	h. deep water habitats in ponds and lakes
limnetic zone	i. more species are found in this biome than in all other biomes combined
profundal zone	j. grassland with fertile soil, in areas with four distinct seasons
estuary	k. lake habitat close to the water surface and to shore
pelagic zone	l. trees drop their leaves in the autumn
benthic zone	m. lake and pond habitats that are close to the water surface, but far from shore
intertidal zone	n. on the ocean bottom
neritic zone	o. marine habitats far from coasts
oceanic zone	p. underwater marine habitats near the coasts

2. Carbon is an essential component of all organic molecules. Most of the inorganic carbon on Earth exists as __________ and is found either in the __________ or dissolved in __________. Carbon moves into the biotic world when ______________________________ ______________________. This carbon becomes available to other organisms as it passes up the food chain. Carbon is returned to the environment by living organisms as __________ during the process of __________. An important part of Earth's carbon supply is also found in fossil fuels such as __________. Human burning of fossil fuels has released so much carbon dioxide that atmospheric carbon dioxide levels are now higher than they have been for 420,000 years. Because atmospheric carbon dioxide traps heat on the planet, this has resulted in __________.

Name ______________________________ Date __________

Ecosystems—continued

3. How do living things obtain water?

____

How do living things return water to the abiotic world?

____

____

4. The difference between primary succession and secondary succession is that

____

______________________________.

During ecological succession, the total biomass of the ecosystem typically__________, and the number of species present in the habitat typically __________.

Ecological succession ends with the __________.

5. What does the intermediate disturbance hypothesis state?

____

____

____

Name ______________________________ Date ____________

Chapter 21: Ecosystems and Environment

Populations

1. Which of the following graphs shows exponential growth and which shows logistic growth?

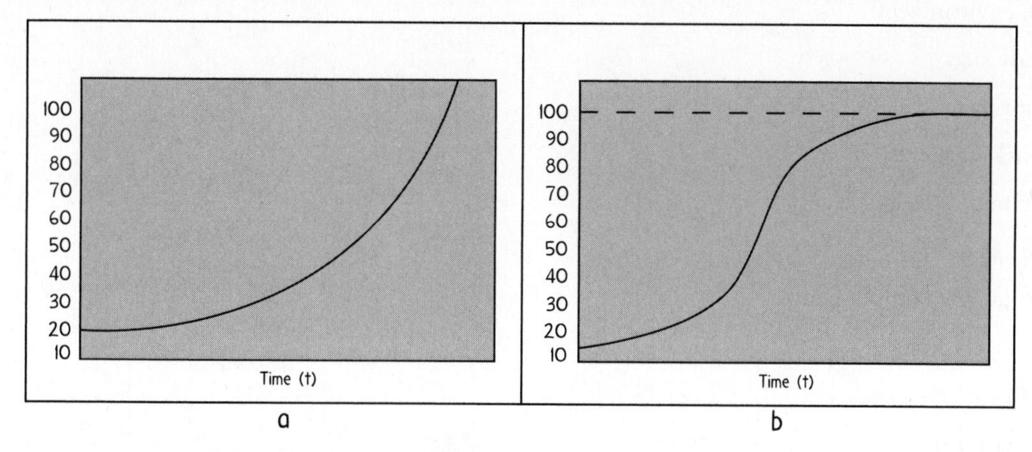

Using words, describe the difference between exponential growth and logistic growth.

____

____

____

What is the carrying capacity in the graph on the right?

____

2. Label the three survivorship curves shown below. Indicate which organisms at right correspond to which survivorship curve.

____

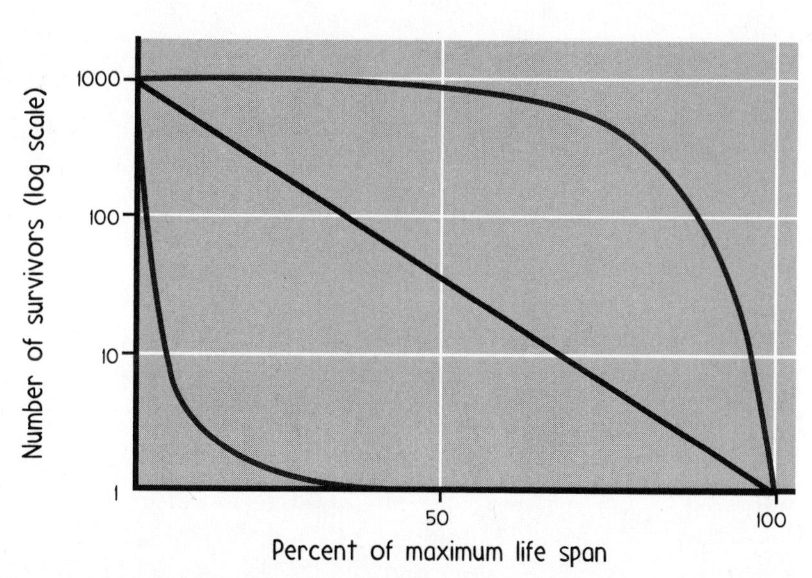

Using words, describe Type I, Type II, and Type III survivorship.

____

____

____

Name ______________________________ Date __________

Populations—continued

3. Of the following traits, indicate which are associated with K-selected populations and which are associated with r-selected populations.

   unstable environment

   large body size

   few offspring

   no parental care

   reach sexual maturity slowly

   long life expectancy

   exponential population growth

   Type III survivorship

   parental care

   stable environment

   short life expectancy

   small body size

   Type I survivorship

Name ______________________________ Date ____________

Chapter 22: Plate Tectonics

Faults

Three block diagrams are illustrated below. Draw arrows on each diagram to show the direction of movement. Answer the questions next to each diagram.

A.

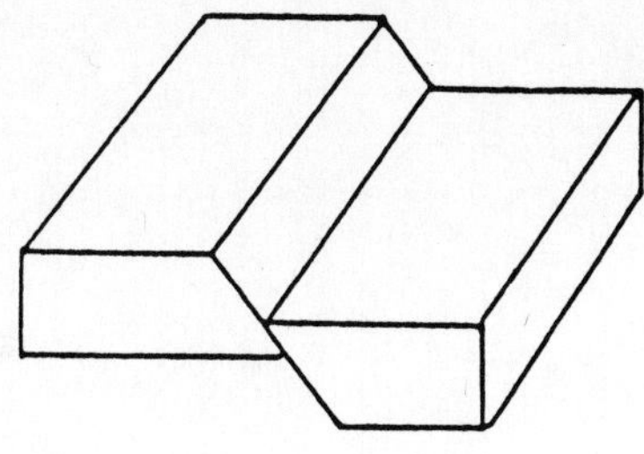

What type of force produced Fault A?

____

Name the fault ______________________________

Where would you expect to find this type of fault?

____

____

B.

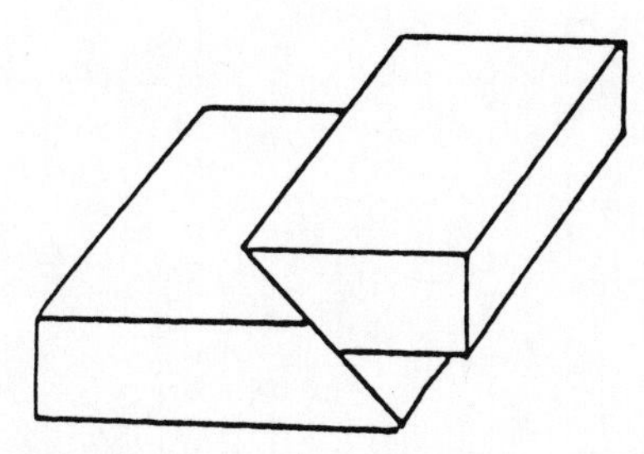

What type of force produced Fault B?

____

Name the fault ______________________________

Where would you expect to find this type of fault?

____

____

C.

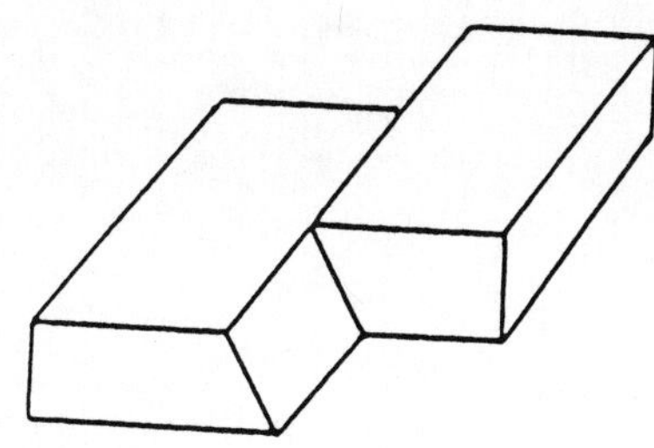

What type of force produced Fault C?

____

Name the fault ______________________________

Where would you expect to find this type of fault?

____

____

Name ______________________________ Date __________

Chapter 22: Plate Tectonics

Structural Geology

Much subsurface information is learned by oil companies when wells are drilled. Some of this information leads to the discovery of oil, and some reveals subsurface structures such as folds and/or faults in Earth's crust.

Four oil wells that have been drilled to the same depth are shown on the cross section below. Each well encounters contacts between different rock formations at the depths shown in the table below. (A contact is simply the boundary between two types or ages of rock). Rock formations are labeled A–F, with youngest A and oldest F.

	Depth to Contact (in meters)			
Contact	Oil Well #1	Oil Well #2	Oil Well #3	Oil Well #4
A-B	200	not encountered	200	not encountered
B-C	400	100	400	100
C-D	600	300	600	300
D-E	800	500	800	500
E-F	1000	700	1000	700

1. In the cross section below, Contacts D–E and E–F are plotted for Oil Wells 1 and 2. Plot the remainder of the data for all four wells, labeling each point you plot.
2. Draw lines to connect the contacts between the rock formations (as is done for Contacts D–E and E–F for Oil Wells 1 and 2).

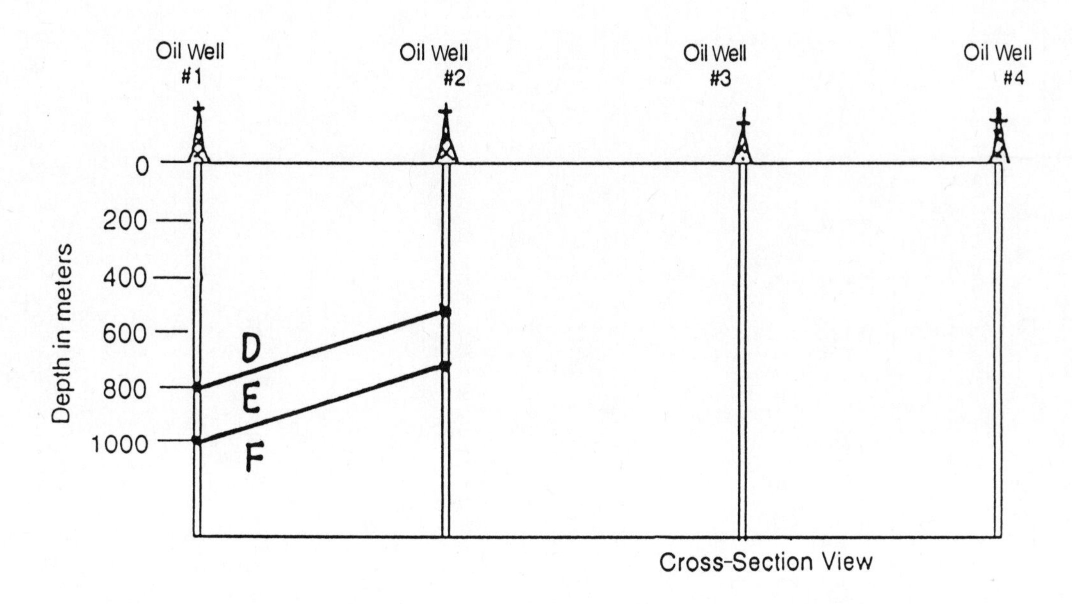

Cross-Section View

Questions

1. What explanation can you offer for no sign of Formation A in Oil Wells 2 and 4?

____

2. What geological structures are revealed? Label them on the cross section. (Hint: Consider the shapes of the contacts.)

____

Name ______________________________ Date ____________

Chapter 22: Plate Tectonics

Plate Boundaries

Draw arrows on the plate boundaries **A, B,** and **C,** to show the relative direction of movement.

Type of plate boundary for **A**? ____________________

What type of force generates this type of boundary?

____

Is this a site of crustal formation, destruction, or crustal transport?

____

A.

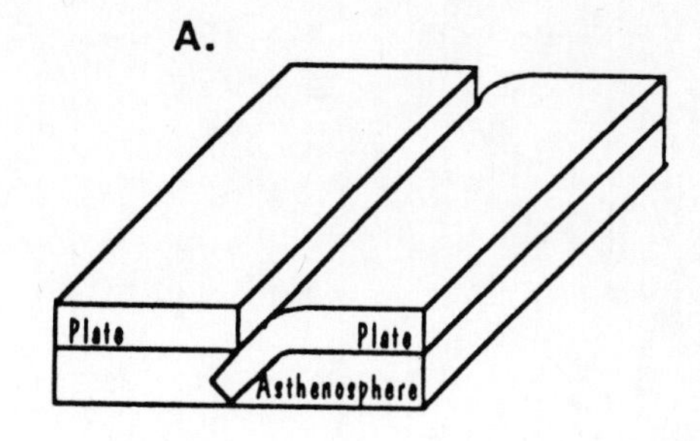

Type of plate boundary for **B**? ____________________

What type of force generates this type of boundary?

____

Is this a site of crustal formation, destruction, or crustal transport?

____

B.

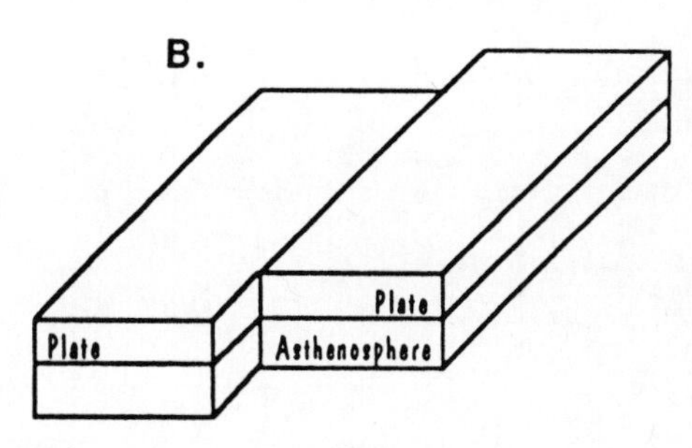

Type of plate boundary for **C**? ____________________

What type of force generates this type of boundary?

____

Is this a site of crustal formation, destruction, or crustal transport?

C.

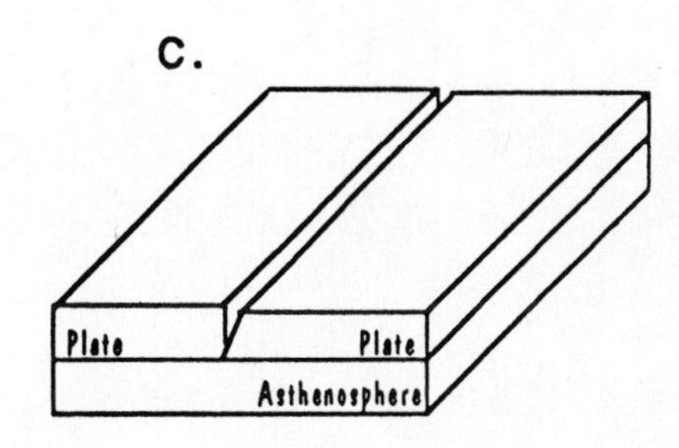

Draw arrows on the transform faults below to indicate relative motion.

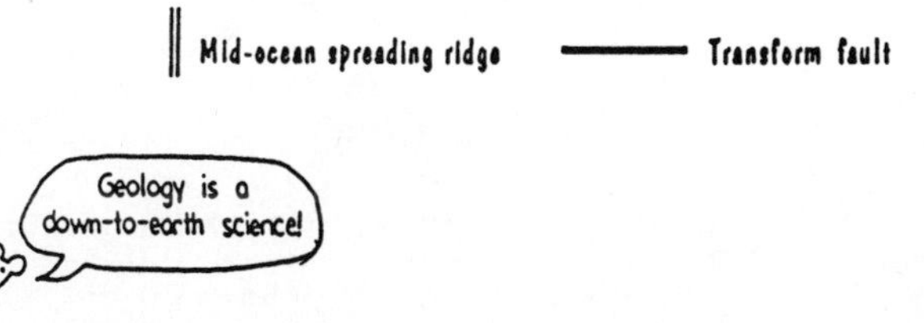

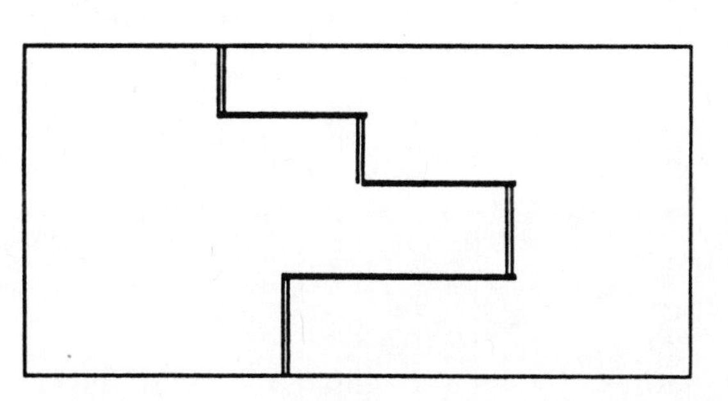

Name ______________________________ Date ____________

Chapter 22: Plate Tectonics

Seafloor Spreading

The rate of seafloor spreading is from 1 to 10 centimeters per year. If we know the distance and age between two points on the ocean floor, we can determine the rate of spreading. Diagrams A, B, and C show stages of sea floor spreading. Spreading begins at A, continues to B where rocks at location **P** begin to spread to the farther-apart positions we see in C. At C newer rock at the ocean crest **S** dated at 10 million years. Using the scale: 1 mm = 50 km, use a ruler on C to find the

1. separation rate of the two continental landmasses in the past 10 million years, in cm/yr: ____________
2. age of the seafloor at **P** in Diagram C (1 cm/yr = 10 km/million years) ____________

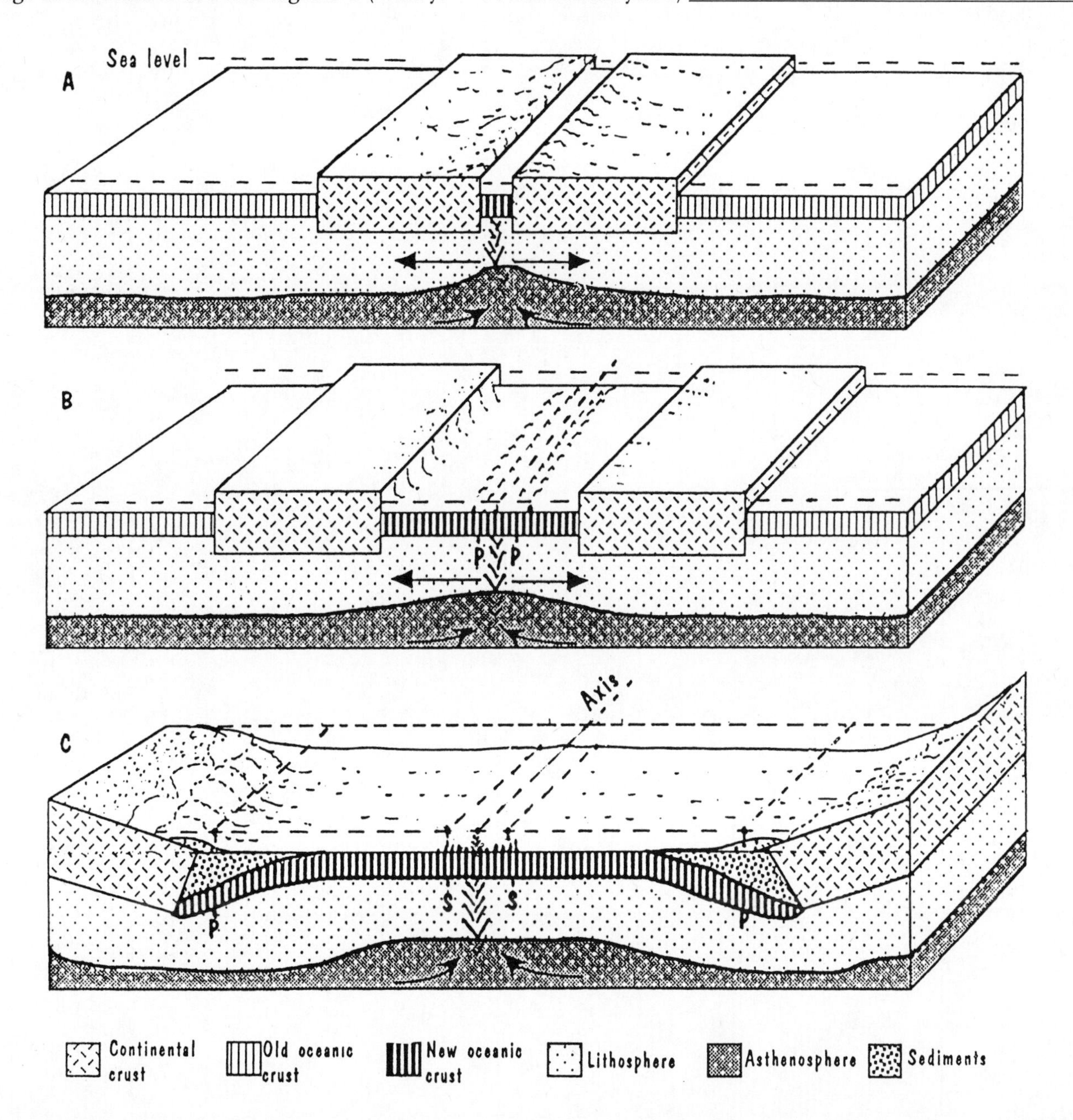

Name ______________________________ Date ____________

Chapter 22: Plate Tectonics

Plate Boundaries and Magma Generation

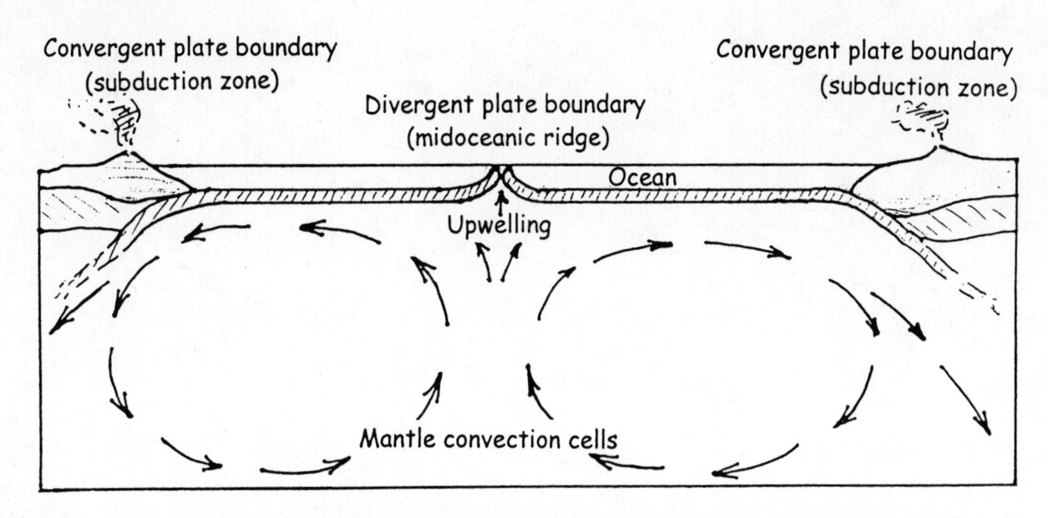

Partial melting occurs in the mantle at divergent and convergent plate boundaries when the melting point of mantle rocks is lowered.

1. What is the mechanism that lowers the melting point of mantle rock at divergent boundaries?

____

____

____

____

____

2. What is the mechanism that lowers the melting point of mantle rock at a subduction zone at a convergent boundary?

____

____

____

____

____

Name ____ Date __________

Chapter 23: Rocks and Minerals

Chemical Structure and Formulas of Minerals

Out of more than 3000 minerals known today, only about two dozen are abundant. Minerals are classified by their chemical composition and atomic structure and are divided into groups.

For each mineral structure diagrammed below, look for a pattern in the structure, count the number of atoms (ions) in each, and fill in the blanks.

The schematic diagrams are simple representations of small mineral structures. Actual mineral structures extend farther and comprise more atoms.

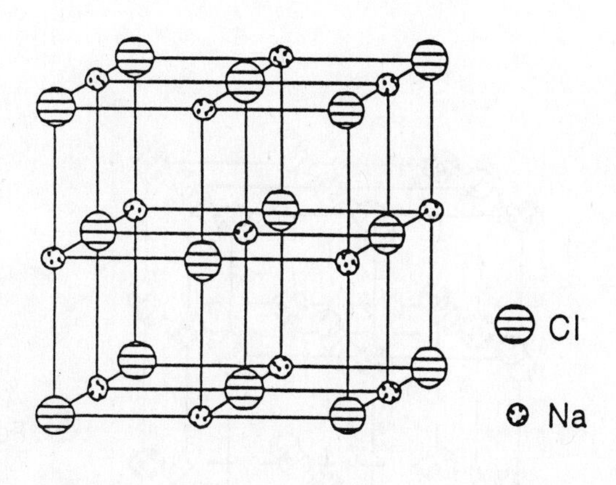

1. Circle pairs of Na and Cl ions in the structure and add any ion(s) needed to complete pairing. This mineral structure contains __________ Na ions and __________ Cl ions. The mineral's formula is __________. This mineral belongs to the __________ group.

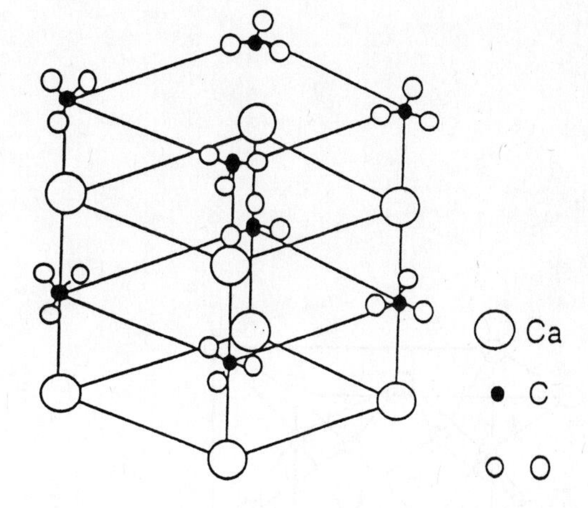

2. This mineral structure contains __________ Ca atoms, __________ C atoms, and __________ O atoms. The mineral's formula is __________. This mineral belongs to the __________ group.

Name ____ Date __________

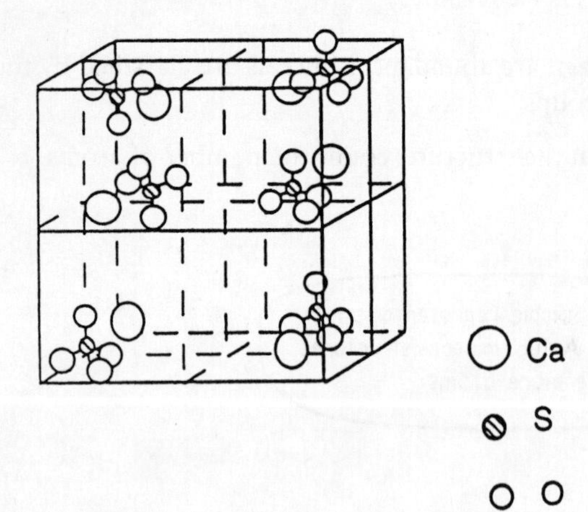

3. This mineral structure contains _________ Ca atoms, _________ S atoms, and _________ O ions. The mineral's formula is _________. This mineral belongs to the _________ group.

4. This mineral structure contains _________ Fe atoms and _________ S atoms. Complete the structure by adding the needed atoms(s). The mineral's formula is _________. This mineral belongs to the _________ group.

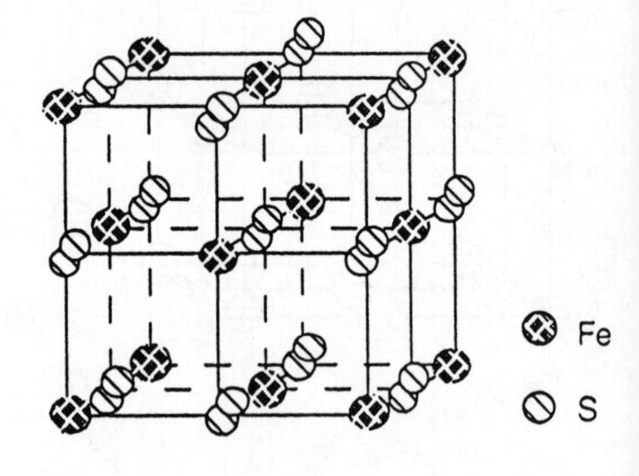

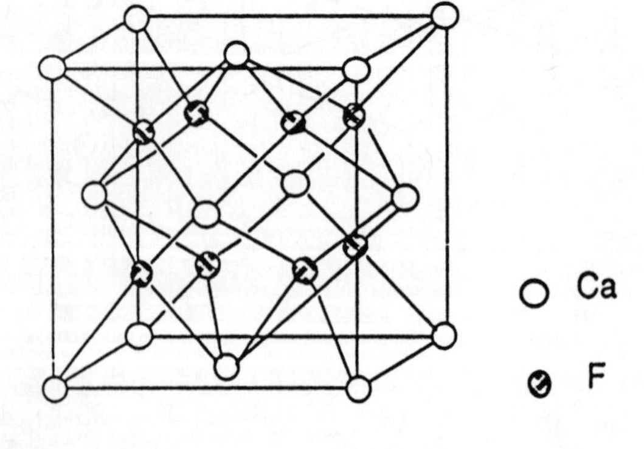

5. Complete the mineral structure so that each Ca atom is linked to two F atoms. Now the mineral structure contains _________ Ca atoms and _________ F atoms. The mineral's formula is _________. This mineral belongs to the _________ group.

Name ______________________________ Date __________

Chapter 23: Rocks and Minerals

The Rock Cycle

Complete the illustration at right, which depicts the different paths in the rock cycle. Insert arrows to show the direction of pathways.

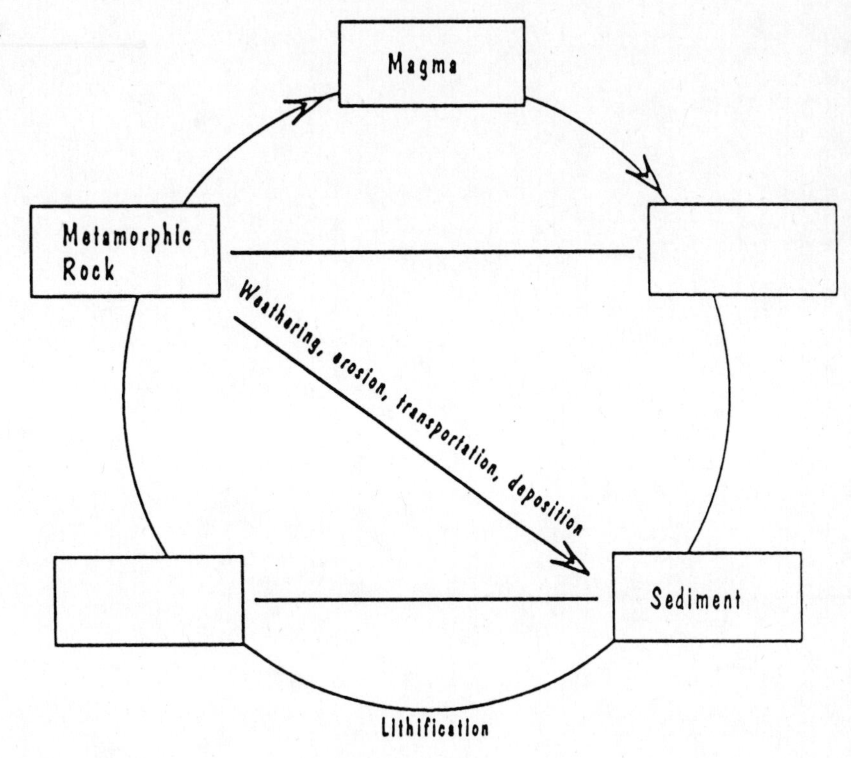

1. Can a rock that has undergone metamorphism turn into sedimentary rock? Why or why not?

____

2. By what process does hot molten magma become rock?

____

3. Can sedimentary rock become metamorphic rock? If so, how?

____

4. Can igneous rock become sedimentary rock? If so, how?

____

5. In what part(s) of the rock cycle does sandstone transform to coal?

____

6. The Big Island of Hawaii is almost 1 million years old. Yet hikers there almost never step on rock that is more than 1000 years old. Explain.

____

Name ______________________________ Date ____________

Chapter 24: Earth's Surface—Land and Water

Groundwater Flow and Contaminant Transport

The occupants of Houses 1, 2, and 3 wish to drill wells for domestic water supply. Note that the locations of all houses are between Lakes A and B, at different elevations.

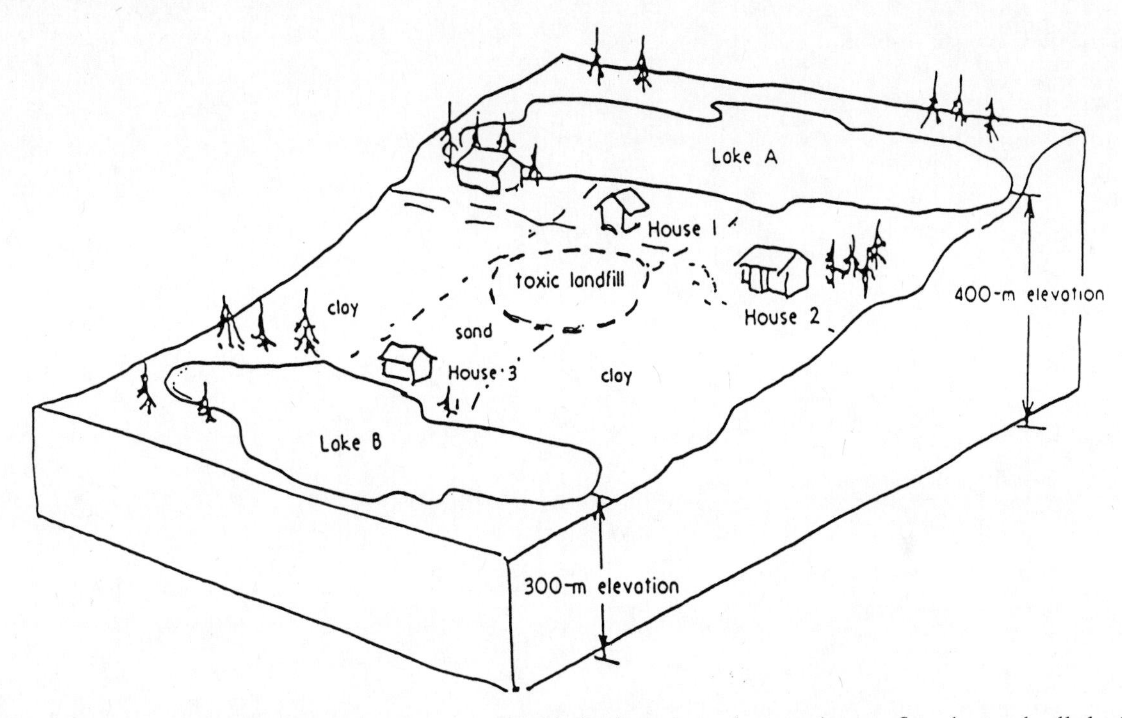

1. Show by sketching dashed lines on the drawing, the likely direction of groundwater flow beneath all the houses.
2. Which of the wells drilled beside Houses 1, 2, and 3 are likely to yield an abundant water supply?

   ____

3. Do any of the three need to worry about the toxic landfill contaminating their water supply? Explain.

   ____

4. Why don't the homeowners simply take water directly from the lakes?

   ____

5. Suggest a potentially better location for the landfill. Defend your choice.

   ____

   ____

   ____

Name ______________________________ Date ____________

Chapter 24: Earth's Surface—Land and Water

Stream Velocity

Let's explore how the average velocity of streams and rivers can change. The volume of water that flows past a given location over any given length of time depends both on the stream velocity and the cross-sectional area of the stream. We say

$$Q = A \times V$$

where Q is the volumetric flow rate (a measure of the volume passing a point per unit time). Also, A is the cross-sectional area of the stream, and V its average velocity.

Consider the stream shown below, with rectangular cross-sectional areas

$$A = \text{width} \times \text{depth}$$

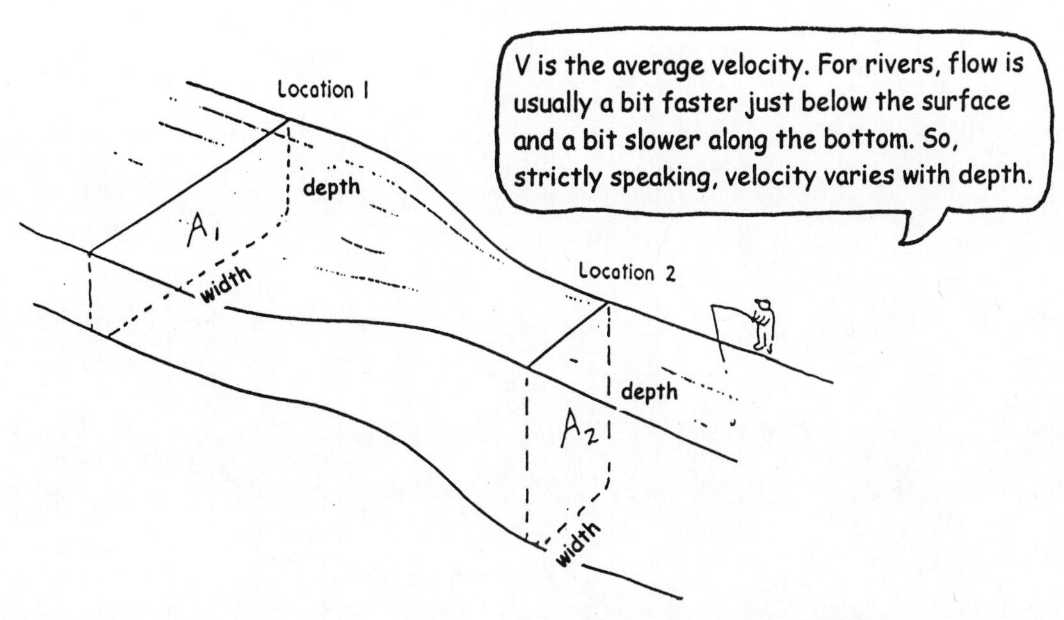

V is the average velocity. For rivers, flow is usually a bit faster just below the surface and a bit slower along the bottom. So, strictly speaking, velocity varies with depth.
Location 1
depth
A₁
width
Location 2
depth
A₂
width

1. The two locations shown have no stream inlets or outlets between them, so Q remains constant. Suppose the cross-sectional areas are also constant ($A_1 = A_2$), with Location 2 deeper but narrower than Location 1. What change, if any, occurs for the stream velocity?

   ____

2. If Q remains constant, what happens to stream velocity at Location 2 if A_2 is less than A_1?

   ____

3. If Q remains constant, what happens to stream velocity at Location 2 if A_2 is greater than A_1?

   ____

4. What happens to stream velocity at Location 2 if area A_2 remains the same, but Q increases (perhaps by an inlet along the way?)

   ____

5. What happens to stream velocity at Location 2 if both A_2 and Q increase?

   ____

Name ____ Date __________

Chapter 25: Weather

Earth's Seasons

1. The warmth of equatorial regions and coldness of polar regions on Earth can be understood by considering light from a flashlight striking a surface. If it strikes perpendicularly, light energy is more concentrated as it covers a smaller area; if it strikes at an angle, the energy spreads over a larger area. So the energy per unit area is less.

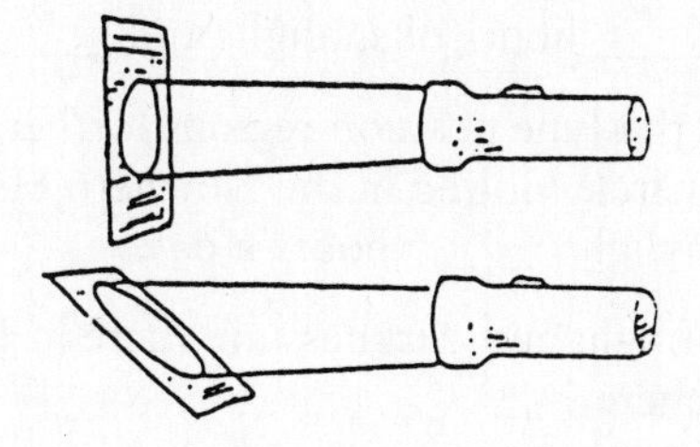

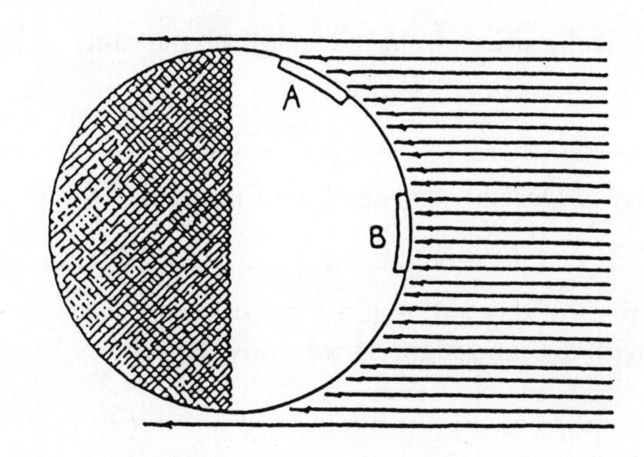

The arrows represent rays of light from the distant Sun incident upon Earth. Two areas of equal size are shown, Area A near the north pole and Area B near the equator. Count the rays that reach each region, and explain why Area B is warmer than Area A.

2. Earth's seasons result from the 23.5-degree tilt of Earth's daily spin axis as it orbits the Sun. When Earth is at the position shown on the right in the sketch below (not shown to scale), the Northern Hemisphere tilts toward the Sun, and sunlight striking it is strong (more rays per area). Sunlight striking the Southern Hemisphere is weak (fewer rays per area). Days in the north are warmer, and daylight lasts longer. You can see this by imagining Earth making its complete daily 24-hour spin.

   Do two things on the sketch: (1) Shade Earth in nighttime darkness for all positions, as is already done in the left position. (2) Label each position with the proper month—March, June, September, or December.

Name ______________________________ Date __________

Earth's Seasons—continued

a. When Earth is in any of the four positions shown, during one 24-hour spin a location at the equator receives sunlight half the time and is in darkness the other half of the time. This means that regions at the equator always get about ______ hours of sunlight and ______ hours of darkness.

b. Can you see that in the June position regions farther north have longer days and shorter nights? Locations north of the Arctic Circle (dotted in the Northern Hemisphere) are always illuminated by the Sun as Earth spins, so they get daylight ______ hours a day.

c. How many hours of light and darkness are there in June at regions south of the Antarctic Circle (dotted line in Southern Hemisphere)?

____

d. Six months later, when Earth is at the December position, is the situation in the Antarctic the same or is it the reverse?

____

e. Why do South America and Australia enjoy warm weather in December instead of in June?

____

3. Earth spins about its polar axis once each 24 hours, which gives us day and night. If Earth's spin was instead only one rotation per year, what difference would there be with day and night as we enjoy them now?

____

____

If the spin of the Earth was the same as its revolution rate around the Sun, would we be like the Moon — one side always facing the body it orbits?

Name ______________________________ Date ____________

Chapter 25: Weather

Short and Long Wavelength

The sine curve is a pictorial representation of a wave—the high points being crests, and the low points troughs. The height of the wave is its *amplitude*. The wavelength is the distance between successive identical parts of the wave (like between crest to crest, or trough to trough). Wavelengths of water waves at the beach are measured in meters, wavelengths of ripples in a pond are measured in centimeters, and the wavelengths of light are measured in billionths of a meter (nanometers).

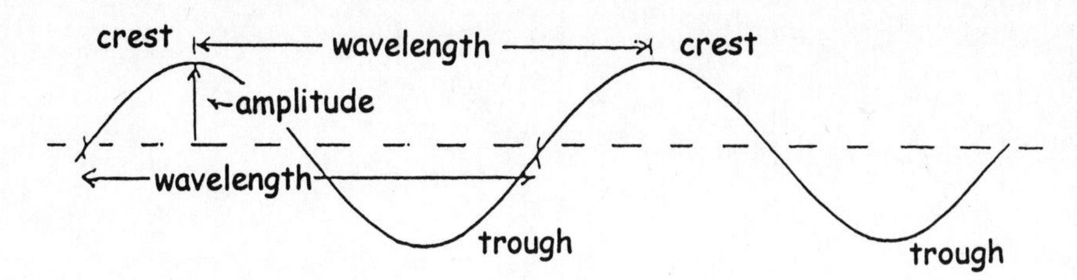

In the boxes below sketch three waves of the same amplitude—Wave A with half the wavelength of Wave B, and Wave C with wavelength twice as long as Wave B.

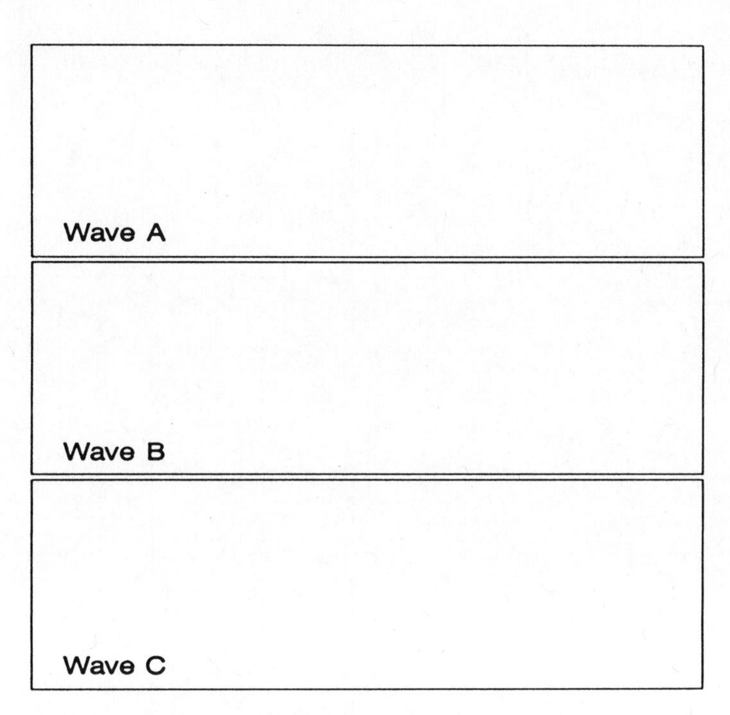

1. If all three waves have the same speed, which has the highest frequency? ____________
2. Compared with solar radiation, terrestrial radiation has a ____________ wavelength.
3. In a florist's greenhouse, ____________ waves are able to penetrate the greenhouse glass, but ____________ waves cannot.
4. Earth's atmosphere is similar to the glass in a greenhouse. If the atmosphere were to contain excess amounts of water vapor and carbon dioxide, the air would be opaque to ____________ waves.

Name ____ Date __________

Chapter 25: Weather

Driving Forces of Air Motion

The primary driving force of Earth's weather is _______. The unequal distribution of solar radiation on Earth's surface creates temperature differences which in turn result in pressure differences in the atmosphere. These pressure differences generate horizontal winds as air moves from _______ pressure to _______ pressure. The weather patterns are not strictly horizontal though, there are other forces affecting the movement of air. Recall from Newton's Second Law that an object moves in the direction of the *net* force acting on it. The forces acting on the movement of air include: 1) pressure gradient force, 2) Coriolis force, 3) centripetal force, and 4) friction.

The greater the pressure difference, the greater the force, and the greater the wind. The "push" caused by the horizontal differences in pressure across a surface is called the *pressure gradient force.* This force is represented by isobars on a weather map. Isobars connect locations on a map that have equal atmospheric pressure. The pressure gradient force is perpendicular to the isobars and strongest where the isobars are closely spaced. So, the steeper the pressure gradient, the _______ the wind.

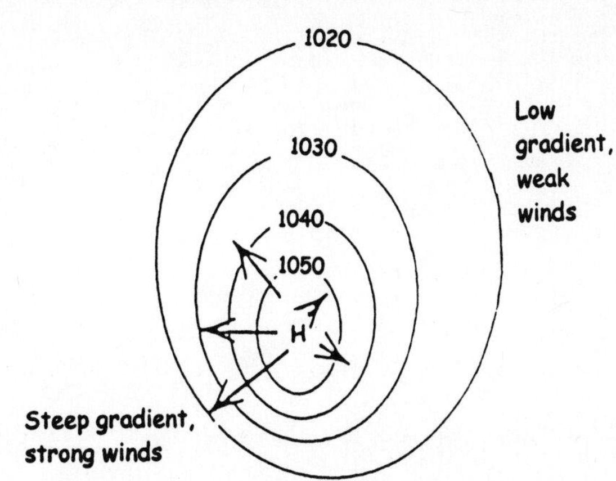

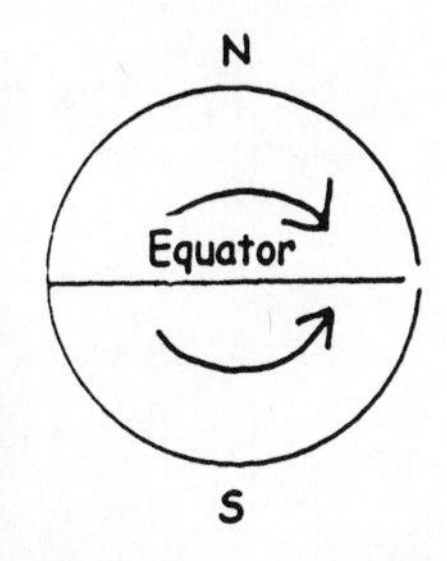

The *Coriolis effect* is a result of Earth's rotation. The Coriolis effect is the deflection of the wind from a _______ path to a _______ path. The Coriolis effect causes the wind to veer to the right of its path in the Northern Hemisphere and to the left of its path in the Southern Hemisphere.

As the wind blows around a low or high pressure center it constantly changes its direction. A change in speed or direction is acceleration. In order to keep the wind moving in a circular path the net force must be directed inward. This __________ force is called *centripetal force.*

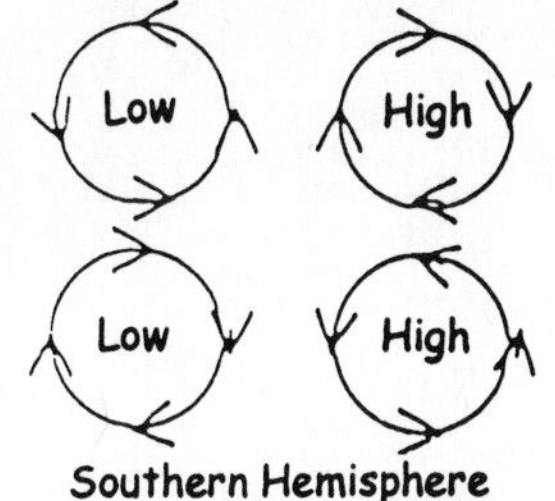

The forces described above greatly influence the flow of upper winds (winds not influenced by surface frictional forces). The interaction of these forces cause the winds in the Northern Hemisphere to rotate __________ around regions of high presssure and __________ around regions of low pressure. In the Southern Hemisphere the situation is reversed—winds rotate __________ around a high pressure zone and __________ around a low pressure zone.

Winds blowing near Earth's surface are slowed by *frictional forces.* In the Northern Hemisphere surface winds blow in a direction __________ into the centers of a low pressure area and __________ out of the centers of a high pressure area. The spiral direction is reversed in the Southern Hemisphere. Draw arrows to show the direction of the pressure gradient force.

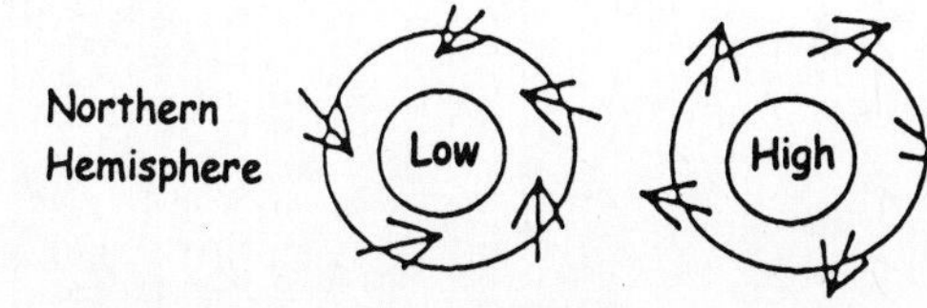

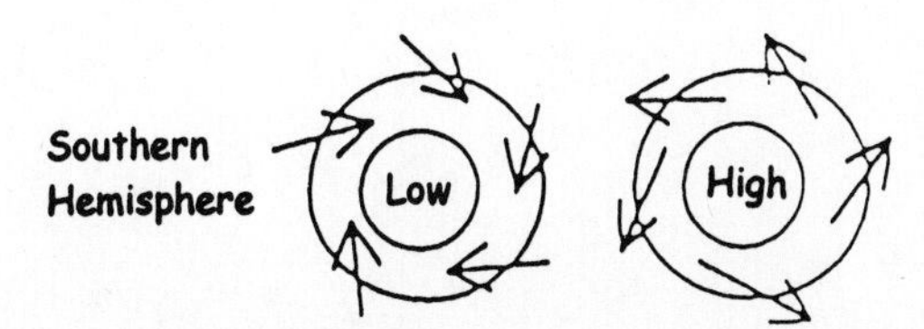

Name ______________________________ Date ____________

Chapter 25: Weather

Air Temperature and Pressure Patterns

Temperature patterns on weather maps are depicted by *isotherms*—lines that connect all points having the same temperature. Each isotherm separates temperatures of higher values from temperatures of lower values.

The weather map to the right shows temperatures in degrees Fahrenheit for various locations. Using 10 degree intervals, connect same value numbers to construct isotherms. Label the temperature value at each end of the isotherm. One isotherm has been completed as an example.

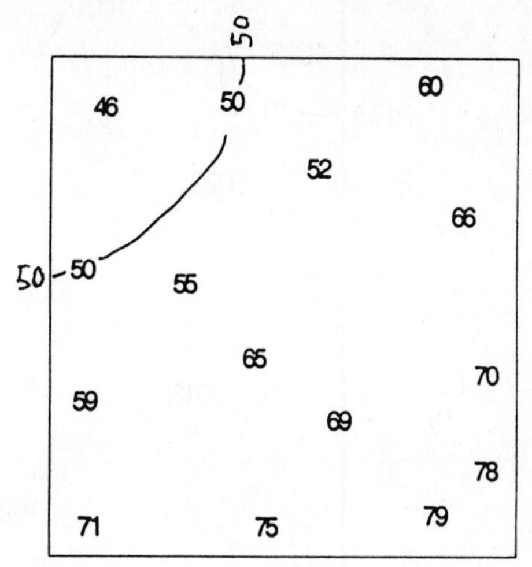

Tips for drawing isotherms:

- Isotherms can never be open ended.
- Isotherms are "closed" if they reach the boundary of plotted data, or make a loop.
- Isotherms can never touch, cross, or fork.
- Isotherms must always appear in sequence; for example, there must be a 60° isotherm between a 50° and 70° isotherm.
- Isotherms should be labeled with their values.

Pressure patterns on weather maps are depicted by *isobars*—lines which connect all points having equal pressure. Each isobar separates stations of higher pressure from stations of lower pressure.

The weather map below shows air pressure in millibar (mb) units at various locations. Using an interval of 4 (for example, 1008, 1012, 1016, etc.), connect equal pressure values to construct isobars. Label the pressure value at each end of the isobar. Two isobars have been completed as an example.

Tips for drawing isobars are similar to those for drawing isotherms.

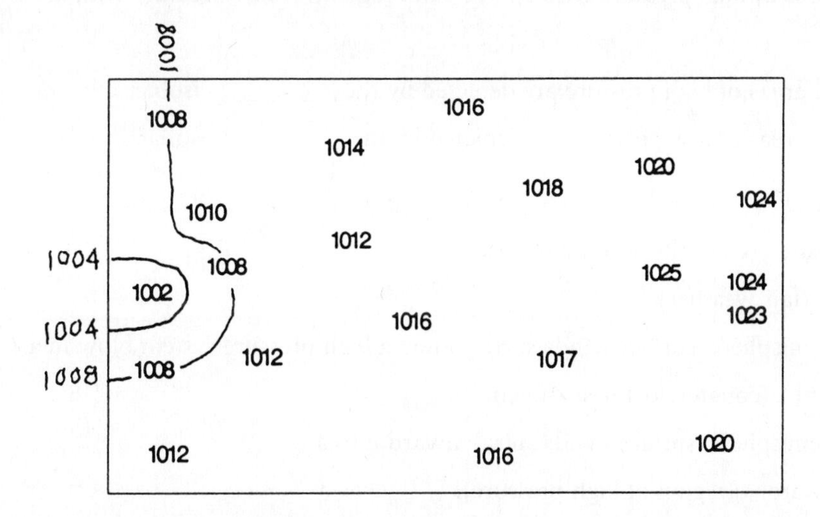

Name ______________________________ Date ____________

Air Temperature and Pressure Patterns—continued

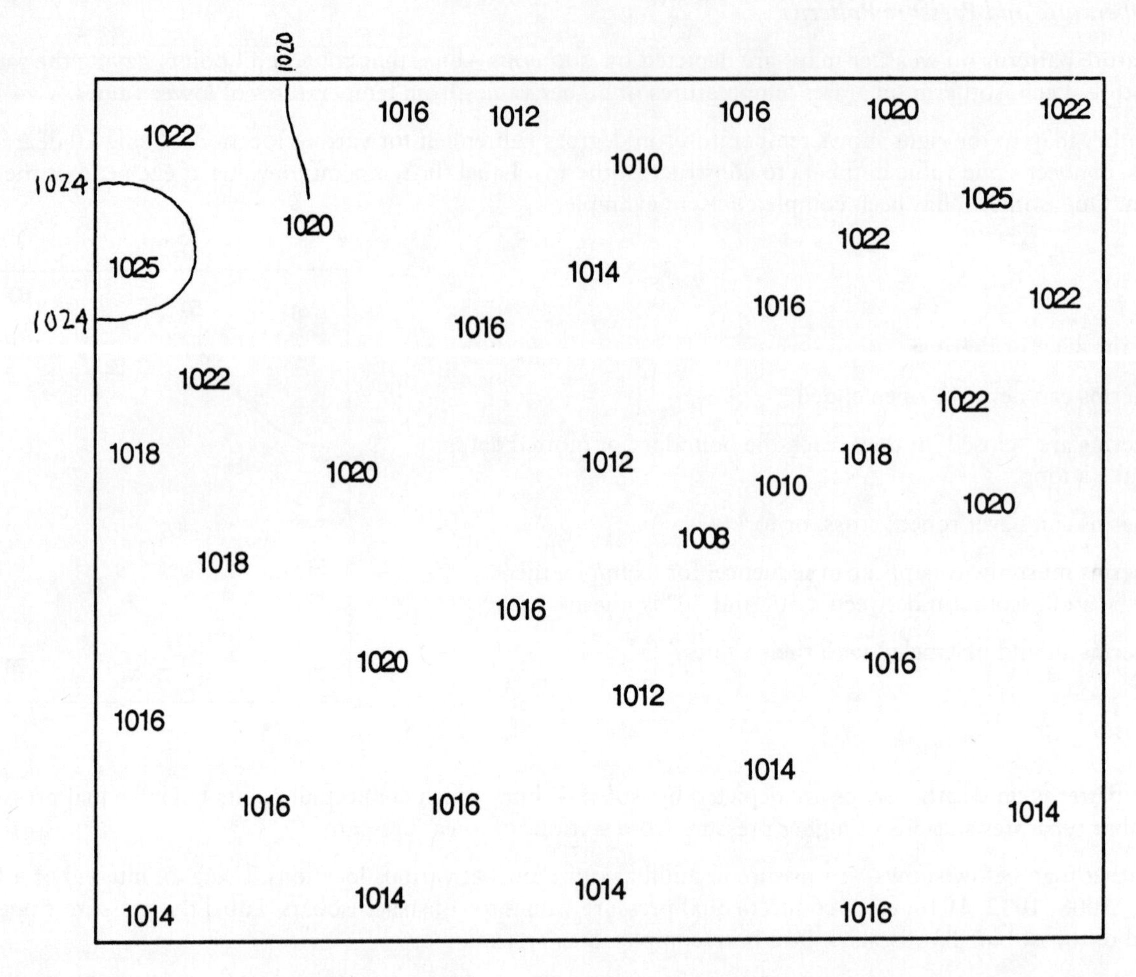

On the map above, use an interval of 4 to draw lines of equal pressure (isobars) to show the pattern of air pressure. Locate and mark regions of high pressure with an "H" and regions of low pressure with an "L".

1. On the map above, areas of high pressure are depicted by the __________ isobar.
2. On the map above, areas of low pressure are depicted by the __________ isobar.

Circle the correct answer.

3. Areas of high pressure are usually accompanied by
   (stormy weather) (fair weather).
4. In the Northern Hemisphere, surface winds surrounding a high pressure system blow in a
   (clockwise direction) (counterclockwise direction).
5. In the Northern Hemisphere, surface winds spiral inward into a
   (region of low pressure) (region of high pressure).

Name ______________________________ Date __________

Chapter 25: Weather

Surface Weather Maps

Station models are used on weather maps to depict weather conditions for individual localities. Weather codes are plotted in, on, and around a central circle that describes the overall appearance of the sky. Jutting from the circle is a wind arrow, its tail in the direction from which the wind comes and its feathers indicating the wind speed. Other weather codes are in standard position around the circle.

Use the simplified station model and weather symbols to complete the statements below.

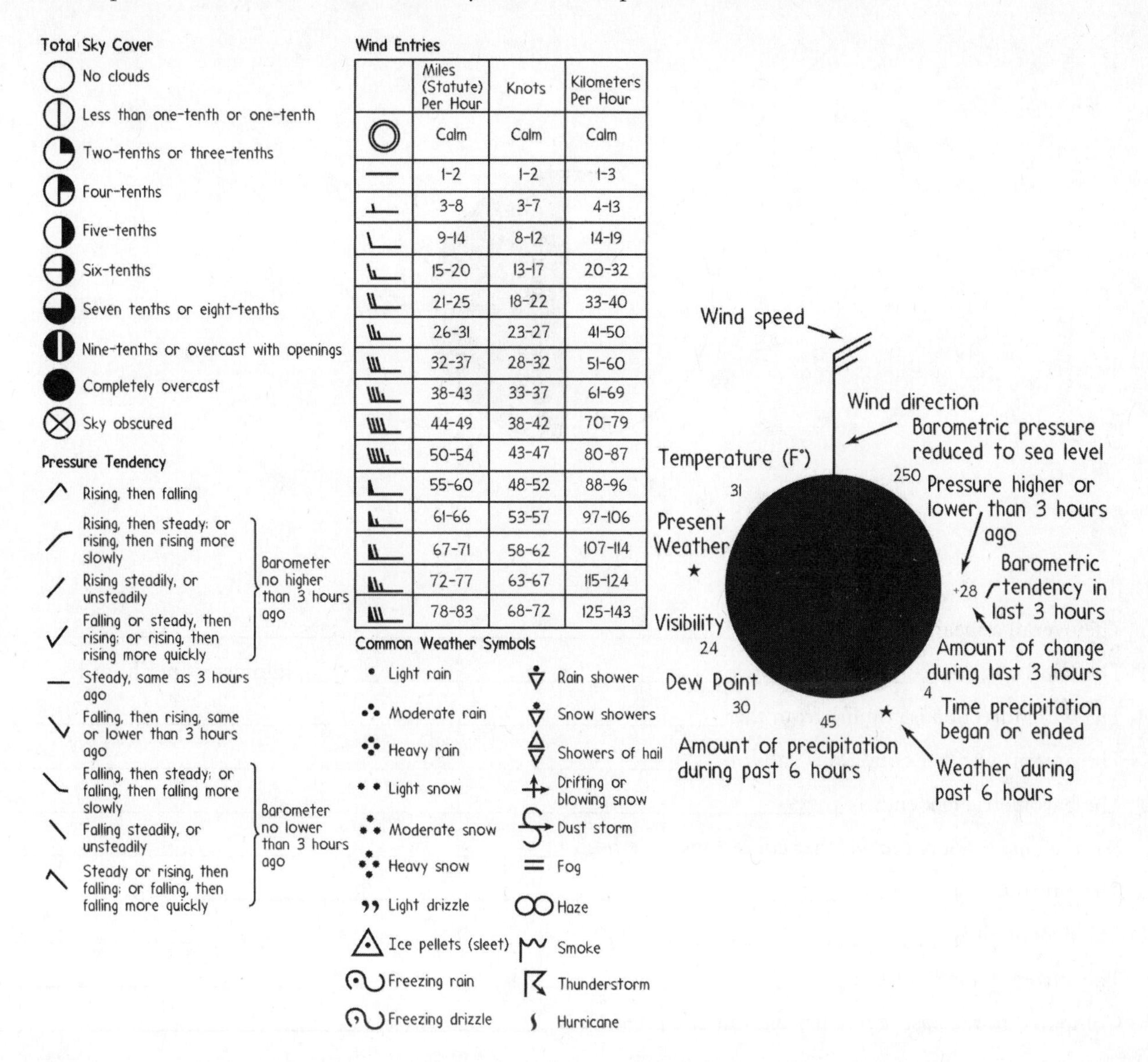

Wind Entries

	Miles (Statute) Per Hour	Knots	Kilometers Per Hour
	Calm	Calm	Calm
	1-2	1-2	1-3
	3-8	3-7	4-13
	9-14	8-12	14-19
	15-20	13-17	20-32
	21-25	18-22	33-40
	26-31	23-27	41-50
	32-37	28-32	51-60
	38-43	33-37	61-69
	44-49	38-42	70-79
	50-54	43-47	80-87
	55-60	48-52	88-96
	61-66	53-57	97-106
	67-71	58-62	107-114
	72-77	63-67	115-124
	78-83	68-72	125-143

1. The overall appearance of the sky is ______________________________.
2. The wind speed is ______________________________ kilometers per hour.
3. The wind direction is coming from the ______________________________.
4. The present weather conditions call for ______________________________.
5. The barometric tendency is ______________________________.
6. For the past 6 hours the weather conditions have been ______________________________.

Name ____ Date __________

Surface Weather Maps—continued

Use the unlabeled station model to answer the questions below.

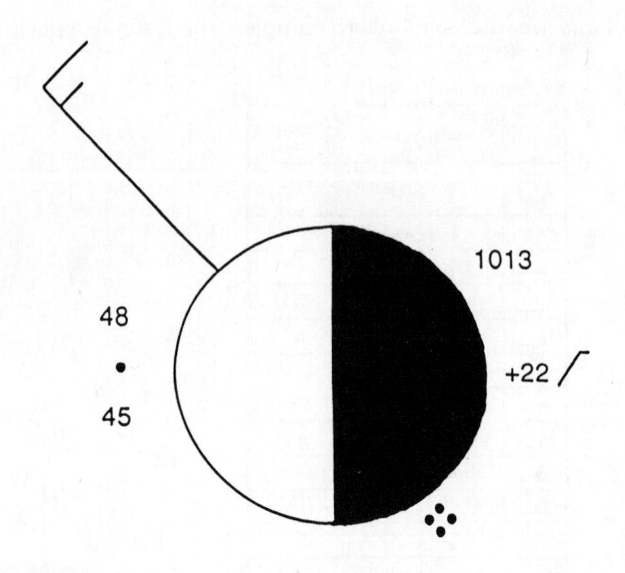

1. The overall appearance of the sky is ________________________________.
2. The wind speed is ________________________ kilometers per hour.
3. The wind direction is coming from the ________________________________.
4. The present weather conditions call for ________________________________.
5. The barometric tendency is ________________________________.
6. For the past 6 hours the weather conditions have been ________________________.
7. The barometric pressure is ________________________________.
8. The dew point is ________________________________.
9. The current temperature is ________________________________.
10. Compared to the past few hours, barometric pressure is ________________________.

Name ______________________________ Date ____________

Chapter 25: Weather

Chilly Winds

Often times it feels colder outside than a thermometer indicates. The apparent difference is called *wind chill*. On November 1, 2001, the National Weather Service implemented a new Wind Chill Temperature index. The new formula uses advances in science, technology, and computer modeling to provide a more accurate, understandable, and useful formula for calculating the dangers from winter winds and freezing temperatures.

For temperatures less than 50°F and wind speeds greater than 3 mph, the new formula used to estimate the temperature we perceive when a cold wind blows is:

$$\text{Wind Chill Temperature (°F)} = 35.74 + 0.6215T - 35.75(V^{0.16}) + 0.4275T(V^{0.16})$$

where V is the wind speed in miles per hour and T is the temperature in degrees Fahrenheit.

Wind Chill Temperature Table

		Temperature (°F)			
V (mph)	$V^{0.16}$	5.00	10.00	15.00	32.00
5.00	1.2937	-4.64	1.24	7.11	
10.00	1.4454				
15.00	1.5423	-12.99	-6.59	-0.19	

1. Using the formula given for Wind Chill Temperature (WCT), complete the above table. The variable $V^{0.16}$ (wind speed raised to 0.16 power) is provided in the table to simplify your calculations.
2. Which has a stronger impact on WCT, changes in wind speed or changes in temperature? Defend your answer with data from your completed table.

____

____

____

3. How does WCT vary when wind speed remains fixed and only temperature changes? Defend your answer.

____

____

____

4. How does WCT vary when temperature stays fixed and only wind speed changes? Defend your answer.

____

____

____

More information can be found on the National Weather Service's Web site:
http://www.nws.noaa.gov/om/windchill

Name ______________________________ Date ____________

Chapter 26: Earth's History

Relative Time—What Came First?

The cross-section below depicts many geologic events. List to the right the sequences of geologic history starting with the oldest event to the youngest event—and where appropriate include tectonic events (such as folding, deposition of strata (beds), subsidence, uplift, erosion, and intrusion).

P G F E D C B A H I J K M N O L

Youngest ____________

____

____

____

____

____

____

____

____

____

____

____

____

____

____

Oldest ____________

Examine the rings in the cross section of a tree and you do more than determine the age of the tree. Relative thicknesses of the rings tells a lot about the climate conditons throughout the trees history. A geologist similarly learns much about Earth's history by examination of rock layers in cross sections of Earth's crust.

Name ______________________________ Date ____________

Chapter 26: Earth's History

Age Relationships

From your investigation of the six geologic regions shown, answer the questions below. The number of each question refers to the same-numbered region.

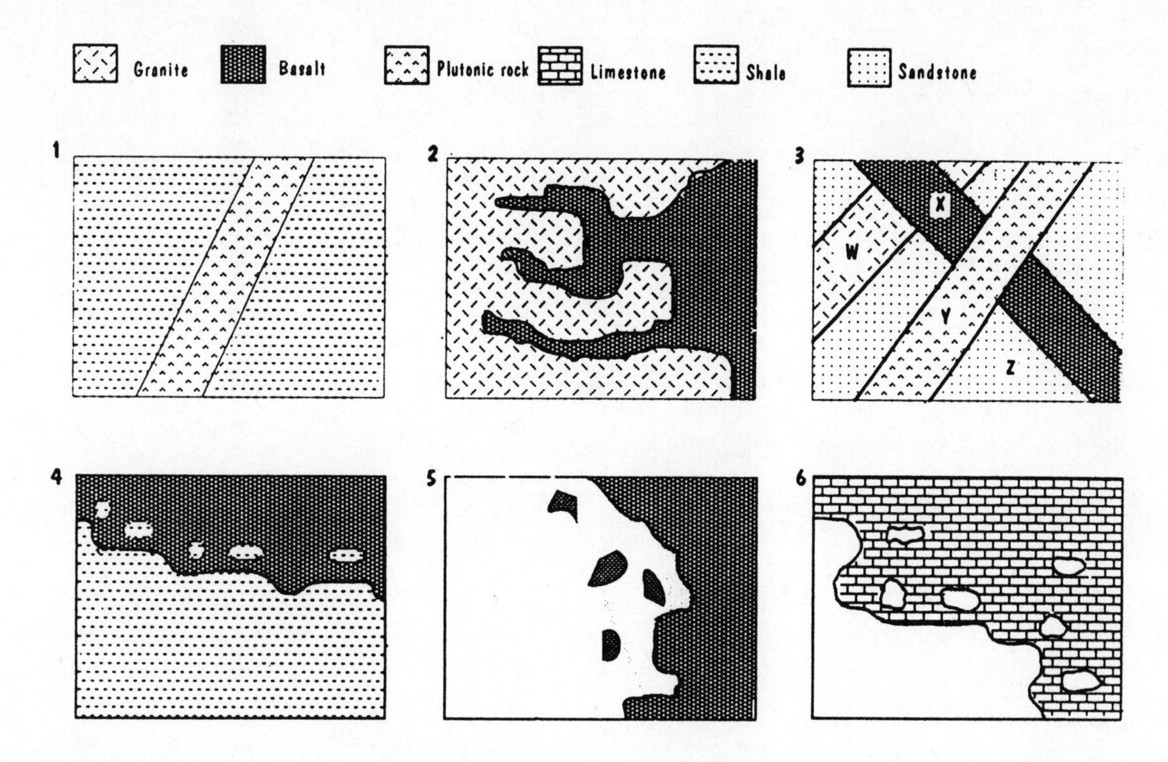

1. The shale has been cut by a dike. The radiometric age of the dike is estimated at 40 million years. Is the shale younger or older? ______________________________

2. Which is older, the granite or the basalt? ______________________________

3. The sandstone bed, Z, has been intruded by dikes. What is the age succession of dikes, going from oldest to youngest? ______________________________

4. Which is older, the shale or the basalt? ______________________________

5. Which is older, the sandstone or the basalt? ______________________________

6. Which is older, the sandstone or the limestone? ______________________________

Name ______________________________ Date ____________

Chapter 26: Earth's History

Unconformities and Age Relationships

The wavy lines in the four regions below represent unconformities. Investigate the regions and answer corresponding questions below.

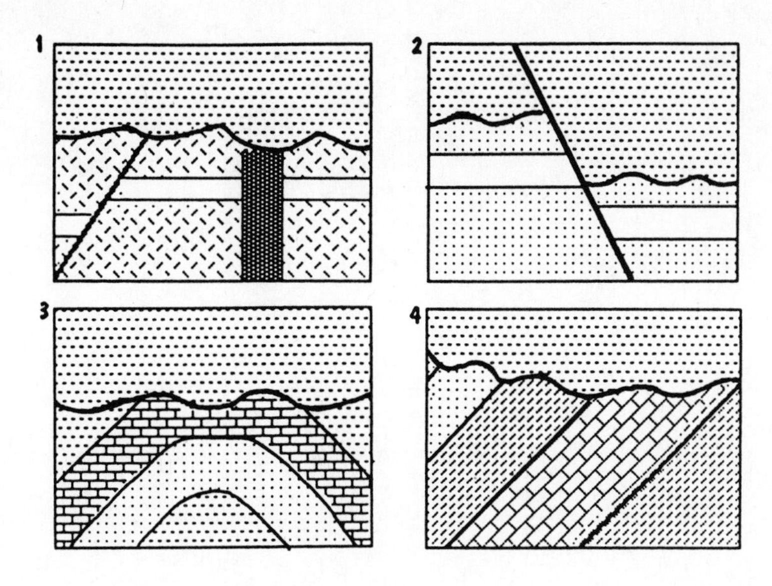

1. Did the faulting and dike occur before or after the unconformity? ______________________________

   What kind of conformity is it? ______________________________

2. Did the faulting occur before or after the unconformity? ______________________________

   What kind of unconformity is represented? ______________________________

3. Did the folding occur before or after the unconformity? ______________________________

   What kind of unconformity is it? ______________________________

4. What kind of unconformity is represented? ______________________________

5. Interestingly, the age of Earth is some 4.5 billion years old—yet the oldest rocks found are some 3.7 billion years old. Why do we find no 4.5-billion year old rocks?

   ____

6. What is the age of the innermost ring in a living redwood tree that is 2000 years old? What is the age of the outermost ring? How does this example relate to the previous question?

   ____

   ____

7. What is the approximate age of the atoms that make up a 3.7-billion year old rock?

   ____

Name ______________________________ Date ____________

Chapter 26: Earth's History

Radiometric Dating

Isotopes Commonly Used for Radiometric Dating

Radioactive Parent	Stable Daughter Product	Currently Accepted Half-life Value
Uranium-238	lead-206	4.5 billion years
Uranium-235	lead-207	704 million years
Potassium-40	argon-40	1.3 billion years
Carbon-14	nitrogen-14	5730 years

1. Consider a radiometric lab experiment wherein 99.98791% of a certain radioactive sample of material remains after one year. What is the decay rate of the sample?

____

2. What is the rate constant?
(Assume that the decay rate is constant for the one year period.)

____

3. What is the half-life?

____

4. Identify the isotope.

____

5. In a sample collected in the field, this isotope was found to be 1/16 of its original amount. What is the age of the sample?

____

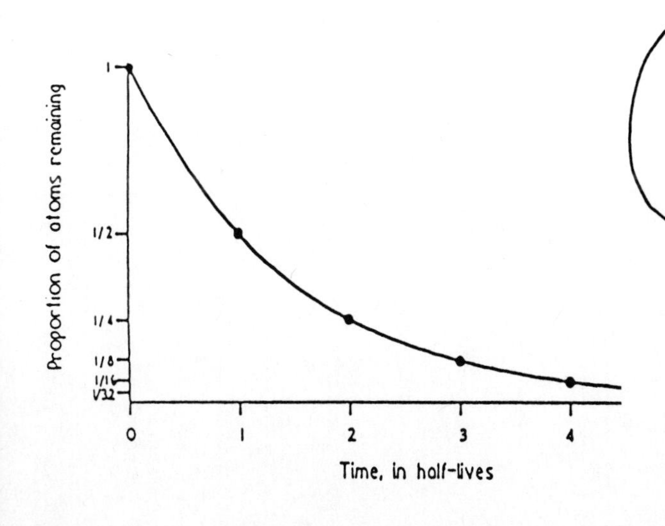

You need to know:

- Decay rate = (amount decayed) / time
- Rate constant K = (decay rate/ starting amount) in units 1/year
- Half-life T = 0.693/K (units in years)

Name ______________________________ Date ____________

Chapter 26: Earth's History

Our Earth's Hot Interior

Scientists faced a major puzzle in the nineteenth century. Volcanoes showed that Earth's interior is semi-molten. Penetration into the crust by bore-holes and mines showed that Earth's temperature increases with depth. Scientists knew that heat flows from the interior to the surface. They assumed that the source of Earth's internal heat was primordial, the afterglow of its fiery birth. Measurements of Earth's rate of cooling indicated a relatively young Earth—some 25 to 30 million years in age, but geological evidence indicated an older Earth. This puzzle wasn't solved until the discovery of radioactivity. Then it was learned that the interior was kept hot by the energy of radioactive decay. We now know the age of Earth is some 4.5 billion years—a much older Earth.

All rock contains trace amounts of radioactive minerals. Radioactive minerals in common granite release energy at the rate 0.03 J/kg · yr. Granite at Earth's surface transfers this energy to the surroundings practically as fast as it is generated, so we don't find granite any warmer than other parts of our environment. But what if a sample of granite were thermally insulated? That is, suppose all the increase of thermal energy due to radioactive decay were contained. Then it would get hotter. How much hotter? Let's figure it out, using 790 J/kg · C° as the specific heat of granite.

Calculations to make:

Let's see now... the relationship between quantity of heat Q, mass, specific heat C and temperature difference is
$Q = cm\Delta T$

1. How many joules are required to increase the temperature of 1-kg of granite by 500 C°?

____

2. How many years would it take radioactivity in a kilogram of granite to produce this many joules?

____

Questions to answer:

1. How many years would it take a thermally insulated 1-kilogram chunk of granite to undergo a 500°C increase in temperature?

____

2. How many years would it take a thermally insulated one-million-kilogram chunk of granite to undergo a 500°C increase in temperature?

____

3. Why does Earth's interior remain molten hot?

____

4. Rock has a higher melting temperature deep in the interior. Why?

____

5. Why doesn't Earth just keep getting hotter until it all melts?

____

An electric toaster stays hot while electric energy is supplied, and doesn't cool until switched off. Similarly, do you think the energy source now keeping Earth hot will one day suddenly switch off like a disconnected toaster - or gradually decrease over a long time?

Name ______________________________ Date __________

Chapter 27: The Solar System

Earth-Moon-Sun Alignment

Here we see a shadow on a wall cast by an apple. Note how the rays define the darkest part of the shadow, the *umbra,* and the lighter part of the shadow, the *penumbra.* The shadows that comprise eclipses of planetary bodies are similarly formed. Below is a diagram of the Sun, Earth, and the orbital path of the Moon (dashed circle). One position of the Moon is shown. Draw the Moon in the appropriate positions on the dashed circle to represent (a) a quarter moon, (b) a half moon, (c) a solar eclipse, and (d) a lunar eclipse. Label your positions. For c and d, extend rays from the top and bottom of the Sun to show umbra and penumbra regions.

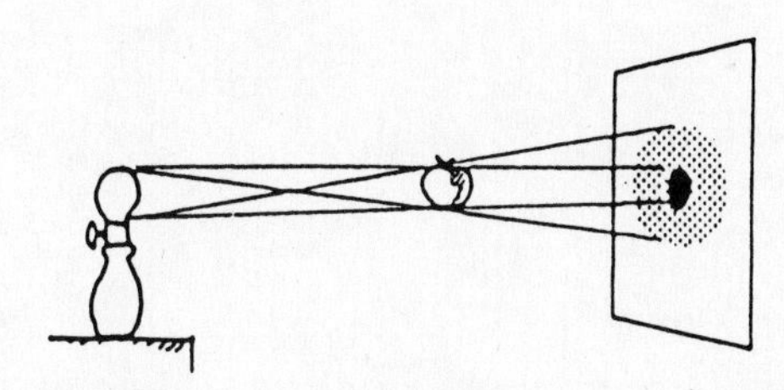

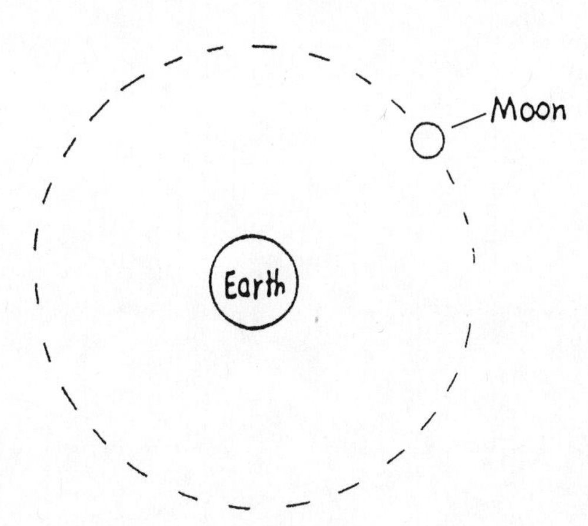

Figure not to scale

The diagram below shows three positions of the Sun, A, B, and C. Sketch the appropriate positions of the Moon in its orbit about Earth for (a) a solar eclipse, and (b) a lunar eclipse. Label your positions. Sketch solar rays similar to the above exercise.

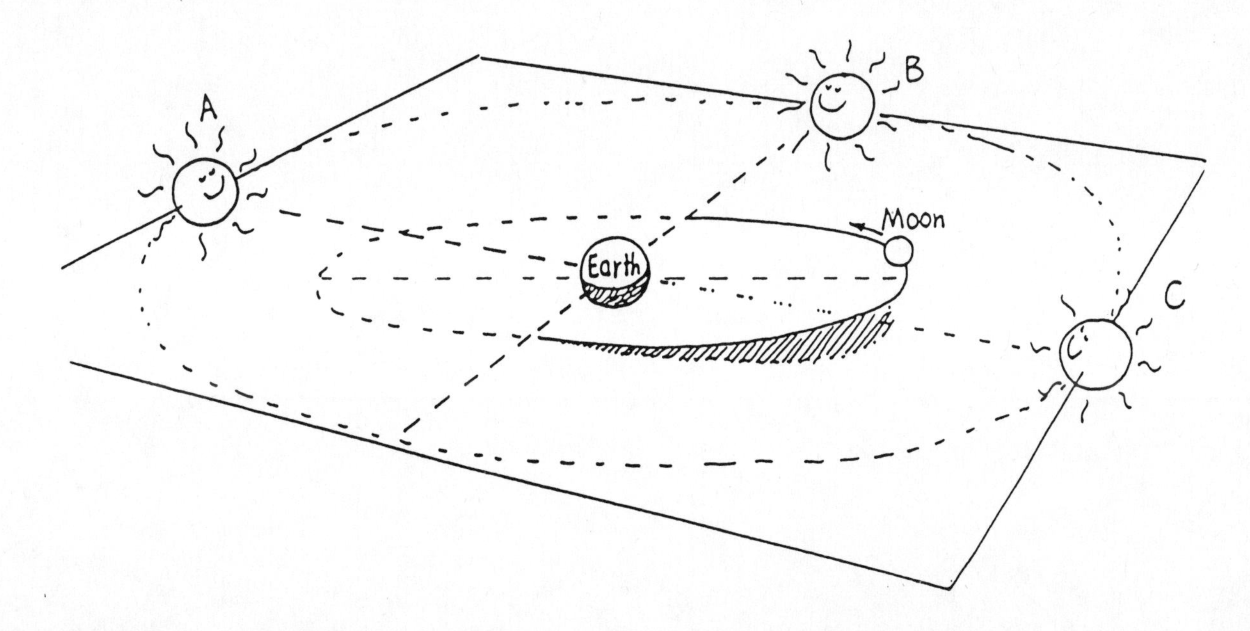

Name ______________________________ Date __________

Chapter 27: The Solar System

Pinhole Image Formation

Look carefully at the round spots of light on the shady ground beneath trees. These are *sunballs,* which are images of the Sun. They are cast by openings between leaves in the trees that act as pinholes. (Did you make a pinhole "camera" back in middle school?) Large sunballs, several centimeters in diameter or so, are cast by openings that are relatively high above the ground, while small ones are produced by closer "pinholes." The interesting point is that the ratio of the diameter of the sunball to its distance from the pinhole is the same as the ratio of the Sun's diameter to its distance from the pinhole. We know the Sun is approximately 150,000,000 km from the pinhole, so careful measurements of the ratio of diameter/distance for a sunball leads you to the diameter of the Sun. That's what this page is about. Instead of measuring sunballs under the shade of trees on a sunny day, make your own easier-to-measure sunball.

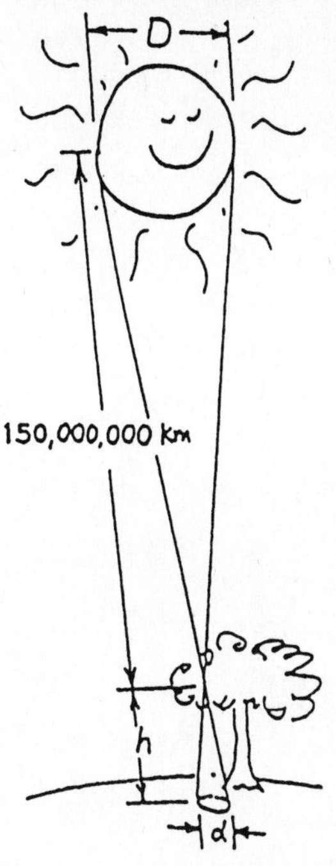

1. Poke a small hole in a piece of card. Perhaps an index card will do, and poke the hole with a sharp pencil or pen. Hold the card in the sunlight and note the circular image that is cast. This is an image of the Sun. Note that its size doesn't depend on the size of the hole in the card, but only on its distance. The image is a circle when cast on a surface perpendicular to the rays—otherwise it's "stretched out" as an ellipse.

2. Try holes of various shapes; say a square hole, or a triangular hole. What is the shape of the image when its distance from the card is large compared with the size of the hole? Does the shape of the pinhole make a difference?

____

____

3. Measure the diameter of a small coin. Then place the coin on a viewing area that is perpendicular to the Sun's rays. Position the card so the image of the sunball exactly covers the coin. Carefully measure the distance between the coin and the small hole in the card. Complete the following:

$$\frac{\text{Diameter of sunball}}{\text{Distance to pinhole}} = \underline{\qquad\qquad}$$

With this ratio, estimate the diameter of the Sun. Show your work on a separate piece of paper.

WHAT SHAPE DO SUNBALLS HAVE DURING A PARTIAL ECLIPSE OF THE SUN?

4. If you did this on a day when the Sun is partially eclipsed, what shape of image would you expect to see?

____

Name ____ Date ____________

Chapter 28: The Universe

Stellar Parallax

Finding distances to objects beyond the solar system is based on the simple phenomenon of **parallax.** Hold a pencil at arm's length and view it against a distant background—each eye sees a different view (try it and see). The displaced view indicates distance. Likewise, when Earth travels around the Sun each year, the position of relatively nearby stars shifts slightly relative to the background stars. By carefully measuring this shift, astronomer types can determine the distance to nearby stars.

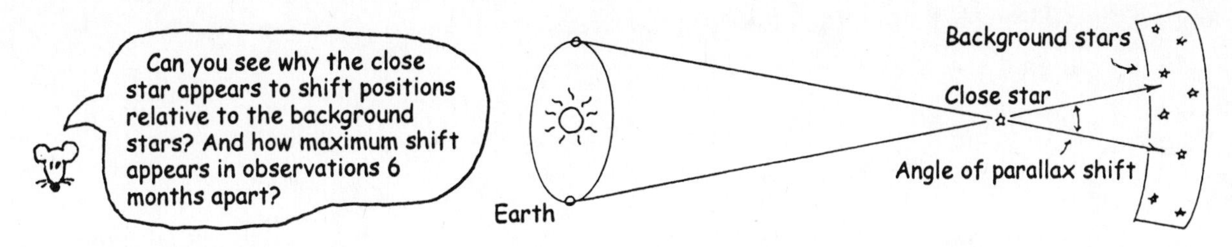
Can you see why the close star appears to shift positions relative to the background stars? And how maximum shift appears in observations 6 months apart?
Earth
Background stars
Close star
Angle of parallax shift

The photographs below show the same section of the evening sky taken at a 6-month interval. Investigate the photos carefully and determine which star appears in a different position relative to the others. Circle the star that shows a parallax shift.

A
B

Below are three sets of photographs, all taken at 6-month intervals. Circle the stars that show a parallax shift in each of the photos.

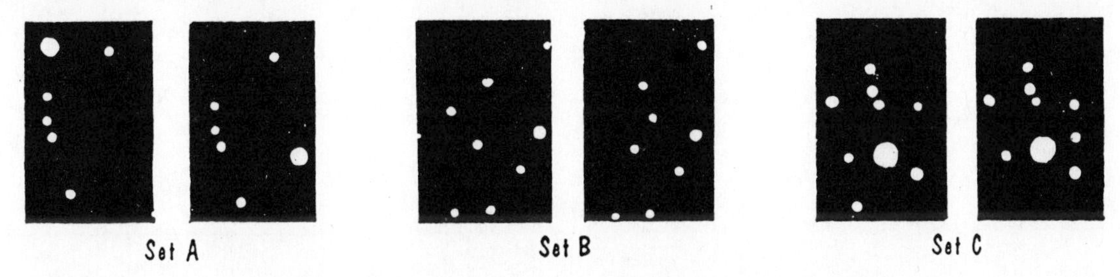
Set A
Set B
Set C

Use a fine ruler and measure the distance of shift in millimeters and place the values below:

Set A ____________ mm   Set B ____________ mm   Set C ____________ mm

Which set of photos indicate the closest star? The most distant "parallaxed" star?

Name ____ Date ____________

Appendix B: Linear and Rotational Motion

Mobile Torques

Apply what you know about torques by making a mobile. Shown below are five horizontal arms with fixed 1- and 2-kg masses attached, and four hangers with ends that fit in the loops of the arms, lettered A through R. You are to figure where the loops should be attached so that when the whole system is suspended from the spring scale at the top, it will hang as a proper mobile, with its arms suspended horizontally. This is best done by working from the bottom upward. Circle the loops where the hangers should be attached. When the mobile is complete, how many kilograms will be indicated on the scale? (Assume the horizontal struts and connecting hooks are practically massless compared to the 1- and 2-kg masses.) On a separate sheet of paper, make a sketch of your completed mobile.

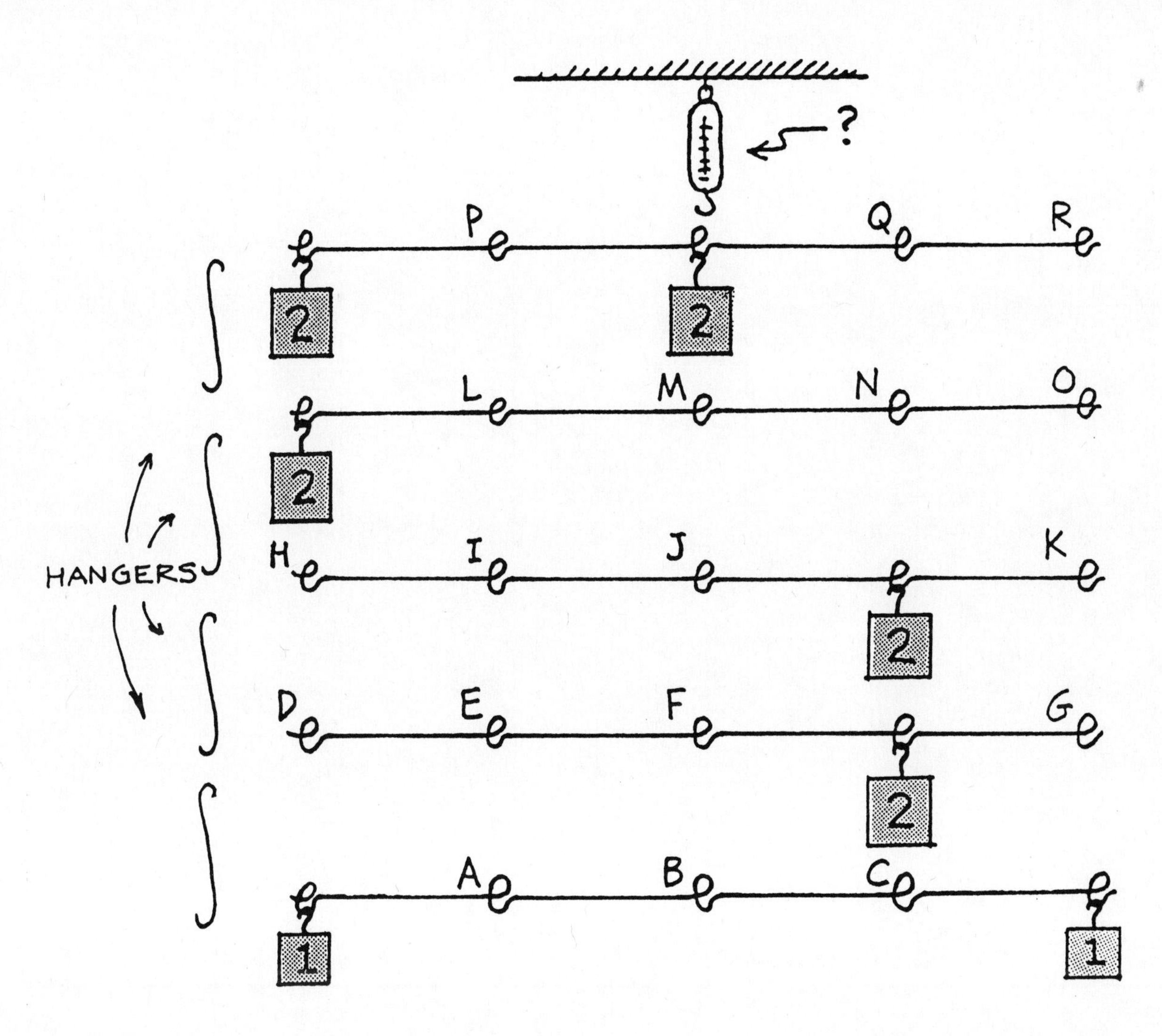

Name ____ Date ____________

Appendix B: Linear and Rotational Motion

Torques and See-Saws

1. Complete the data for the three see-saws in equilibrium.

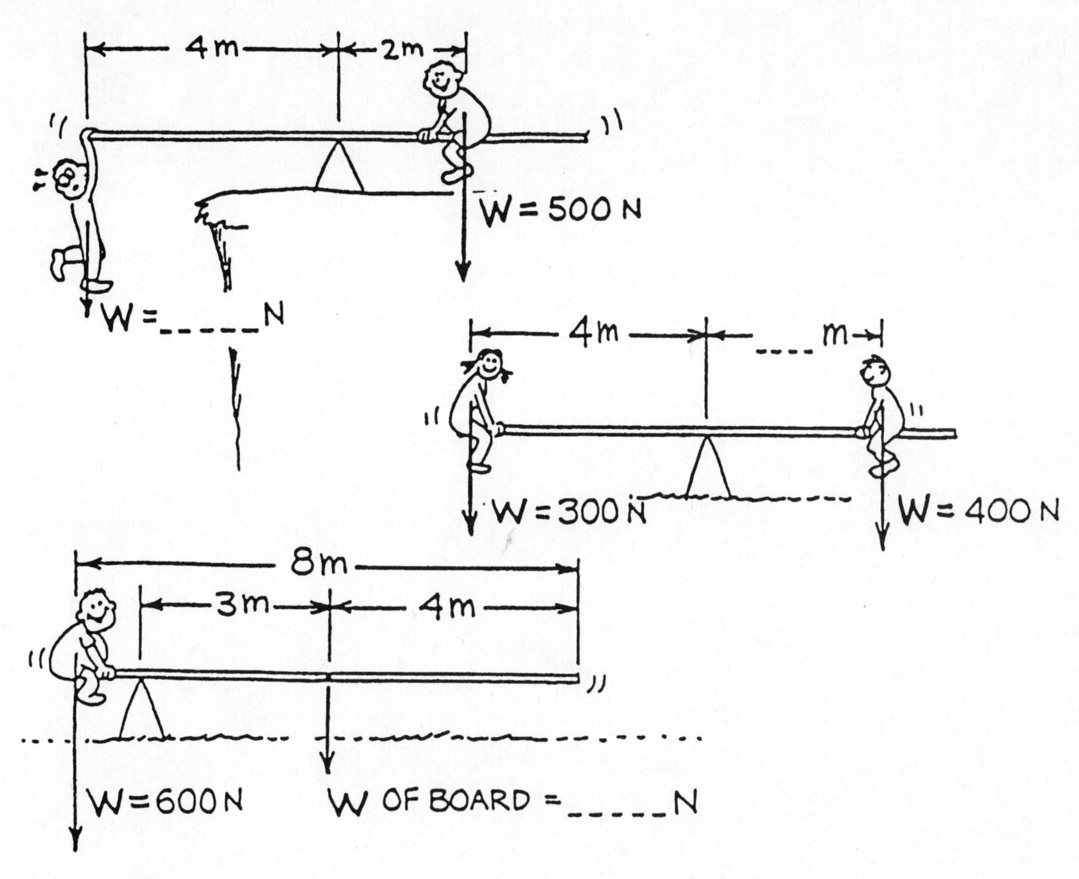

2. The broom balances at its CG. If you cut the broom in half at the CG and weigh each part of the broom, which end would weigh more?

____

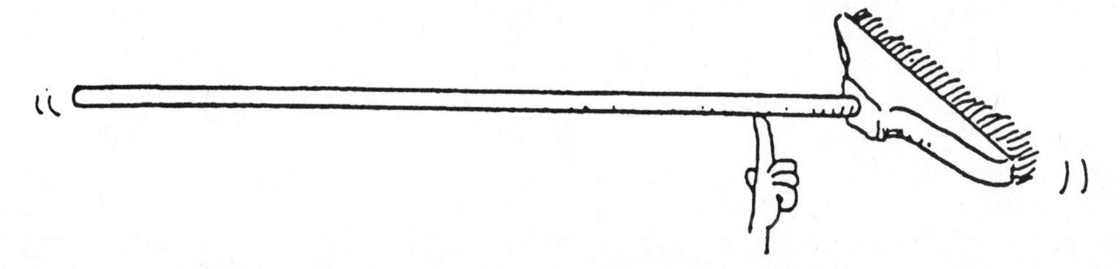

Explain why each end has or does not have the same weight? (Hint: Compare this to one of the see-saw systems above.)

____

____

Name ______________________________ Date ____________

Appendix E: Physics of Fluids

Archimedes' Principle

1. Consider a balloon filled with 1 liter of water (1000 cm^3) in equilibrium in a container of water, as shown in Figure 1.

WATER DOES NOT SINK IN WATER!

   a. What is the mass of the 1 liter of water?

   ____

   b. What is the weight of the 1 liter of water?

   ____

   c. What is the weight of water displaced by the balloon?

   ____

   d. What is the buoyant force on the balloon?

   ____

   e. Sketch a pair of vectors in Figure 1: one for the weight of the balloon and the other for the buoyant force that acts on it. How do the size and directions of your vectors compare?

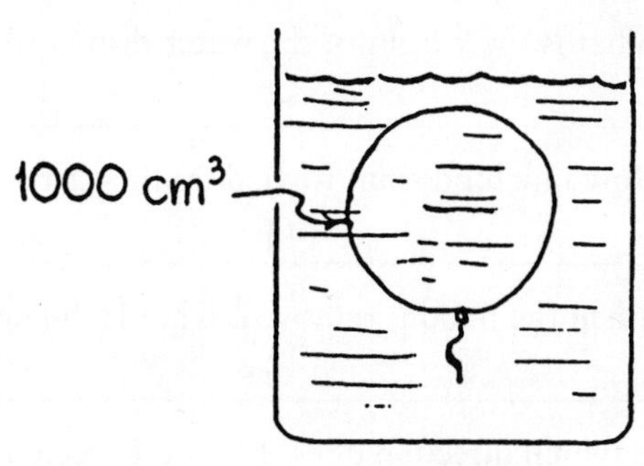

Figure 1

____

2. As a thought experiment, pretend we could remove the water from the balloon but still have it remain the same size of 1 liter. The inside the balloon is a vacuum.

   a. What is the mass of the liter of nothing?

   ____

   b. What is the weight of the liter of nothing?

   ____

ANYTHING THAT DISPLACES 9.8 N OF WATER EXPERIENCES 9.8 N OF BUOYANT FORCE.

   c. What is the weight of water displaced by the massless balloon?

   ____

   d. What is the buoyant force on the massless balloon?

   ____

IF YOU PUSH 9.8 N OF WATER ASIDE THE WATER PUSHES BACK ON YOU WITH 9.8 N!

   e. In which direction would the massless balloon be accelerated?

   ____

Name ______________________________ Date ______________

Archimedes' Principle—continued

3. Assume the balloon is replaced by a 0.5-kilogram piece of wood that has exactly the same volume (1000 cm^3), as shown in Figure 2. The wood is held in the same submerged position beneath the surface of the water.

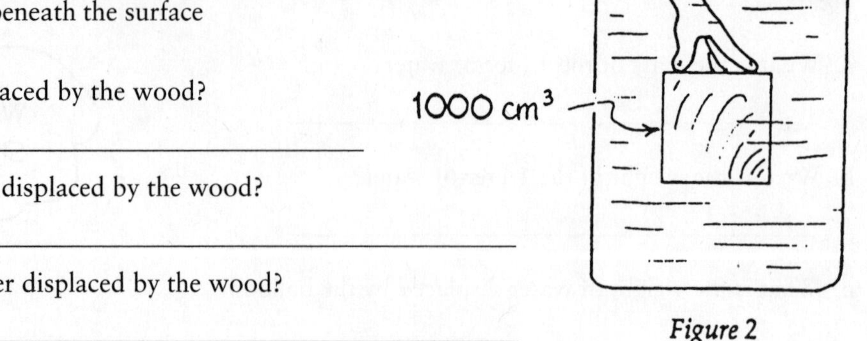

Figure 2

   a. What volume of water is displaced by the wood?

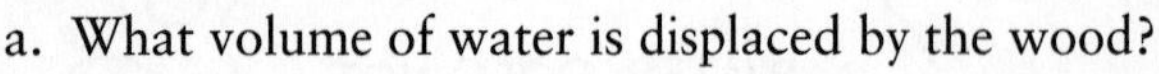

   b. What is the mass of the water displaced by the wood?

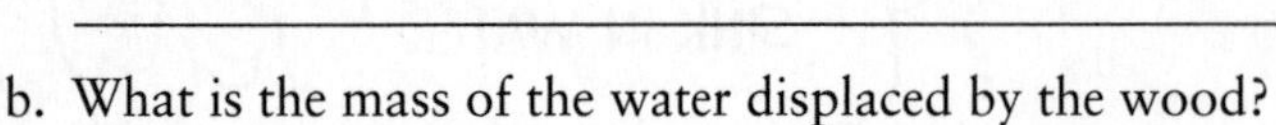

   c. What is the weight of the water displaced by the wood?

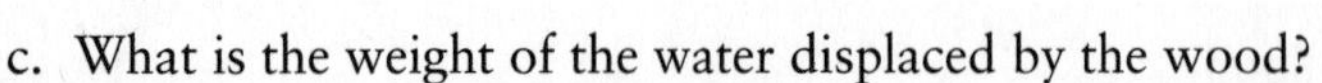

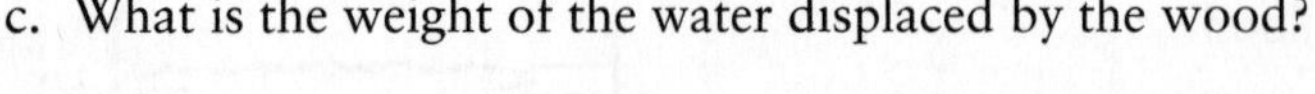

   d. How much buoyant force does the surrounding water exert on the wood?

   ____

   e. When the hand is removed, what is the net force on the wood?

   ____

   f. In which direction does the wood accelerate when released? ______________

4. Repeat parts *a* through *f* in the previous question for a 5-kg rock that has the same volume (1000 cm^3), as shown in Figure 3. Assume the rock is suspended by a string in the container of water.

   a. ______________
   b. ______________
   c. ______________
   d. ______________
   e. ______________
   f. ______________

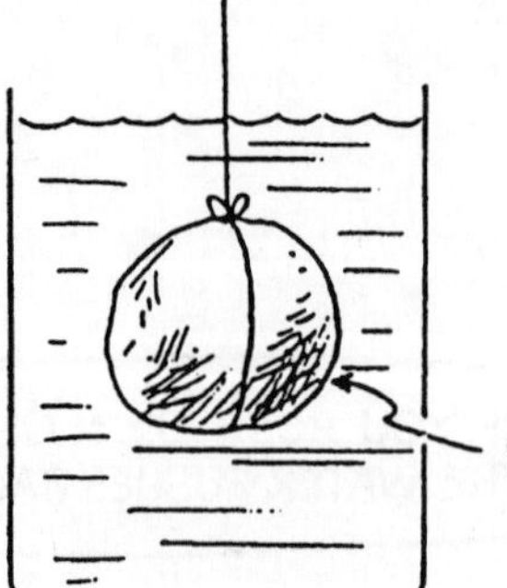

1000 cm³

Figure 3

WHEN THE WEIGHT OF AN OBJECT IS GREATER THAN THE BUOYANT FORCE EXERTED ON IT, IT SINKS!

Name ______________________________ Date ____________

Appendix E: Physics of Fluids

More on Archimedes' Principle

1. The water lines for the first three cases are shown. Sketch in the appropriate water lines for cases *d* and *e*, and make up your own for case *f*.

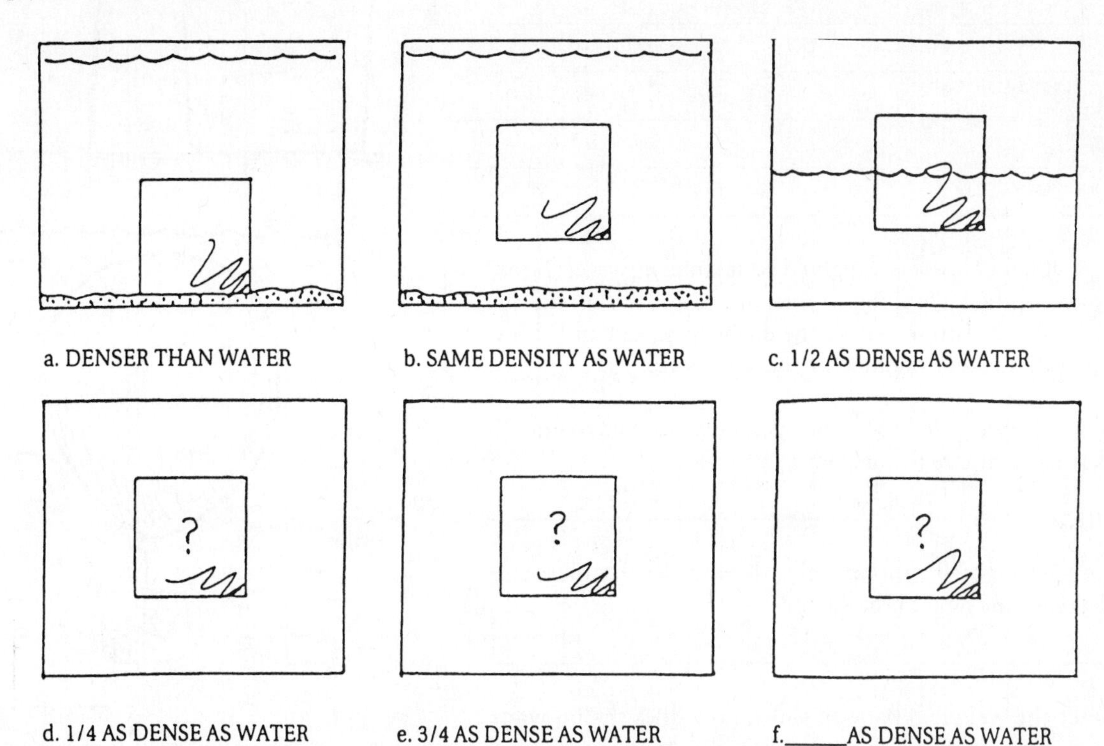

a. DENSER THAN WATER

b. SAME DENSITY AS WATER

c. 1/2 AS DENSE AS WATER

d. 1/4 AS DENSE AS WATER

e. 3/4 AS DENSE AS WATER

f.______AS DENSE AS WATER

2. If the weight of a ship is 100 million N, then the water it displaces weighs ______________. If cargo weighing 1000 N is put on board, then the ship will sink down until an extra ______________ of water is displaced.

3. The first two sketches below show the water line for an empty and a loaded ship. Draw in the appropriate water line for the third sketch.

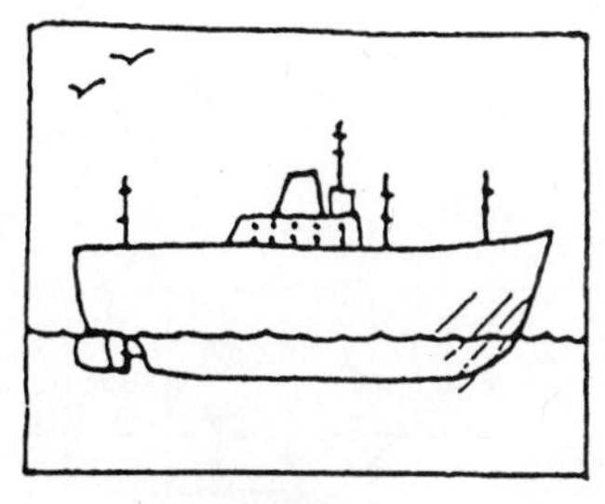

a. SHIP EMPTY

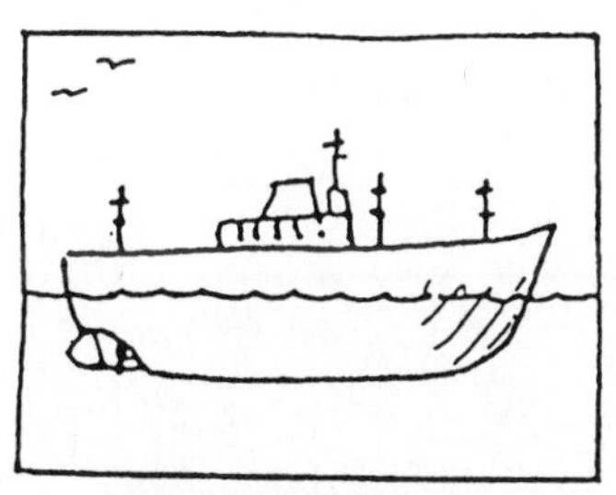

b. SHIP LOADED WITH 50 TONS OF IRON

c. SHIP LOADED WITH 50 TONS OF STYROFOAM

More on Archimedes' Principle—continued

4. Here is a glass of ice water with an ice cube floating in it. Draw the water line after the ice cube melts. (Will the water line rise, fall, or remain the same?)

____

____

____

____

____

5. The air-filled balloon is weighted so it sinks in water. Near the surface, the balloon has a certain volume. Draw the balloon at the bottom (inside the dashed square) and show whether it is bigger, smaller, or the same size.

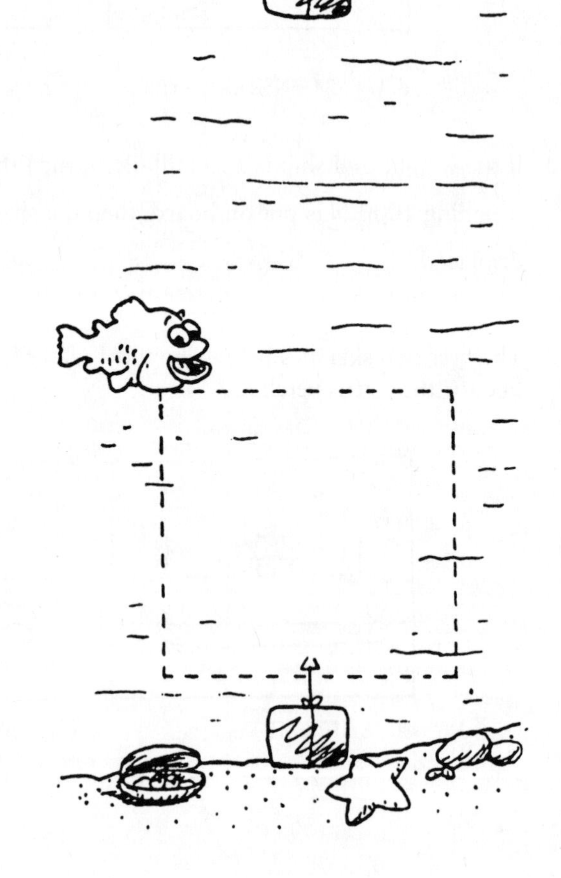

   a. Since the weighted balloon sinks, how does its overall density compare to the density of water?

   ____

   b. As the weighted balloon sinks, does its density increase, decrease, or remain the same?

   ____

   c. Since the weighted balloon sinks, how does the buoyant force on it compare to its weight?

   ____

   d. As the weighted balloon sinks deeper, does the buoyant force on it increase, decrease, or remain the same?

   ____

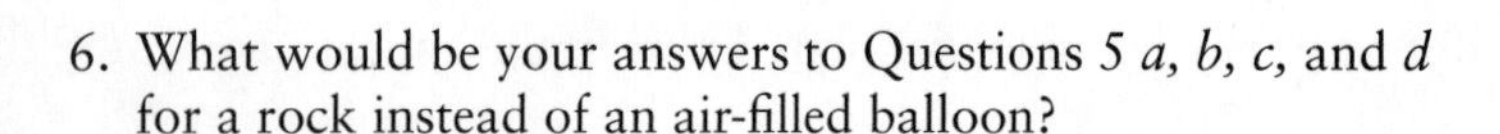

6. What would be your answers to Questions 5 *a, b, c,* and *d* for a rock instead of an air-filled balloon?

   a. ____

   b. ____

   c. ____

   d. ____

Name ____ Date ____________

Appendix E: Physics of Fluids

Gases

1. A principle difference between a liquid and a gas is that when a liquid is under pressure, its volume

   (increases) (decreases) (doesn't change noticeably)

   and its density

   (increases) (decreases) (doesn't change noticeably).

   When a gas is under pressure, its volume

   (increases) (decreases) (doesn't change noticeably)

   and its density

   (increases) (decreases) (doesn't change noticeably).

2. The sketch shows the launching of a weather balloon at sea level. Make a sketch of the same weather balloon when it is high in the atmosphere. In words, what is different about its size and why?

   ____

   ____

   ____

   ____

HIGH-ALTITUDE SIZE
GROUND-LEVEL SIZE

3. A hydrogen-filled balloon that weighs 10-N must displace ________ N of air in order to float in air. If it displaces less than ________ N it will be buoyed up with less than ________ N and sink. If it displaces more than ________ N of air it will move upward.

4. Why is the cartoon at right more humorous to physics types than to non-physics types? What physics has occurred?

RATS TO YOU TOO, DANIEL BERNOULLI!

   ____

   ____

   ____

   ____

Chapter 2: Describing Motion

Vectors and Equilibrium

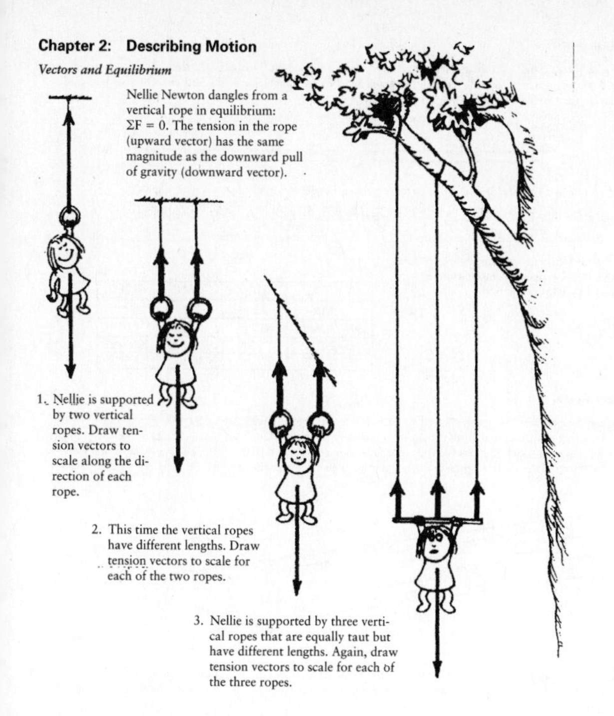

Nellie Newton dangles from a vertical rope in equilibrium: $\Sigma F = 0$. The tension in the rope (upward vector) has the same magnitude as the downward pull of gravity (downward vector).

1. Nellie is supported by two vertical ropes. Draw tension vectors to scale along the direction of each rope.

2. This time the vertical ropes have different lengths. Draw tension vectors to scale for each of the two ropes.

3. Nellie is supported by three vertical ropes that are equally taut but have different lengths. Again, draw tension vectors to scale for each of the three ropes.

Circle the correct answer:

4. We see that tension in a rope is [dependent on] [independent of] the length of the rope. So the length of a vector representing rope tension is [dependent on] [independent of] the length of the rope.

Rope tension depends on the angle the rope makes with the vertical, as Practice Pages for Chapter 3 will show!

Chapter 2: Describing Motion

Free Fall Speed

1. Aunt Minnie gives you \$10 per second for 4 seconds. How much money do you have after 4 seconds? \$40

2. A ball dropped from rest picks up speed at 10 m/s per second. After it falls for 4 seconds, how fast is it going? 40 m/s
3. You have \$20, and Uncle Harry gives you \$10 each second for 3 seconds. How much money do you have after 3 seconds? \$50
4. A ball is thrown straight down with an initial speed of 20 m/s. After 3 seconds, how fast is it going? 50 m/s
5. You have \$50 and you pay Aunt Minnie \$10/second. When will your money run out? 5 s
6. You shoot an arrow straight up at 50 m/s. When will it run out of speed? 5 s
7. What will be the arrow's speed 5 seconds after you shoot it? 0 m/s
8. What will its speed be 6 seconds after you shoot it? 7 seconds? 10 m/s 20 m/s

Free Fall Distance

1. Speed is one thing; distance another. *Where* is the arrow you shoot up at 50 m/s when it runs out of speed? 125 m
2. How high will the arrow be 7 seconds after being shot up at 50 m/s? 105 m
3. a. Aunt Minnie drops a penny into a wishing well and it falls for 3 seconds before hitting the water. How fast is it going when it hits? 30 m/s

   b. What is the penny's average speed during its 3 second drop? 15 m/s

   c. How far down is the water surface? 45 m
4. Aunt Minnie didn't get her wish, so she goes to a deeper wishing well and throws a penny straight down into it at 10 m/s. How far does this penny go in 3 seconds? 75 m

FROM REST, $v = 10t$, $d = 5t^2$

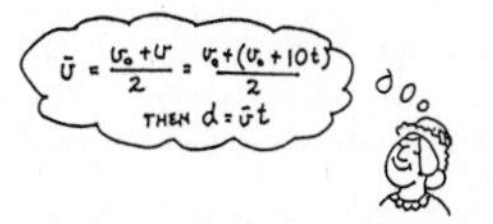

Acceleration of Free Fall

A rock dropped from the top of a cliff picks up speed as it falls. Pretend that a speedometer and odometer are attached to the rock to show readings of speed and distance at 1 second intervals. Both speed and distance are zero at time = zero (see sketch). Note that after falling 1 second the speed reading is 10 m/s and the distance fallen is 5 m. The readings for succeeding seconds of fall are not shown and are left for you to complete. Draw the position of the speedometer pointer and write in the correct odometer reading for each time. Use $g = 10\ m/s^2$ and neglect air resistance.

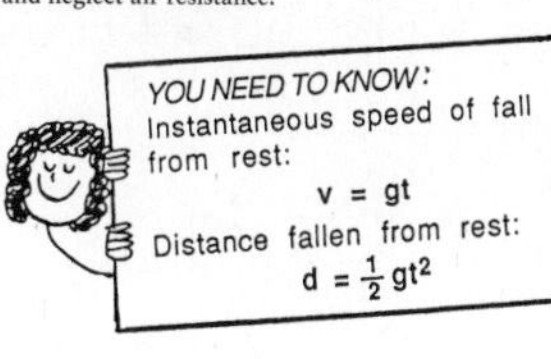

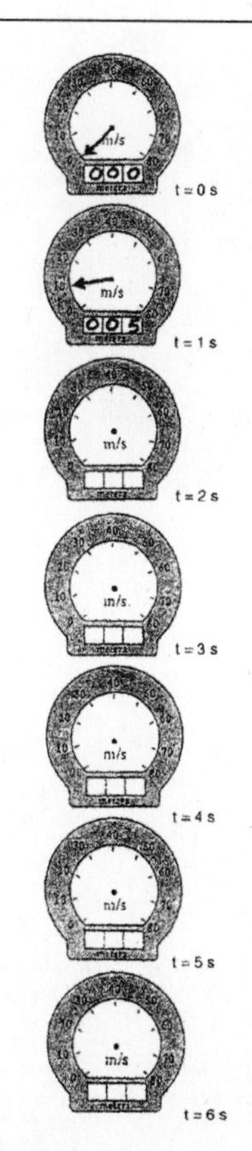

1. The speedometer reading increases by the same amount, 10 m/s, each second. This increase in speed per second is called ACCELERATION.
2. The distance fallen increases as the square of the TIME.
3. If it takes 7 seconds to reach the ground, then its speed at impact is 70 m/s, the total distance fallen is 245 m, and its acceleration of fall just before impact is 10 m/s^2.

Chapter 3: Newton's Laws of Motion

Newton's First Law and Friction

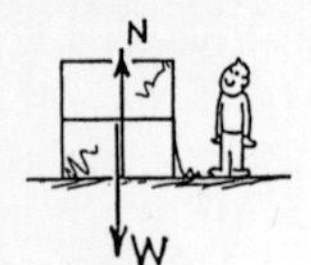

1. A crate filled with video games rests on a horizontal floor. Only gravity and the support force of the floor act on it, as shown by the vectors for weight **W** and normal force **N**.

   a. The net force on the crate is **(zero)** (greater than zero).

   b. Evidence for this is NO ACCELERATION.

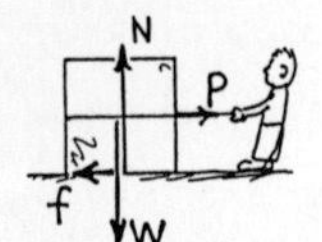

2. A slight pull **P** is exerted on the crate, not enough to move it. A force of friction **f** now acts,

   a. which is (less than) **(equal to)** (greater than) **P**.

   b. Net force on the crate is **(zero)** (greater than zero).

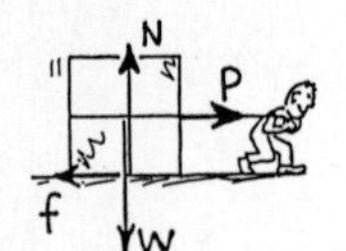

3. Pull **P** is increased until the crate begins to move. It is pulled so that it moves with constant velocity across the floor.

   a. Friction **f** is (less than) **(equal to)** (greater than) **P**.

   b. Constant velocity means acceleration is **(zero)** (greater than zero).

   c. Net force on the crate is (less than) **(equal to)** (greater than) zero.

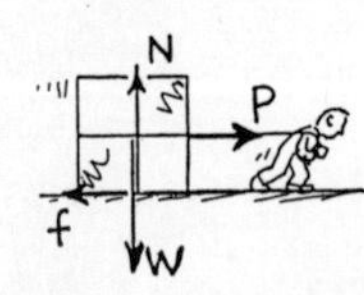

4. Pull **P** is further increased and is now greater than friction **f**.

   a. Net force on the crate is (less than) (equal to) **(greater than)** zero.

   b. The net force acts toward the right, so acceleration acts toward the (left) **(right)**.

5. If the pulling force P is 150 N and the crate doesn't move, what is the magnitude of f? 150 N
6. If the pulling force P is 200 N and the crate doesn't move, what is the magnitude of f? 200 N
7. If the force of sliding friction is 250 N, what force is necessary to keep the crate sliding at constant velocity? 250 N
8. If the mass of the crate is 50 kg and sliding friction is 250 N, what is the acceleration of the crate when the pulling force is 250 N? $0\ m/s^2$ 300 N? $1\ m/s^2$ 500 N? $5\ m/s^2$

Nonaccelerated Motion

1. The sketch shows a ball rolling at constant velocity along a level floor. The ball rolls from the first position shown to the second in 1 second. The two positions are 1 meter apart. Sketch the ball at successive 1-second intervals all the way to the wall (Neglect resistance.)

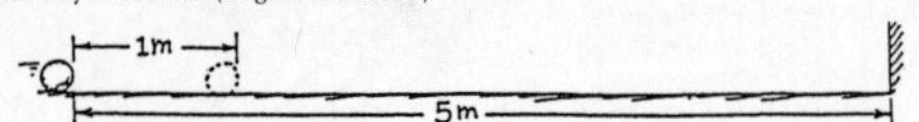

   a. Did you draw successive ball positions evenly spaced, farther apart, or close together? Why?
   EVENLY SPACED – EQUAL DISTANCE IN EQUAL TIME → CONSTANT v

   b. The ball reaches the wall with a speed of 1 m/s and takes a time of 5 seconds.

2. Table 1 shows data of sprinting speeds of some animals. Make whatever computations are necessary to complete the table.

Table 1

ANIMAL	DISTANCE	TIME	SPEED
CHEETAH	75 m	3 s	25 m/s
GREYHOUND	160 m	10 s	16 m/s
GAZELLE	1 km	0.01 h	100 km/h
TURTLE	30 cm	30 s	1 cm/s

Accelerated Motion

3. An object starting from rest gains a speed $v = at$ when it undergoes uniform acceleration. The distance it covers is $d = 1/2\ at^2$. Uniform acceleration occurs for a ball rolling down an inclined plane. The plane below is tilted so a ball picks up a speed of 2 m/s each second; then its acceleration is $a = 2\ m/s^2$. The positions of the ball are shown at 1-second intervals. Complete the six blank spaces for distance covered, and the four blank spaces for speeds.

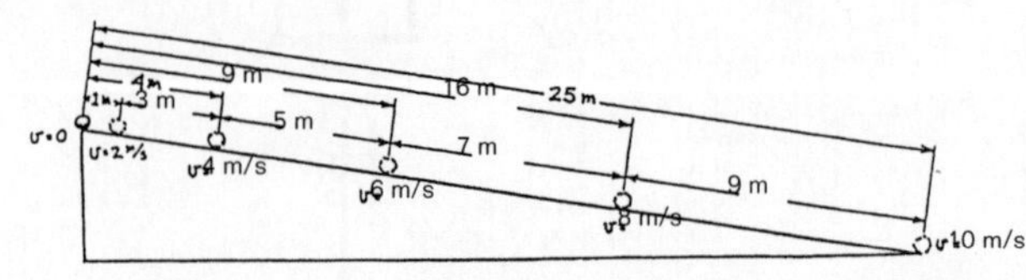

   a. Do you see that the total distance from the starting point increases as the square of the time? This was discovered by Galileo. If the incline were to continue, predict the ball's distance from the starting point for the next 3 seconds.
   YES; DISTANCE INCREASES AS SQUARE OF TIME; 36 m, 49 m, 64 m.

   b. Note the increase of distance between ball positions with time. Do you see an odd-integer pattern (also discovered by Galileo) for the increase? If the incline were to continue, predict the successive distances between ball positions for the next 3 seconds.
   YES; 11 m, 13 m, 15 m

Chapter 3: Newton's Laws of Motion

A Day at the Races with Newton's Second Law: $a = \frac{F}{m}$

In each situation below, Cart A has a mass of **1 kg**. The mass of Cart B varies as indicated. Circle the correct answer (A, B, or Same for both).

1. Cart A is pulled with a force of **1 N**. Cart B also has a mass of **1 kg** and is pulled with a force of **2 N**. Which undergoes the greater acceleration?

   A **(B)** Same for both

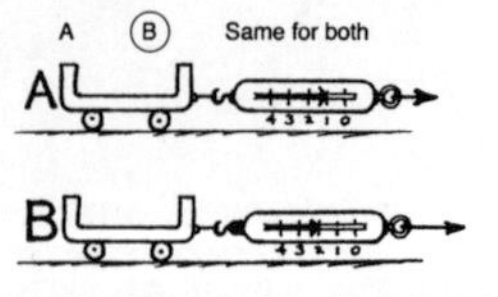

2. Cart A is pulled with a force of **1 N**. Cart B has a mass of **2 kg** and is also pulled with a force of **1 N**. Which undergoes the greater acceleration?

   (A) B Same for both

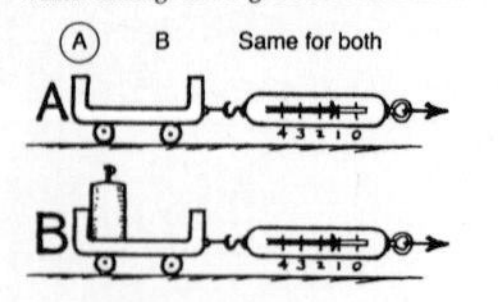

3. Cart A is pulled with a force of **1 N**. Cart B has a mass of **2 kg** and is pulled with a force of **2 N**. Which undergoes the greater acceleration?

   A B **(Same for both)**

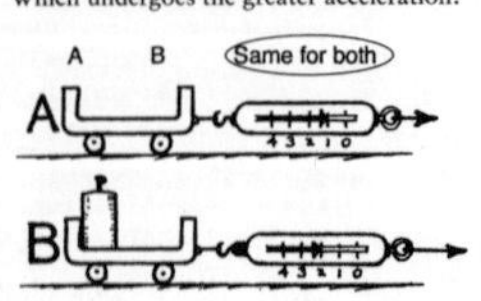

4. Cart A is pulled with a force of **1 N**. Cart B has a mass of **3 kg** and is pulled with a force of **3 N**. Which undergoes the greater acceleration?

   A B **(Same for both)**

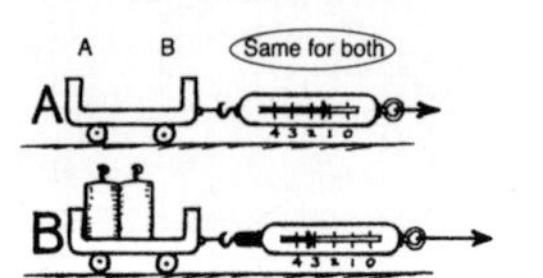

5. This time Cart A is pulled with a force of **4 N**. Cart B has a mass of **4 kg** and is pulled with a force of **4 N**. Which undergoes the greater acceleration?

   (A) B Same for both

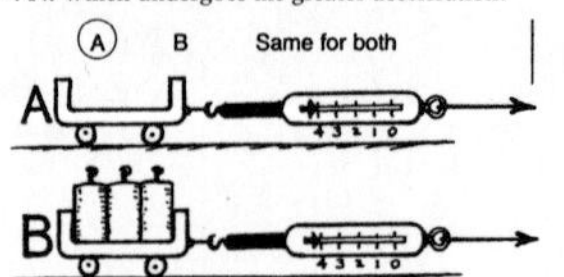

6. Cart A is pulled with a force of **2 N**. Cart B has a mass of **4 kg** and is pulled with a force of **3 N**. Which undergoes the greater acceleration?

   (A) B Same for both

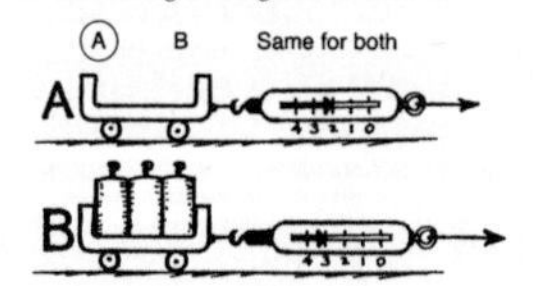

Dropping Masses and Accelerating Cart

1. Consider the simple case of a 1-kg cart being pulled by a 10 N applied force. According to Newton's Second Law, acceleration of the cart is

$$a = \frac{F}{m} = \frac{10\text{ N}}{1\text{ kg}} = 10\text{ m/s}^2$$

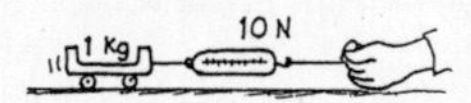

This is the same as the acceleration of free fall, g—because a force equal to the cart's weight accelerates it.

2. Now consider the acceleration of the cart when a second mass is also accelerated. This time the applied force is due to a 10-N iron weight attached to a string draped over a pulley. Will the cart accelerate as before, at 10 m/s²? The answer is no, because the mass being accelerated is the mass of the cart *plus* the mass of the piece of iron that pulls it. Both masses accelerate. The mass of the 10-N iron weight is 1 kg—so the total mass being accelerated (cart + iron) is 2 kg. Then,

$$a = \frac{F}{m} = \frac{10\text{ N}}{2\text{ kg}} = 5\text{ m/s}^2$$

The pulley changes only the direction of the force.

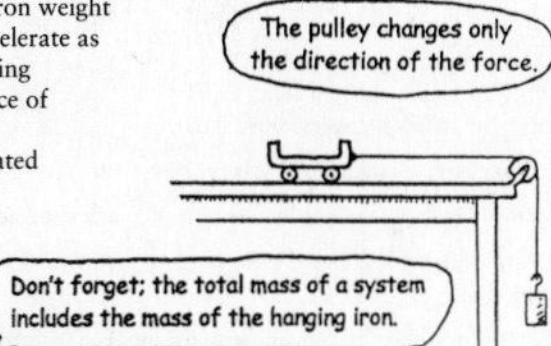

Don't forget: the total mass of a system includes the mass of the hanging iron.

Note this is half the acceleration due to gravity alone, g. So the acceleration of 2 kg produced by the weight of 1 kg is $g/2$.

a. Find the acceleration of the 1-kg cart when two identical 10-N weights are attached to the string.

$$a = \frac{F}{m} = \frac{\text{unbalanced force}}{\text{total mass}} = \frac{20\text{ N}}{3\text{ kg}} = \underline{6.7}\text{ m/s}^2.$$

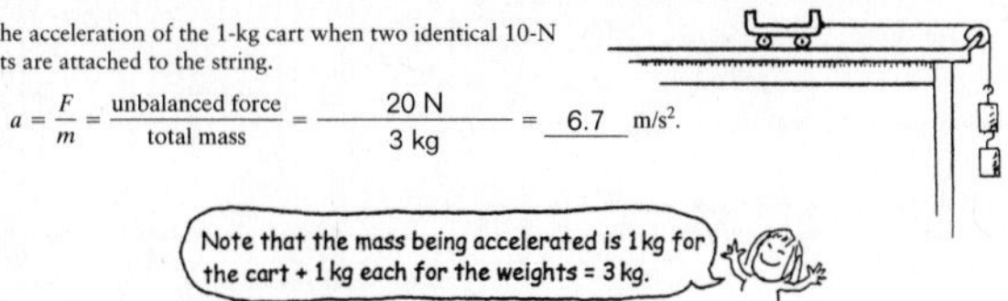

Note that the mass being accelerated is 1 kg for the cart + 1 kg each for the weights = 3 kg.

Dropping Masses and Accelerating Cart—continued

b. Find the acceleration of the 1-kg cart when three identical 10-N weights are attached to the string.

$$a = \frac{F}{m} = \frac{\text{unbalanced force}}{\text{total mass}} = \frac{30\text{ N}}{4\text{ kg}} = \underline{7.5}\text{ m/s}^2.$$

c. Find the acceleration of the 1-kg cart when four identical 10-N weights (not shown) are attached to the string.

$$a = \frac{F}{m} = \frac{\text{unbalanced force}}{\text{total mass}} = \frac{40\text{ N}}{5\text{ kg}} = \underline{8.0}\text{ m/s}^2.$$

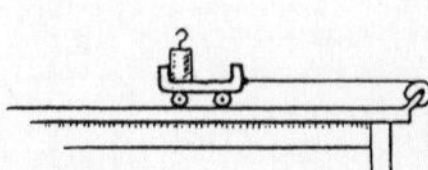

d. This time, 1 kg of iron is added to the cart, and only one iron piece dangles from the pulley. Find the acceleration of the cart.

$$a = \frac{F}{m} = \frac{\text{unbalanced force}}{\text{total mass}} = \frac{10\text{ N}}{3\text{ kg}} = \underline{3.4}\text{ m/s}^2.$$

The force due to gravity on a mass m is mg. So gravitational force on 1 kg is (1 kg)(10 m/s²) = 10 N.

e. Find the acceleration of the cart when it carries two pieces of iron and only one iron piece dangles from the pulley.

$$a = \frac{F}{m} = \frac{\text{unbalanced force}}{\text{total mass}} = \frac{10\text{ N}}{4\text{ kg}} = \underline{2.5}\text{ m/s}^2.$$

Dropping Masses and Accelerating Cart—continued

f. Find the acceleration of the cart when it carries three pieces of iron and only one iron piece dangles from the pulley.

$$a = \frac{F}{m} = \frac{\text{unbalanced force}}{\text{total mass}} = \frac{10\text{ N}}{5\text{ kg}} = \underline{2.0}\text{ m/s}^2.$$

g. Find the acceleration of the cart when it carries three pieces of iron and four iron pieces dangle from the pulley.

$$a = \frac{F}{m} = \frac{\text{unbalanced force}}{\text{total mass}} = \frac{40\text{ N}}{8\text{ kg}} = \underline{5.0}\text{ m/s}^2.$$

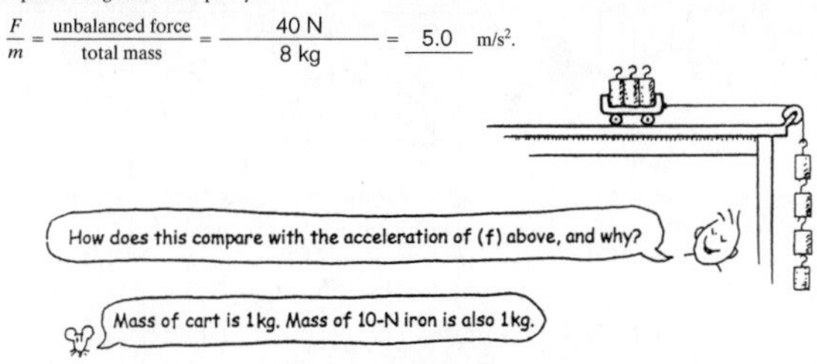

How does this compare with the acceleration of (f) above, and why?
Mass of cart is 1 kg. Mass of 10-N iron is also 1 kg.

h. Draw your own combination of masses and find the acceleration.

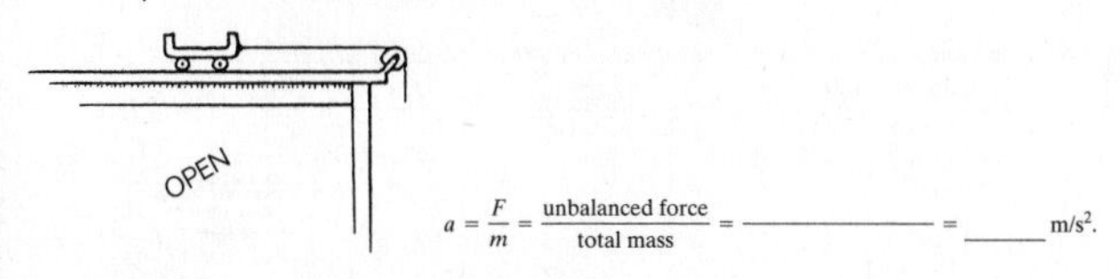

$$a = \frac{F}{m} = \frac{\text{unbalanced force}}{\text{total mass}} = \underline{\qquad\qquad} = \underline{\qquad}\text{ m/s}^2.$$

Chapter 3: Newton's Laws of Motion

Mass and Weight

Learning physics is learning the connections among concepts in nature, and also learning to distinguish between closely related concepts. Velocity and acceleration are often confused. Similarly, in this chapter, we find that mass and weight are often confused. They aren't the same! Please review the distinction between mass and weight in your textbook. To reinforce your understanding of this distinction, circle the correct answers below.

Comparing the concepts of mass and weight, one is basic—fundamental—depending only on the internal makeup of an object and the number and kind of atoms that compose it. The concept that is fundamental is [mass] [weight].

The concept that additionally depends on location in a gravitational field is [mass] [weight].

[Mass] [Weight] is a measure of the amount of matter in an object and only depends on the number and kind of atoms that compose it.

We can correctly say that [mass] [weight] is a measure of an object's "laziness."

[Mass] [Weight] is related to the gravitational force acting on the object.

[Mass] [Weight] depends on an object's location, whereas [mass] [weight] does not.

In other words, a stone would have the same [mass] [weight] whether it is on the surface of Earth or on the surface of the Moon. However, its [mass] [weight] depends on its location.

On the Moon's surface, where gravity is only about 1/16 of Earth's gravity, [mass] [weight] [both the mass and the weight] of the stone would be the same as on Earth.

While mass and weight are not the same, they are [directly proportional] [inversely proportional] to each other. In the same location, twice the mass has [twice] [half] the weight.

The Standard International (SI) unit of mass is the [kilogram] [newton], and the SI unit of force is the [kilogram] [newton].

In the United States, it is common to measure the mass of something by measuring its gravitational pull to Earth, its weight. The common unit of weight in the U.S. is the [pound] [kilogram] [newton].

When I step on a scale, two forces act on it: a downward pull of gravity, and an upward support force. These equal and opposite forces effectively compress a spring inside the scale that is calibrated to show weight. When in equilibrium, my weight = *mg*.

Chapter 3: Newton's Laws of Motion

Converting Mass to Weight

Objects with mass also have weight (although they can be weightless under special conditions). If you know the mass of something in **kilograms** and want its weight in **newtons**, at Earth's surface, you can take advantage of the formula that relates weight and mass:

$$\text{Weight} = \text{mass} \times \text{acceleration due to gravity}$$

$$W = mg.$$

This is in accord with Newton's Second Law, written as $F = ma$. When the force of gravity is the only force, the acceleration of any object of mass m will be g, the acceleration of free fall. Importantly, g acts as a proportionality constant, 9.8 N/kg, which is equivalent to 9.8 m/s^2.

Sample Question:

How much does a 1-kg bag of nails weigh on Earth?

$W = mg = (1\text{ kg})(9.8\text{ m/s}^2) = 9.8\text{ m/s}^2 = 9.8\text{ N}.$

or simply, $W = mg = (1\text{ kg})(9.8\text{ N/kg}) = 9.8\text{ N}.$

From $F = ma$, we see that the unit of force equals the units [kg × m/s^2]. Can you see the units [m/s^2] = [N/kg]?

Answer the following questions:

Felicia the ballet dancer has a mass of 45.0 kg.

1. What is Felicia's weight in newtons at Earth's surface? 441 N
2. Given that 1 kilogram of mass corresponds to 2.2 pounds at Earth's surface, what is Felicia's weight in pounds on Earth? 99 LB
3. What would be Felicia's mass on the surface of Jupiter? 45.0 kg
4. What would be Felcia's weight on Jupiter's surface, where the acceleration due to gravity is 25.0 m/s^2? 1125 N

Different masses are hung on a spring scale calibrated in newtons. The force exerted by gravity on 1 kg = 9.8 N.

5. The force exerted by gravity on 5 kg = 49 N.
6. The force exerted by gravity on 10 kg = 98 N.

Make up your own mass and show the corresponding weight:

The force exerted by gravity on ______ kg = ______ N.

9.8 N — 1 kg

By whatever means (spring scales, measuring balances, etc.), find the mass of your integrated science book. Then complete the table.

OBJECT	MASS	WEIGHT
MELON	1 kg	9.8 N
APPLE	0.1 kg	1 N
BOOK		
A FRIEND	60 kg	588 N

Chapter 3: Newton's Laws of Motion

Bronco and Newton's Second Law

Bronco skydives and parachutes from a stationary helicopter. Various stages of fall are shown in positions *a* through *f*. Using Newton's Second Law:

$$a = \frac{F_{NET}}{m} = \frac{W - R}{m}$$

Find Bronco's acceleration at each position (answer in the blanks to the right). You need to know that Bronco's mass m is 100 kg so his weight is a constant 1000 N. Air resistance R varies with speed and cross-sectional area as shown.

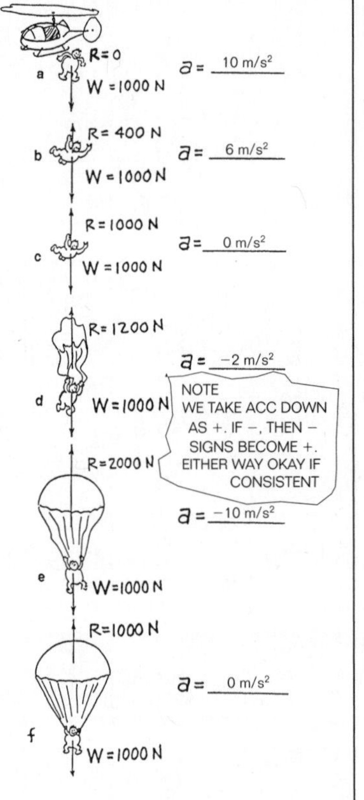

Circle the correct answers:

1. When Bronco's speed is least, his acceleration is (least) (most)
2. In which position(s) does Bronco experience a downward acceleration? (a) (b) (c) (d) (e) (f)
3. In which position(s) does Bronco experience an upward acceleration? (a) (b) (c) (d) (e) (f)
4. When Bronco experiences an upward acceleration, his velocity is (still downward) (upward also).
5. In which position(s) is Bronco's velocity constant? (a) (b) (c) (d) (e) (f)
6. In which position(s) does Bronco experience terminal velocity? (a) (b) (c) (d) (e) (f)
7. In which position(s) is terminal velocity greatest? (a) (b) (c) (d) (e) (f)
8. If Bronco were heavier, his terminal velocity would be (greater) (less) (the same).

Chapter 3: Newton's Laws of Motion

Newton's Third Law

Your thumb and finger pull on each other when you stretch a rubber band between them. This pair of forces, thumb on finger and finger on thumb, make up an action-reaction pair of forces, both of which are equal in magnitude and oppositely directed. Draw the reaction vector and state in words the reaction force for each of the examples a through g. Then make up your own example in h.

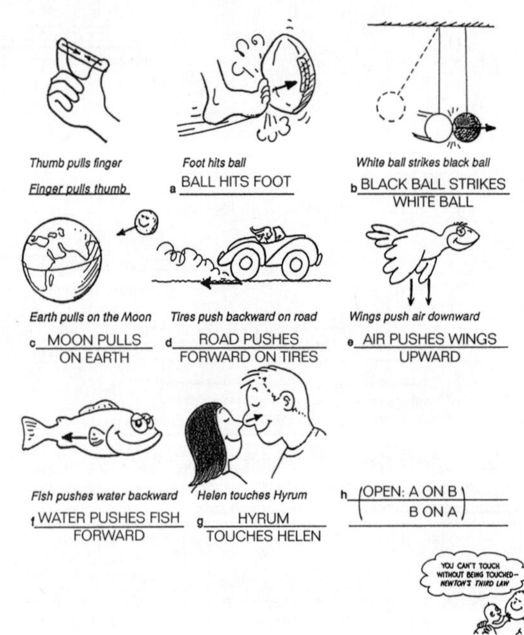

YOU CAN'T TOUCH WITHOUT BEING TOUCHED— NEWTON'S THIRD LAW

Thumb pulls finger — *Finger pulls thumb*

Foot hits ball — a BALL HITS FOOT

White ball strikes black ball — b BLACK BALL STRIKES WHITE BALL

Earth pulls on the Moon — c MOON PULLS ON EARTH

Tires push backward on road — d ROAD PUSHES FORWARD ON TIRES

Wings push air downward — e AIR PUSHES WINGS UPWARD

Fish pushes water backward — f WATER PUSHES FISH FORWARD

Helen touches Hyrum — g HYRUM TOUCHES HELEN

h (OPEN: A ON B / B ON A)

Nellie and Newton's Third Law

Nellie Newton holds an apple weighing 1 newton at rest on the palm of her hand. *Circle the correct answers.*

1. To say the weight (W) of the apple is 1 N is to say that a downward gravitational force of 1 N is exerted on the apple by
   (Earth) (her hand).
2. Nellie's hand supports the apple with normal force N, which acts in a direction opposite to W. We can say N
   (equals W) (has the same magnitude as W).
3. Since the apple is at rest, the net force on the apple is
   (zero) (nonzero).
4. Since N is equal and opposite to W, we (can) (cannot) say that N and W comprise an action-reaction pair. The reason is because action and reaction (act on the same object) (act on different objects), and here we see N and W (both acting on the apple) (acting on different objects).

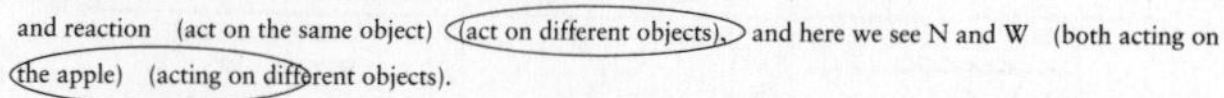

5. In accord with the rule, "If ACTION is A acting on B, then REACTION is B acting on A," if we say action is Earth pulling down on the apple, reaction is
   (the apple pulling up on Earth) (N, Nellie's hand pushing up on the apple.)
6. To repeat for emphasis, we see that N and W are equal and opposite to each other
   (and comprise an action-reaction pair) (but do *not* comprise an action-reaction pair).

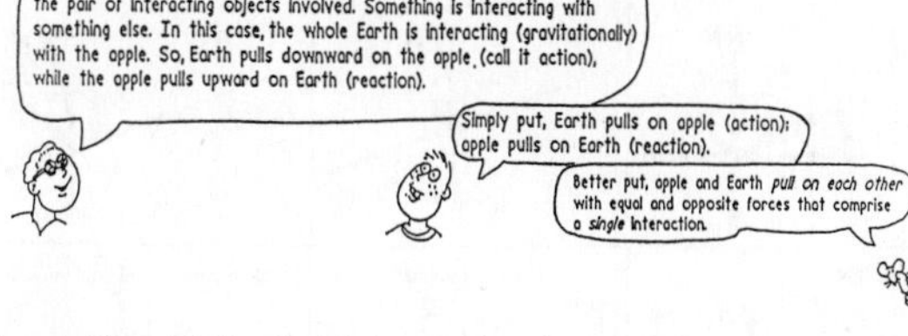

To identify a pair of action-reaction forces in any situation, first identify the pair of interacting objects involved. Something is interacting with something else. In this case, the whole Earth is interacting (gravitationally) with the apple. So, Earth pulls downward on the apple (call it action), while the apple pulls upward on Earth (reaction).
Simply put, Earth pulls on apple (action); apple pulls on Earth (reaction).
Better put, apple and Earth pull on each other with equal and opposite forces that comprise a single interaction.

7. Another pair of forces is N [shown] and the downward force of the apple against Nellie's hand [not shown]. This pair of forces (is) (isn't) an action-reaction pair.
8. Suppose Nellie now pushes upward on the apple with the force of 2 N. The apple (is still in equilibrium) (accelerates upward), and compared with W, the magnitude of N is (the same) (twice) (not the same, and not twice).
9. Once the apple leaves Nellie's hand, N is (zero) (still twice the magnitude of W), and the net force on the apple is (zero) (only W) (still W − N, which is a negative force).

Vectors and the Parallelogram Rule

1. When vectors A and B are at an angle to each other, they add to produce the resultant C by the *parallelogram rule*. Note that C is the diagonal of a parallelogram where A and B are adjacent sides. Resultant C is shown in the first two diagrams, *a* and *b*. Construct the resultant C in diagrams *c* and *d*. Note that in diagram *d* you form a rectangle (a special case of a parallelogram).

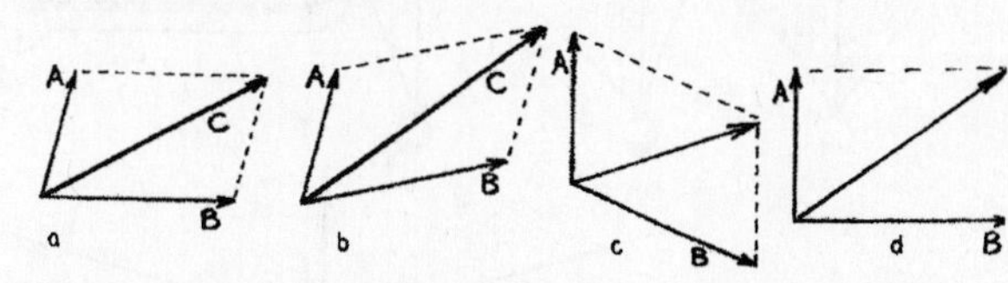

2. Below we see a top view of an airplane being blown off course by wind in various directions. Use the parallelogram rule to show the resulting speed and direction of travel for each case. In which case does the airplane travel fastest across the ground? d Slowest? c

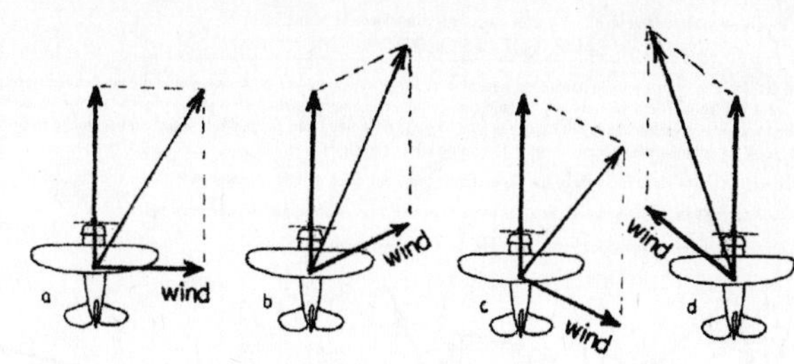

3. To the right we see top views of three motorboats crossing a river. All have the same speed relative to the water, and all experience the same water flow.

   Construct resultant vectors showing the speed and direction of the boats.

   a. Which boat takes the shortest path to the opposite shore? a

   b. Which boat reaches the opposite shore first? b

   c. Which boat provides the fastest ride? c

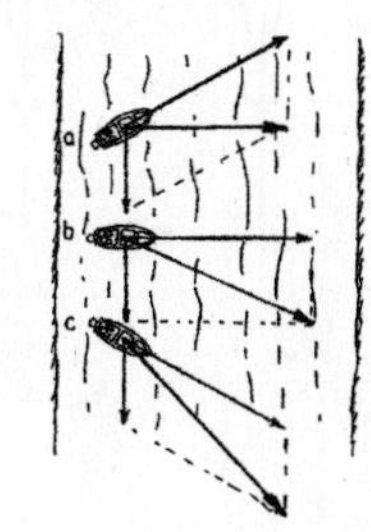

Vectors

Use the parallelogram rule to carefully construct the resultants for the eight pairs of vectors.

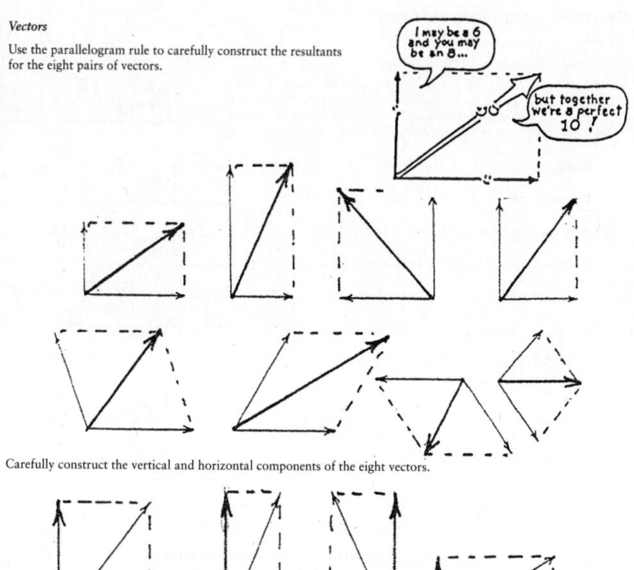

I may be a 6 and you may be an 8...
but together we're a perfect 10!

Carefully construct the vertical and horizontal components of the eight vectors.

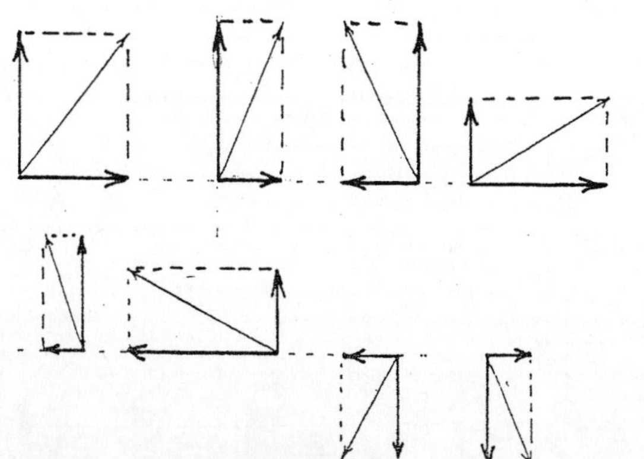

Chapter 3: Newton's Laws of Motion

Force Vectors and the Parallelogram Rule

1. The heavy ball is supported in each case by two strands of rope. The tension in each strand is shown by the vectors. Use the parallelogram rule to find the resultant of each vector pair.

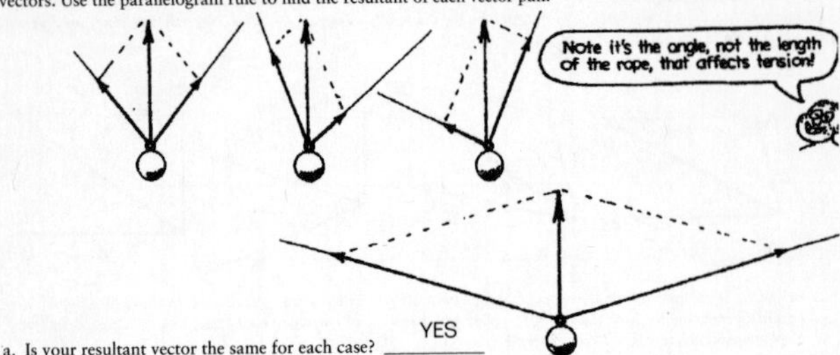

Note it's the angle, not the length of the rope, that affects tension!

a. Is your resultant vector the same for each case? YES

b. How do you think the resultant vector compares to the weight of the ball? SAME (BUT OPPOSITE DIRECTION)

2. Now let's do the opposite of what we've done above. More often, we know the weight of the suspended object, but we don't know the rope tensions. In each case below, the weight of the ball is shown by the vector W. Each dashed vector represents the resultant of the pair of rope tensions. Note that each is equal and opposite to vectors W (they must be; otherwise the ball wouldn't be at rest).

a. Construct parallelograms where the ropes define adjacent sides and the dashed vectors are the diagonals.

b. How do the relative lengths of the sides of each parallelogram compare to rope tensions?

c. Draw rope-tension vectors, clearly showing their relative magnitudes.

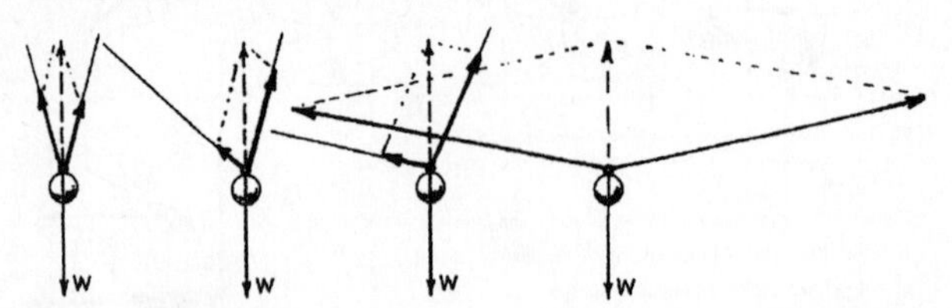

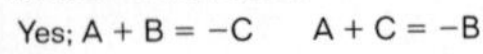

3. A lantern is suspended as shown. Draw vectors to show the relative tensions in ropes A, B, and C. Do you see a relationship between your vectors A + B and vector C? Between vectors A + C and vector B?

Yes; A + B = −C   A + C = −B

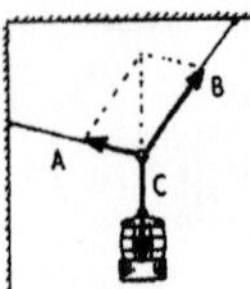

Force-Vector Diagrams

In each case, a rock is acted on by one or more forces. Draw an accurate vector diagram showing all forces acting on the rock, and no other forces. Use a ruler, and do it in pencil so you can correct mistakes. The first two are done as examples. Show by the parallelogram rule in 2 that the vector sum of **A** + **B** is equal and opposite to **W** (that is, **A** + **B** = −**W**). Do the same for 3 and 4. Draw and label vectors for the weight and normal forces in 5 to 10, and for the appropriate forces in 11 and 12.

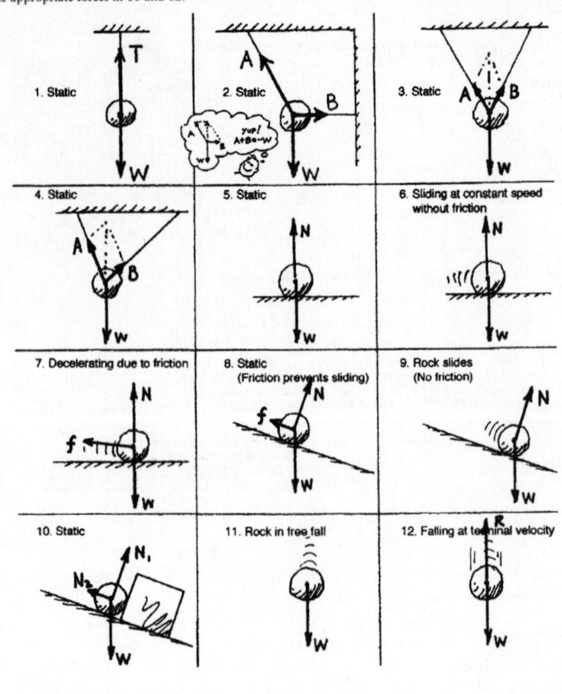

Chapter 4: Momentum and Energy

Momentum

1. A moving car has momentum. If it moves twice as fast, its momentum is TWICE as much.

2. Two cars, one twice as heavy as the other, move down a hill at the same speed. Compared to the lighter car, the momentum of the heavier car is TWICE as much.

3. The recoil momentum of a gun that kicks is
(more than) (less than) (the same as) [circled]
the momentum of the gases and bullet it fires.

4. If a man firmly holds a gun when fired, then the momentum of the bullet and expelled gases is equal to the recoil momentum of the
(gun alone) (gun-man system) [circled] (man alone).

5. Suppose you are traveling in a bus at highway speed on a nice summer day and the momentum of an unlucky bug is suddenly changed as it splatters onto the front window.

a. Compared to the force that acts on the bug, how much force acts on the bus?
(more) (the same) [circled] (less)

b. The time of impact is the same for both the bug and the bus. Compared to the impulse on the bug, this means the impulse on the bus is
(more) (the same) [circled] (less).

c. Although the momentum of the bus is very large compared to the momentum of the bug, the change in momentum of the bus, compared to the *change* of momentum of the bug is
(more) (the same) [circled] (less).

d. Which undergoes the greater acceleration?
(bus) (both the same) (bug) [circled]

e. Which, therefore, suffers the greater damage?
(bus) (both the same) (the bug of course!) [circled]

Chapter 4: Momentum and Energy

Systems

Momentum conservation (and Newton's Third Law) apply to *systems* of bodies. Here we identify some systems.

1. When the compressed spring is released, Blocks A and B will slide apart. There are three systems to consider here, indicated by the closed dashed lines below—System A, System B, and System A+B. Ignore the vertical forces of gravity and the support force of the table.

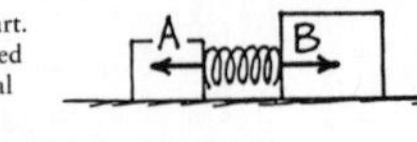

a. Does an external force act on System A? (yes) [circled] (no)
Will the momentum of System A change? (yes) [circled] (no)

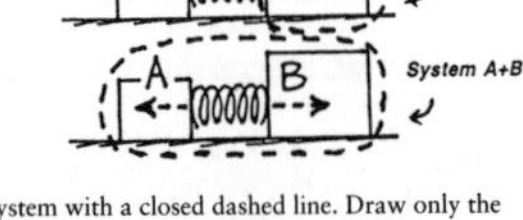

b. Does an external force act on System B? (yes) [circled] (no)
Will the momentum of System B change? (yes) [circled] (no)

c. Does an external force act on System A+B? (yes) (no) [circled]
Will the momentum of System A+B change? (yes) (no) [circled]

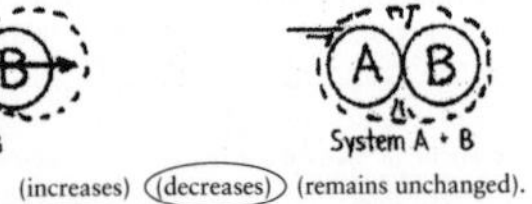

2. Billiard ball A collides with billiard ball B at rest. Isolate each system with a closed dashed line. Draw only the external force vectors that act on each system.

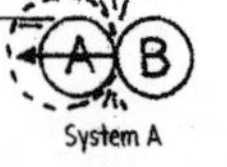

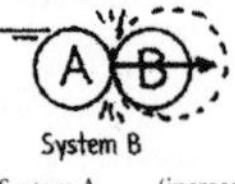

System A + B

a. Upon collision, the momentum of System A (increases) (decreases) [circled] (remains unchanged).
b. Upon collision, the momentum of System B (increases) [circled] (decreases) (remains unchanged).
c. Upon collision, the momentum of System A+B (increases) (decreases) (remains unchanged) [circled].

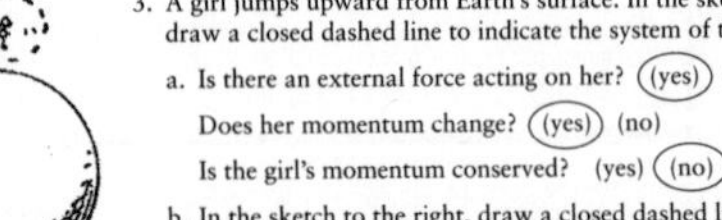

3. A girl jumps upward from Earth's surface. In the sketch to the left, draw a closed dashed line to indicate the system of the girl.

a. Is there an external force acting on her? (yes) [circled] (no)
Does her momentum change? (yes) [circled] (no)
Is the girl's momentum conserved? (yes) (no) [circled]

b. In the sketch to the right, draw a closed dashed line to indicate the system [girl + Earth]. Is there an external force due to the interaction between the girl and Earth that acts on the system? (yes) (no) [circled]
Is the momentum of the system conserved? (yes) [circled] (no)

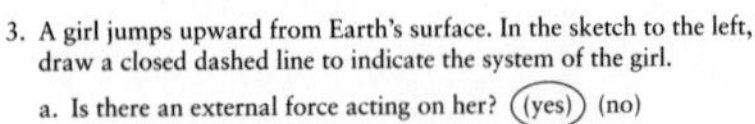

4. A block strikes a blob of jelly. Isolate three systems with a closed dashed line and show the external force on each. In which system is momentum conserved? SYSTEM AT RIGHT

5. A truck crashes into a wall. Isolate three systems with a closed dashed line and show the external force on each. In which system is momentum conserved? AT RIGHT

Chapter 4: Momentum and Energy

Impulse–Momentum

Bronco Brown wants to put $Ft = \Delta mv$ to the test and try bungee jumping. Bronco leaps from a high cliff and experiences free fall for 3 seconds. Then the bungee cord begins to stretch, reducing his speed to zero in 2 seconds. Fortunately, the cord stretches to its maximum length just short of the ground below.

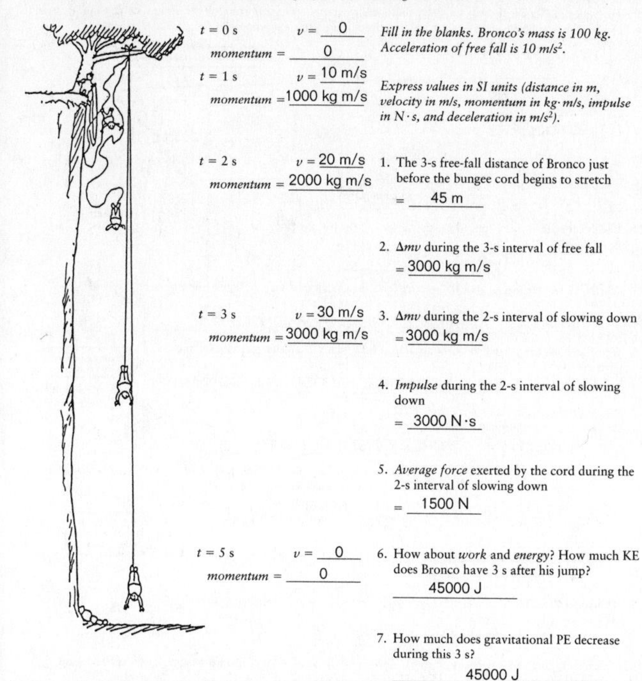

Fill in the blanks. Bronco's mass is 100 kg. Acceleration of free fall is 10 m/s².

Express values in SI units (distance in m, velocity in m/s, momentum in kg·m/s, impulse in N·s, and deceleration in m/s²).

$t = 0$ s v = 0 momentum = 0

$t = 1$ s v = 10 m/s momentum = 1000 kg m/s

$t = 2$ s v = 20 m/s momentum = 2000 kg m/s

$t = 3$ s v = 30 m/s momentum = 3000 kg m/s

$t = 5$ s v = 0 momentum = 0

1. The 3-s free-fall distance of Bronco just before the bungee cord begins to stretch = 45 m
2. Δmv during the 3-s interval of free fall = 3000 kg m/s
3. Δmv during the 2-s interval of slowing down = 3000 kg m/s
4. *Impulse* during the 2-s interval of slowing down = 3000 N·s
5. *Average force* exerted by the cord during the 2-s interval of slowing down = 1500 N
6. How about *work* and *energy*? How much KE does Bronco have 3 s after his jump? 45000 J
7. How much does gravitational PE decrease during this 3 s? 45000 J
8. What two kinds of PE are changing during the slowing-down interval? GRAVITATIONAL AND ELASTIC

Chapter 4: Momentum and Energy

Conservation of Momentum

Granny whizzes around the rink and is suddenly confronted with Ambrose at rest directly in her path. Rather than knock him over, she picks him up and continues in motion without "braking." Consider both Granny and Ambrose as two parts of one system. Since no outside forces act on the system, the momentum of the system before collision equals the momentum of the system after collision.

a. Complete the before-collision data in the table below.

BEFORE COLLISION	
Granny's mass	80 kg
Granny's speed	3 m/s
Granny's momentum	240 kg m/s
Ambrose's mass	40 kg
Ambrose's speed	0 m/s
Ambrose's momentum	0
Total momentum	240 kg m/s

b. After collision, does Granny's speed increase or decrease? DECREASE

c. After collision, does Ambrose's speed increase or decrease? INCREASE

d. After collision, what is the total mass of Granny + Ambrose? 120 kg

e. After collision, what is the total momentum of Granny + Ambrose? 240 kg m/s

f. Use the conservation of momentum law to find the speed of Granny and Ambrose together after collision. (Show your work in the space below.)

$$Mv + mv' = (M + m)V$$
$$(80\text{ kg})(3\text{ m/s}) + 0 = (80\text{ kg} + 40\text{ kg})V$$
$$240\text{ kg m/s} = (120\text{ kg})V$$
$$V = 2\text{ m/s}$$

New speed = 2 m/s

Chapter 4: Momentum and Energy

Work and Energy

1. How much work (energy) is needed to lift an object that weighs 200 N to a height of 4 m? 800 J
2. How much power is needed to lift the 200-N object to a height of 4 m in 4 s? 200 W
3. What is the power output of an engine that does 60,000 J of work in 10 s? 6 kW
4. The block of ice weighs 500 newtons.
   a. Neglecting friction, how much force is needed to push it up the incline? 250 N
   b. How much work is required to push it up the incline compared with lifting the block vertically 3 m? SAME (250 × 6 = 500 × 3)

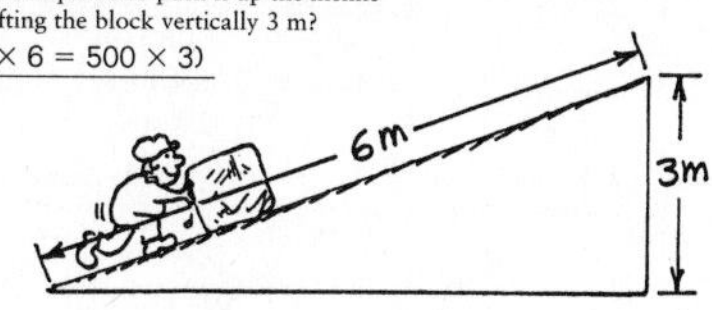

5. All the ramps are 5 m high. We know that the KE of the block at the bottom of the ramp will be equal to the loss of PE (conservation of energy). Find the speed of the block at ground level in each case. [Hint: Do you recall from earlier chapters how long it takes something to fall a vertical distance of 5 m from a position of rest (assume g = 10 m/s²)? And how much speed a falling object acquires in this time? This gives you the answer to Case 1. Discuss with your classmates how energy conservation gives you the answers to Cases 2 and 3.]

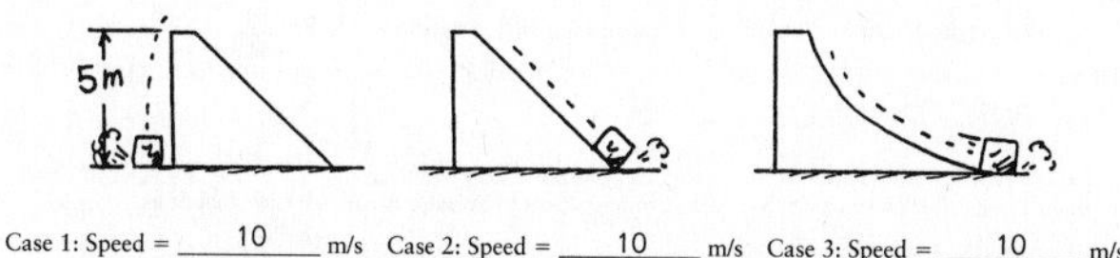

Case 1: Speed = 10 m/s Case 2: Speed = 10 m/s Case 3: Speed = 10 m/s

SAME SPEED BECAUSE ΔKE SAME; BUT **TIME** IS DIFFERENT!

Work and Energy—continued

6. Which block gets to the bottom of the incline first? Assume there is no friction. (Be careful!) Explain your answer.
   BLOCK A GETS TO BOTTOM FIRST. IT HAS MORE ACCELERATION (STEEPER) AND LESS SLIDING DISTANCE—(HOWEVER, BOTH HAVE SAME **SPEED** AT BOTTOM—BUT WE'RE ASKED FOR **TIME**)

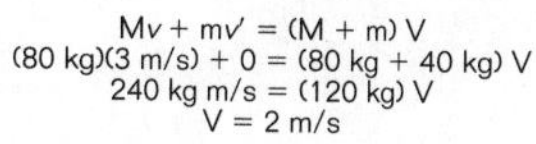

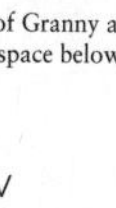

7. The KE and PE of a block freely sliding down a ramp are shown in only one place in the sketch. Fill in the missing information.

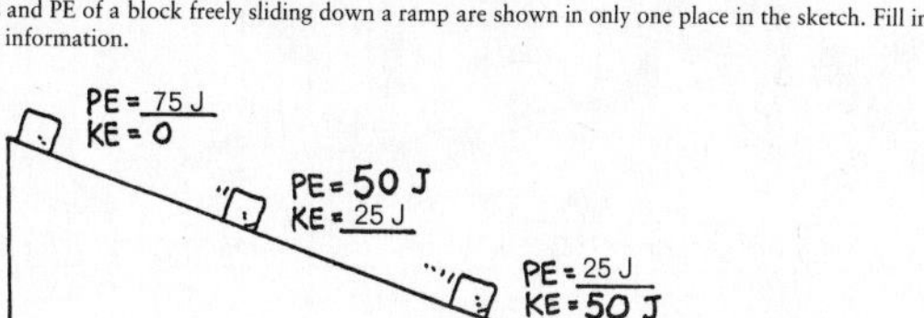

8. A big metal bead slides due to gravity along an upright friction-free wire. It starts from rest at the top of the wire as shown in the sketch. How fast is it traveling as it passes
   Point B? 10 m/s
   Point D? 10 m/s
   Point E? 10 m/s
   At what point does it have the maximum speed? C

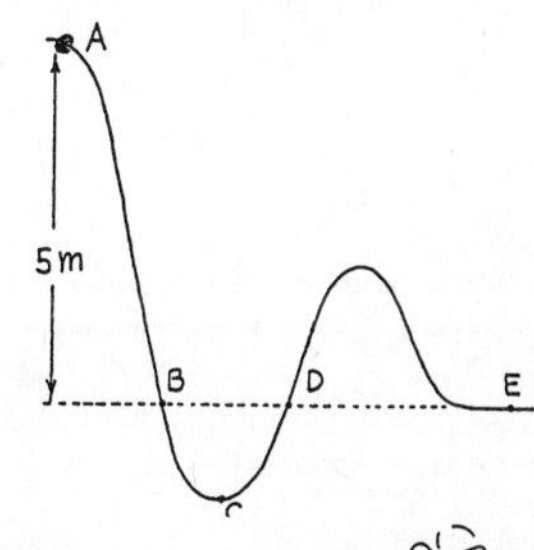

9. Rows of wind-powered generators are used in various windy locations to generate electric power. Does the power generated affect the speed of the wind? Would locations behind the "windmills" be windier if they weren't there? Discuss this in terms of energy conservation with your classmates.
   YES! BY CONS OF ENERGY, ENERGY GAINED BY WINDMILLS IS TAKEN FROM KE OF WIND—SO WIND MUST SLOW DOWN. LOCATIONS BEHIND WOULD BE A BIT WINDIER WITHOUT THE WINDMILLS!

Chapter 4: Momentum and Energy

Conservation of Energy

Fill in the blanks for the six systems shown:

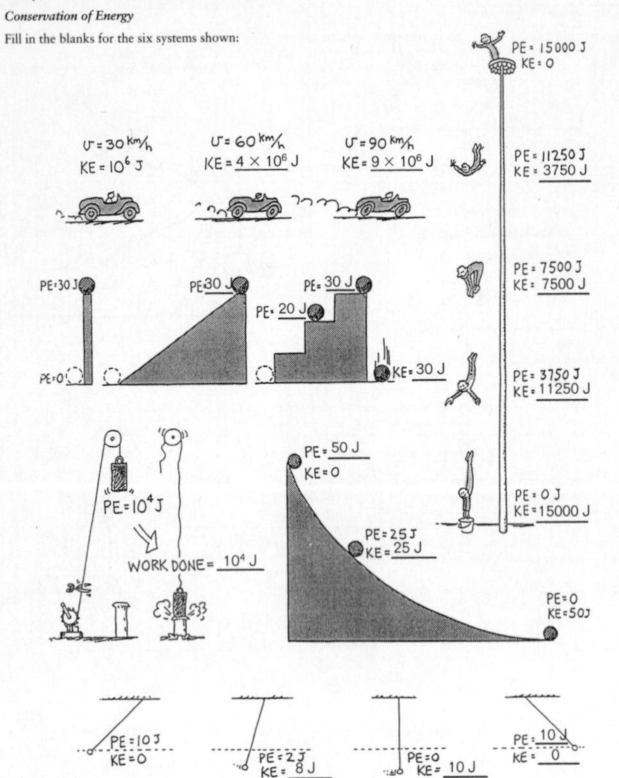

Chapter 4: Momentum and Energy

Momentum, Impulse, and Kinetic Energy

A Honda Civic and a Lincoln Town Car are initially at rest on a horizontal parking lot at the edge of a steep cliff. For simplicity, we assume that the Town Car has twice as much mass as the Civic. Equal constant forces are applied to each car and they accelerate across equal distances (we ignore the effects of friction). When they reach the far end of the lot the force is suddenly removed, whereupon they sail through the air and crash to the ground below. (The cars are beat up to begin with, and this is a scientific experiment!)

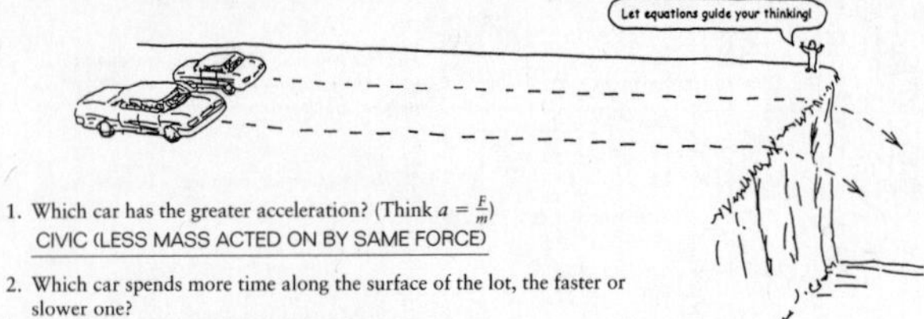

1. Which car has the greater acceleration? (Think $a = \frac{F}{m}$.)
   CIVIC (LESS MASS ACTED ON BY SAME FORCE)
2. Which car spends more time along the surface of the lot, the faster or slower one?
   TOWN CAR (SLOWER)
3. Which car has the larger impulse imparted to it by the applied force? (Think Impulse = Ft.) Defend your answer.
   TOWN CAR. SAME FORCE APPLIED OVER LONGER TIME.
4. Which car has the greater momentum at the cliff's edge? (Think $Ft = \Delta mv$.) Defend your answer.
   TOWN CAR. MORE IMPULSE, MORE CHANGE IN MOMENTUM
5. Which car has the greater work done on it by the applied force? (Think $W = Fd$) Defend your answer in terms of the distance traveled.
   SAME ON EACH—FORCE × DISTANCE SAME
6. Which car has the greater kinetic energy at the edge of the cliff? (Think $W = \Delta KE$) Does your answer follow from your explanation of 5? Does it contradict your answer to 3? Why or why not?
   SAME, BECAUSE SAME WORK. NO CONTRADICTION BECAUSE GREATER MOMENTUM OF TOWN CAR DUE TO MASS
7. Which car spends more time in the air, from the edge of the cliff to the ground below?
   BOTH THE SAME
8. Which car lands farthest horizontally from the edge of the cliff onto the ground below?
   CIVIC

Impulse = Δmomentum
Ft = Δmv

Work = Fd = ΔKE = Δ½mv²

Making the distinction between momentum and kinetic energy is high-level physics.

Challenge: Suppose the slower car crashes a horizontal distance of 10 m from the ledge. At what horizontal distance does the faster car hit? 14.1 M

(FROM $KE_{CIVIC} = KE_{TC}$, CIVIC HAS $\sqrt{2}\,v$ . . .1.41 TIMES FASTER)

Chapter 5: Gravity

The Inverse-Square Law—Weight

1. Paint spray travels radially away from the nozzle of the can in straight lines. Like gravity, the strength (intensity) of the spray obeys an inverse-square law. Complete the diagram by filling in the blank spaces.

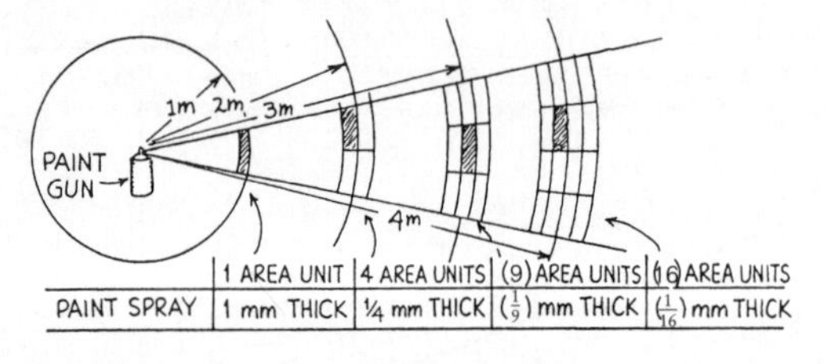

PAINT SPRAY	1 AREA UNIT	4 AREA UNITS	(9) AREA UNITS	(16) AREA UNITS
	1 mm THICK	¼ mm THICK	($\frac{1}{9}$) mm THICK	($\frac{1}{16}$) mm THICK

2. A small light source located 1 m in front of an opening of area 1 m² illuminates a wall behind. If the wall is 1 m behind the opening (2 m from the light source), the illuminated area covers 4 m². How many square meters will be illuminated if the wall is

   5 m from the source? 25 m²

   10 m from the source? 100 m²

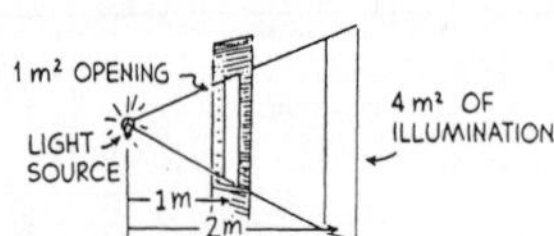

3. If we stand on a weighing scale and find that we are pulled toward Earth with a force of 500 N, then we weigh 500 N. Strictly speaking, we weigh 500 N relative to Earth. How much does Earth weigh? If we tip the scale upside down and repeat the weighing process, we can say that we and Earth are still pulled together with a force of 500 N, and, therefore, relative to us, the whole 6,000,000,000,000,000,000,000,000-kg Earth weighs 500 N! Weight, unlike mass, is a relative quantity.

We are pulled to Earth with a force of 500 N, so we weigh 500 N.

Earth is pulled toward us with a force of 500 N, so it weighs 500 N.

Chapter 5: Gravity

Ocean Tides

1. Consider two equal-mass blobs of water, A and B, initially at rest in the Moon's gravitational field. The vector shows the gravitational force of the Moon on A.

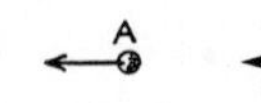

   a. Draw a force vector on B due to the Moon's gravity.
   b. Is the force on B more or less than the force on A? LESS
   c. Why? FARTHER AWAY
   d. The blobs accelerate toward the Moon. Which has the greater acceleration? (A) (B) — circled: (A)
   e. Because of the different accelerations, with time (A gets farther ahead of B) (A and B gain identical speeds) and the distance between A and B (increases) (stays the same) (decreases). — circled: A gets farther ahead of B; increases
   f. If A and B were connected by a rubber band, with time the rubber band would (stretch) (not stretch). — circled: stretch
   g. This (stretching) (nonstretching) is due to the (difference) (nondifference) in the Moon's gravitational pulls. — circled: difference
   h. The two blobs will eventually crash into the Moon. To orbit around the Moon instead of crashing into it, the blobs should move (away from the Moon) (tangentially). Then their accelerations will consist of changes in (speed) (direction). — circled: tangentially; direction

2. Now consider the same two blobs located on opposite sides of Earth.

Earth
A
B

   a. Because of differences in the Moon's pull on the blobs, they tend to (spread away from each other) (approach each other). This produces ocean tides! — circled: spread away from each other
   b. If Earth and the Moon were closer, gravitational force between them would be (more) (the same) (less), and the difference in gravitational forces on the near and far parts of the ocean would be (more) (the same) (less). — circled: more; more
   c. Because Earth's orbit about the Sun is slightly elliptical, Earth and the Sun are closer in December than in June. Taking the Sun's tidal force into account, on a world average, ocean tides are greater in (December) (June) (no difference). — circled: December

Projectile Motion

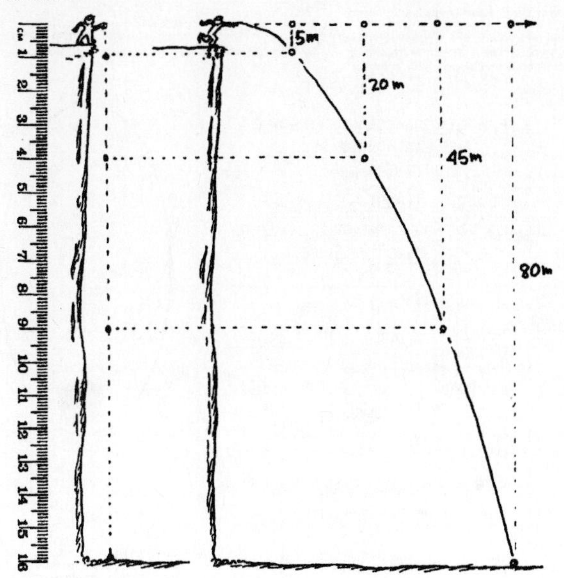

1. Above left: Use the scale 1 cm : 5 m and draw the positions of the dropped ball at 1-second intervals. Neglect air drag and assume $g = 10$ m/s². Estimate the number of seconds the ball is in the air.
   ___4___ seconds.

2. Above right: The four positions of the thrown ball with *no gravity* are at 1-second intervals. At 1 cm : 5 m, carefully draw the positions of the ball *with* gravity. Neglect air drag and assume $g = 10$ m/s². Connect your positions with a smooth curve to show the path of the ball. How is the motion in the vertical direction affected by motion in the horizontal direction?
   VERTICAL MOTION AFFECTED BY GRAVITY—HORIZONTAL MOTION DOESN'T AFFECT VERTICAL MOTION

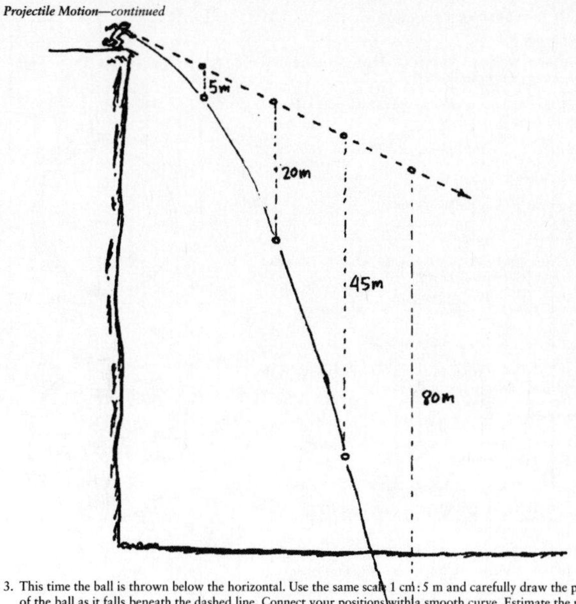

3. This time the ball is thrown below the horizontal. Use the same scale 1 cm : 5 m and carefully draw the positions of the ball as it falls beneath the dashed line. Connect your positions with a smooth curve. Estimate the number of seconds the ball remains in the air. ___3.5___ seconds

4. Suppose that you are an accident investigator and you are asked to figure out whether or not the car was speeding before it crashed through the rail of the bridge and into the mudbank as shown. The speed limit on the bridge is 55 mph = 24 m/s. What is your conclusion?

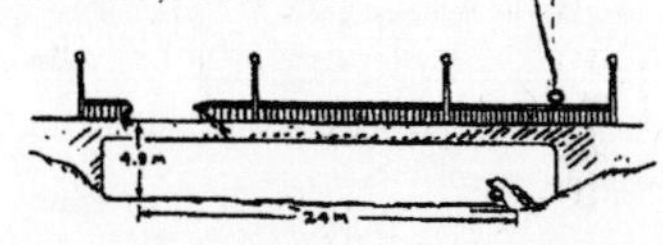

CAR COVERS 24 M IN 1 SEC (5 M DROP!), SO IT'S GOING 24 M/S **AFTER** CRASHING THRU RAIL. SO IT MUST HAVE BEEN GOING FASTER **BEFORE** HITTING RAIL. SO DRIVER WAS SPEEDING!

Tossed-Ball Vectors

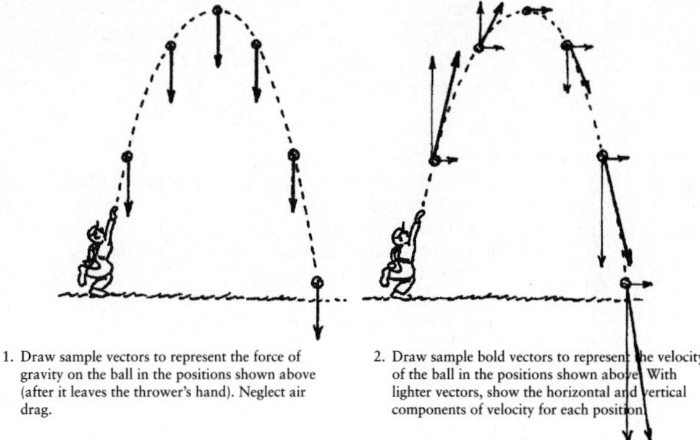

1. Draw sample vectors to represent the force of gravity on the ball in the positions shown above (after it leaves the thrower's hand). Neglect air drag.

2. Draw sample bold vectors to represent the velocity of the ball in the positions shown above. With lighter vectors, show the horizontal and vertical components of velocity for each position.

3. a. Which velocity component in the previous question remains constant? Why?
   HORIZONTAL, BECAUSE NO FORCE ACTS HORIZONTALLY.

   b. Which velocity component changes along the path? Why?
   VERTICAL, BECAUSE THE FORCE OF GRAVITY IS VERTICAL.

4. It is important to distinguish between force and velocity vectors. Force vectors combine with other force vectors, and velocity vectors combine with other velocity vectors. Do velocity vectors combine with force vectors? ___NO!___

A ball tossed upward has initial velocity components 30 m/s vertical and 5 m/s horizontal. The position of the ball is shown at 1-second intervals. Air resistance is negligible, and $g = 10$ m/s². Fill in the boxes, writing in the values of velocity *components* ascending, and your calculated *resultant velocities* descending.

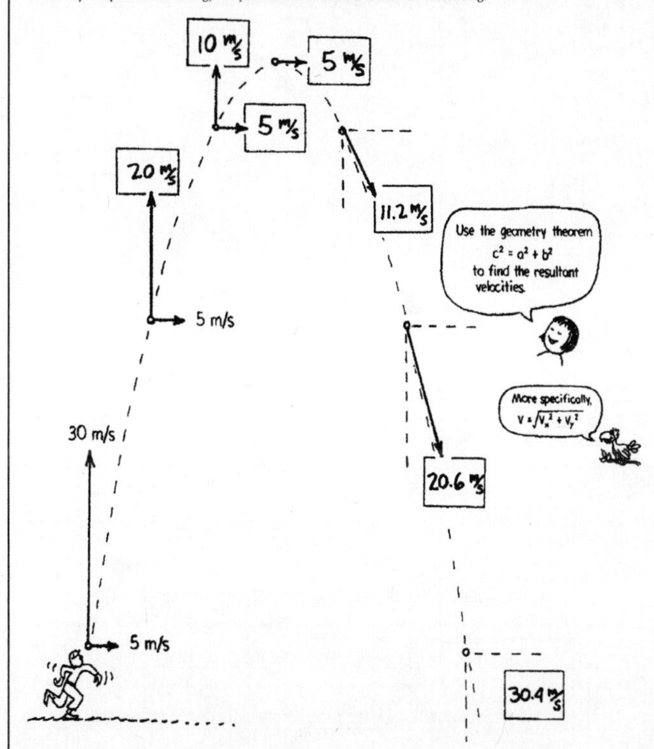

Chapter 5: Gravity

Circular and Elliptical Orbits

I. *Circular Orbits*

1. Figure 1 shows "Newton's Mountain," so high that its top is above the drag of the atmosphere. The cannonball is fired and hits the ground as shown.
   a. Draw the path the cannonball might take if it were fired a little bit faster.
   b. Repeat for a still greater speed, but still less than 8 km/s.
   c. Draw the orbital path it would take if its speed were 8 km/s.
   d. What is the shape of the 8-km/s curve?
   CIRCLE
   e. What would be the shape of the orbital path if the cannonball were fired at a speed of about 9 km/s?
   ELLIPSE

Figure 1

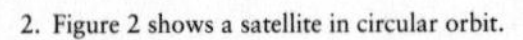

2. Figure 2 shows a satellite in circular orbit.
   a. At each of the four positions draw a vector that represents the gravitational *force* exerted on the satellite.
   b. Label the force vectors *F*.
   c. Draw at each position a vector to represent the *velocity* of the satellite at that position and label it *V*.
   d. Are all four *F* vectors the same length? Why or why not?
   YES; SAME DISTANCE, SAME FORCE

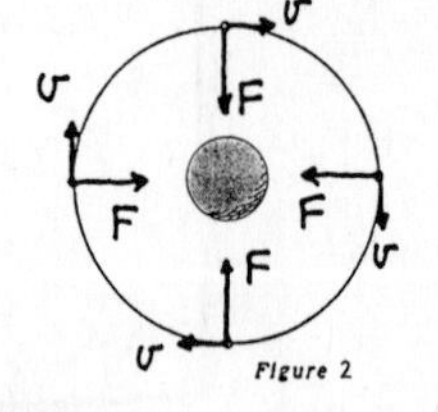

Figure 2

   e. Are all four *V* vectors the same length? Why or why not?
   YES—IN CIRCULAR ORBIT $F \perp v$ SO NO COMPONENT OF F ALONG v.
   f. What is the angle between your *F* and *V* vectors? 90°
   g. Is there any component of *F* along *V*? NO, ($F \perp v$)
   h. What does this tell you about the work the force of gravity does on the satellite?
   NO WORK BECAUSE NO COMPONENT OF F ALONG PATH
   i. Does the KE of the satellite in Figure 2 remain constant, or does it vary? CONSTANT
   j. Does the PE of the satellite remain constant, or does it vary?
   CONSTANT

Circular and Elliptical Orbits—continued

II. *Elliptical Orbits*

3. Figure 3 shows a satellite in elliptical orbit.
   a. Repeat the procedure you used for the circular orbit, drawing vectors *F* and *V* for each position, including proper labeling. Show equal magnitudes with equal lengths, and greater magnitudes with greater lengths, but don't bother making the scale accurate.
   b. Are your vectors *F* all the same magnitude? Why or why not?
   NO, FORCE DECREASES WHEN DISTANCE FROM EARTH INCREASES
   c. Are your vectors *V* all the same magnitude? Why or why not?
   NO, WHEN KE DECREASES, SPEED DECREASES. WHEN KE INCREASES (CLOSER TO EARTH) SPEED INCREASES.

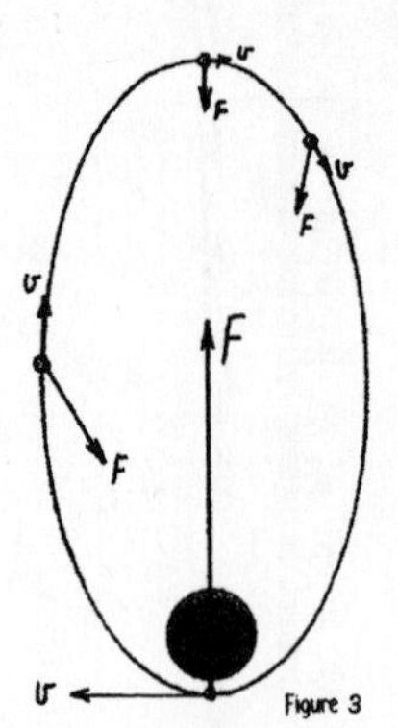

Figure 3

   d. Is the angle between vectors *F* and *V* everywhere the same, or does it vary?
   IT VARIES
   e. Are there places where there is a component of *F* along *V*?
   YES (EVERYWHERE EXCEPT AT THE APOGEE AND PERIGEE)
   f. Is work done on the satellite when there is a component of *F* along and in the same direction of *V*, and if so, does this increase or decrease the KE of the satellite?
   YES, THIS INCREASES KE OF SATELLITE
   g. When there is a component of *F* along and opposite to the direction of *V*, does this increase or decrease the KE of the satellite?
   THIS DECREASES KE OF SATELLITE
   h. What can you say about the sum KE + PE along the orbit?
   CONSTANT (IN ACCORD WITH CONSERVATION OF ENERGY)

Be very, very careful when placing both velocity and force vectors on the same diagram. Not a good practice, for one may construct the resultant of the vectors—ouch!

Chapter 5: Gravity

Mechanics Overview

1. The sketch shows the elliptical path described by a satellite about Earth. In which of the marked positions, A–D, (put S for "same everywhere") does the satellite experience the maximum
   a. gravitational force? A
   b. speed? A
   c. velocity? A
   d. momentum? A
   e. kinetic energy? A
   f. gravitational potential energy? C
   g. total energy (KE + PE)? S
   h. acceleration? A

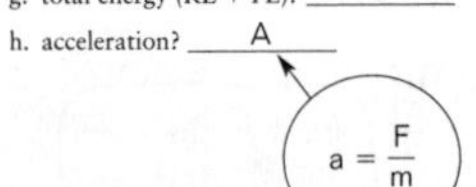

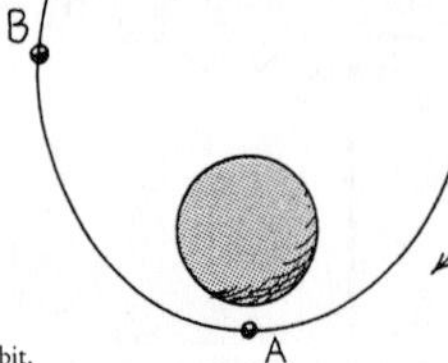

2. Answer the above questions for a satellite in circular orbit.
   a. S b. S c. S d. S e. S f. S g. S h. S

3. In which position(s) is there momentarily no work done on the satellite by the force of gravity? Why?
   A AND C, BECAUSE NO FORCE COMPONENTS ALONG PATH

4. Work changes energy. Let the equation for work, $W = Fd$, guide your thinking on these questions. Defend your answers in terms of $W = Fd$.
   a. In which position will a several-minutes thrust of rocket engines do the most work on the satellite and give it the greatest change in kinetic energy?
   A, BECAUSE d GREATEST DURING THRUST—$F \times d$ IS MORE WORK
   b. In which position will a several-minutes thrust of rocket engines do the most work on the *exhaust gases* and give the *exhaust gases* the greatest change in kinetic energy?
   C, WHERE SATELLITE ROCKET IS SLOWEST
   c. In which position will a several-minutes thrust of rocket engines give the satellite the least boost in kinetic energy?
   C, BECAUSE RELATIVE TO PLANET, MOST ENERGY IS GIVEN TO THE EXHAUST GASES.

Chapter 6: Heat

Temperature Mix

1. You apply heat to 1 L of water and raise its temperature by 10°C. If you add the same quantity of heat to 2 L of water, how much will the temperature rise? To 3 L of water?

Record your answers on the blanks in the drawing at the right. (Hint: Heat transferred is directly proportional to its temperature change, $Q = mc\Delta T$.)

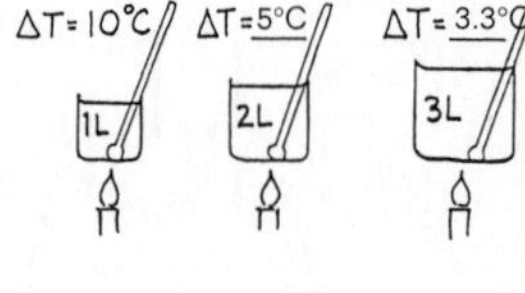

2. A large bucket contains 1 L of 20°C water.

   a. What will be the temperature of the mixture when 1 L of 20°C water is added?
   STILL 20°C
   b. What will be the temperature of the mixture when 1 L of 40°C water is added?
   30°C
   c. If 2 L of 40°C water were added, would the temperature of the mixture be greater or less than 30°C?
   GREATER

$2\,(40°C - T) = 1\,(T - 20°C)$
$T = 33.3°C$

3. A red-hot iron kilogram mass is put into 1 L of cool water. Mark each of the following statements true (T) of false (F). (Ignore heat transfer to the container.)
   a. The increase in the water temperature is equal to the decrease in the iron's temperature.
   F
   NOTE DISTINCTION!
   b. The quantity of heat gained by the water is equal to the quantity of heat lost by the iron.
   T
   c. The iron and the water will both reach the same temperature.
   T
   THERMAL EQUILIBRIUM
   d. The final temperature of the iron and water is about halfway between the initial temperatures of each.
   F

4. *True or False:* When Queen Elizabeth throws the last sip of her tea over Queen Mary's rail, the ocean gets a little warmer. T (UNLESS IT WAS ICE TEA!)

Chapter 6: Heat

Absolute Zero

A mass of air is contained so that the volume can change but the pressure remains constant. Table 1 shows air volumes at various temperatures when the air is heated slowly.

1. Plot the data in Table 1 on the graph and connect the points.

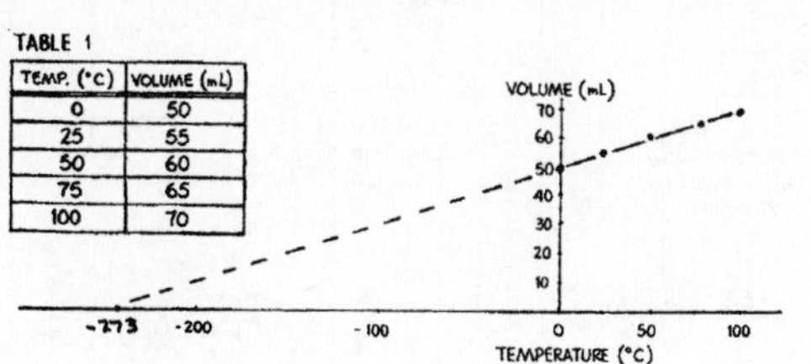

TABLE 1

TEMP. (°C)	VOLUME (mL)
0	50
25	55
50	60
75	65
100	70

2. The graph shows how the volume of air varies with temperature at constant pressure. The straightness of the line means that the air expands uniformly with temperature. From your graph, you can predict what will happen to the volume of air when it is cooled.

   Extrapolate (extend) the straight line of your graph to find the temperature at which the volume of the air would become zero. Mark this point on your graph. Estimate this temperature: −273°C

3. Although air would liquify before cooling to this temperature, the procedure suggests that there is a lower limit to how cold something can be. This is the absolute zero of temperature.

   Careful experiments show that absolute zero is −273 °C.

4. Scientists measure temperature in *kelvins* instead of degrees Celsius, where the absolute zero of temperature is 0 kelvins. If you relabeled the temperature axis on the graph in Question 1 so that it shows temperature in kelvins, would your graph look like the one below? YES

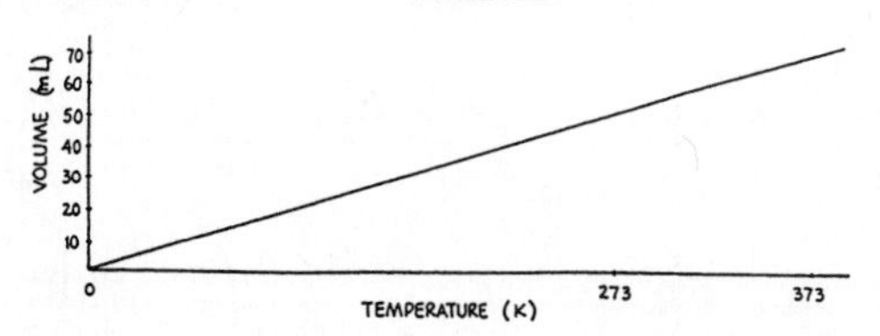

Chapter 6: Heat

Thermal Expansion

1. Steel expands by about 1 part in 100,000 for each 1°C increase in temperature.

   a. How much longer will a piece of steel 1000 mm long (1 meter) be when its temperature is increased by 10°C? 0.1 mm

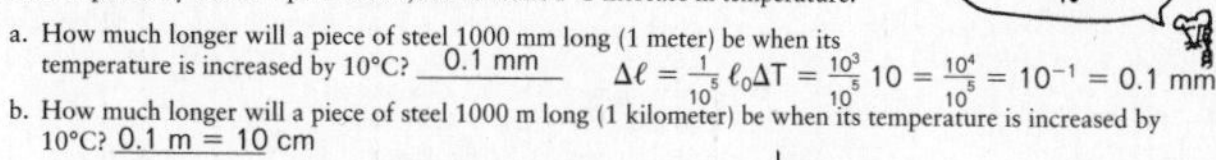

$\Delta \ell = \frac{1}{10^5}\ell_0\Delta T = \frac{10^3}{10^5}10 = \frac{10^4}{10^5} = 10^{-1} = 0.1$ mm

   b. How much longer will a piece of steel 1000 m long (1 kilometer) be when its temperature is increased by 10°C? 0.1 m = 10 cm

   c. You place yourself between a wall and the end of a 1-m steel rod when the opposite end is securely fastened as shown. No harm comes to you if the temperature of the rod is increased a few degrees. Discuss the consequences of doing this with a rod many meters long?

   $\Delta\ell$ IS SMALL FOR SMALL ℓ_0, BUT CAN BE FATALLY LARGE (YOUR BODY WIDTH!) FOR LARGE ℓ_0.

2. The Eiffel Tower in Paris is 298 meters high. On a cold winter night, it is shorter than on a hot summer day. What is its change in height for a 30°C temperature difference?

   $\Delta\ell = \frac{298}{10^5} \cdot 30 = 0.09$ m = 9 cm

3. Consider a gap in a piece of metal. Does the gap become wider or narrower when the metal is heated? [Consider the piece of metal made up of 11 blocks—if the blocks are individually heated, each is slightly larger. Make a sketch of them, slightly enlarged, beside the sketch shown.]

   GAP IS WIDER (AS MUCH IF IT WERE ALL METAL)

4. The equatorial radius of Earth is about 6370 km. Consider a 40,000-km long steel pipe that forms a giant ring that fits snugly around the equator of the earth. Suppose people all along its length breathe on it so as to raise its temperature 1°C. The pipe gets longer. It is also no longer snug. How high does it stand above the ground? (Hint: Concentrate on the radial distance.)

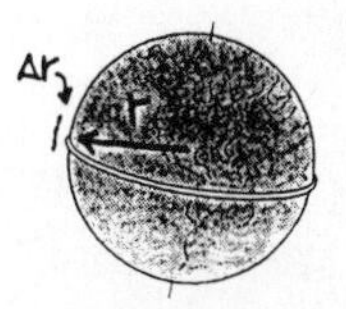

$\Delta r = \frac{6370}{10^5}$ km · 10 = 0.637 km = 63.7 m! WOW!

Thermal Expansion—continued

5. A weight hangs above the floor from the copper wire. When a candle is moved along the wire and heats it, what happens to the height of the weight above the floor? Why?

   HEIGHT DECREASES AS WIRE LENGTHENS

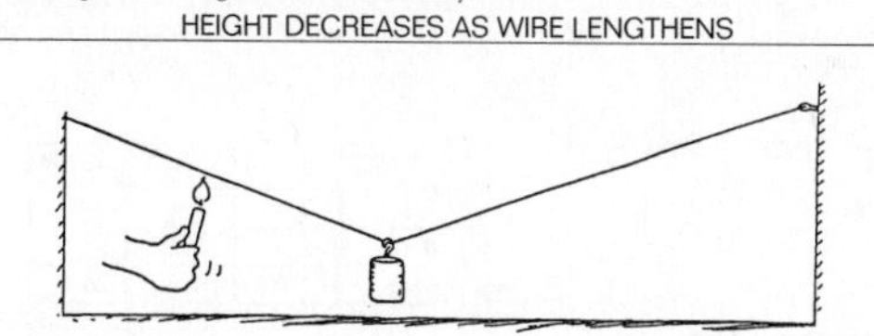

6. The levels of water at 0°C and 1°C are shown below in the first two flasks. At these temperatures there is microscopic slush in the water. There is slightly more slush at 0°C than at 1°C. As the water is heated, some of the slush collapses as it melts, and the level of the water falls in the tube. That's why the level of water is slightly lower in the 1°C tube. Make rough estimates and sketch in the appropriate levels of water at the other temperatures shown. What is important about the level when the water reaches 4°C?

   SINCE WATER IS MOST DENSE AT 4°C, WATER LEVEL IS LOWEST AT 4°C

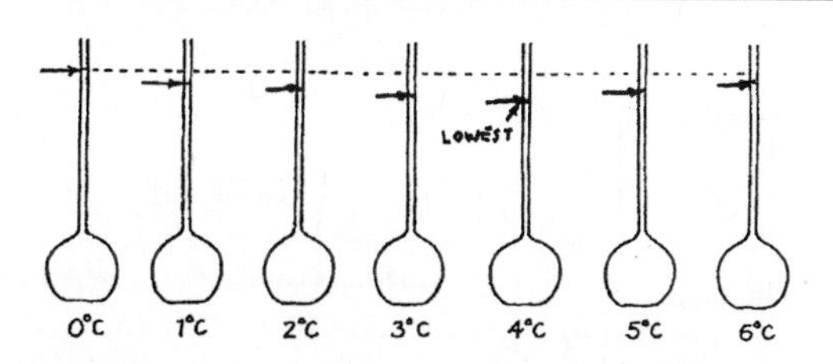

7. The diagram at right shows an ice-covered pond. Mark the probable temperatures of water at the top and bottom of the pond.

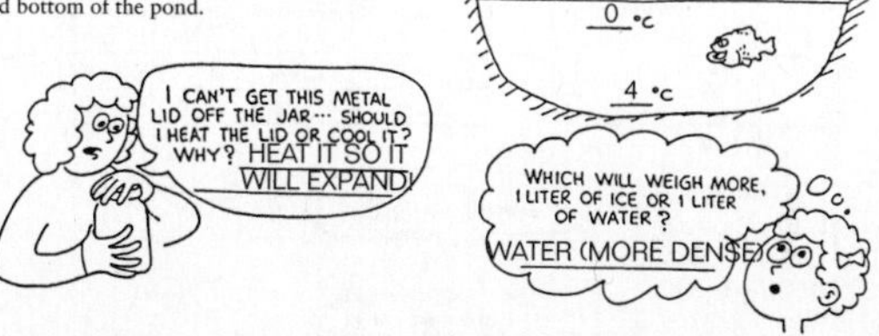

Chapter 6: Heat

Transmission of Heat

1. The tips of both brass rods are held in the gas flame. *Mark the following true (T) or false (F).*

   a. Heat is conducted only along Rod A. F

   b. Heat is conducted only along Rod B. F

   c. Heat is conducted equally along both Rod A and Rod B. T

   d. The idea that "heat rises" applies to heat transfer by *convection*, not by *conduction*. ____

2. Why does a bird fluff its feathers to keep warm on a cold day?

   FLUFFED FEATHERS TRAP AIR THAT INSULATES.

3. Why does a down-filled sleeping bag keep you warm on a cold night? Why is it useless if the down is wet?

   AS IN 2, WHEN WATER TAKES PLACE OF TRAPPED AIR, INSULATION IS REDUCED.

4. What does *convection* have to do with the holes in the shade of the desk lamp?

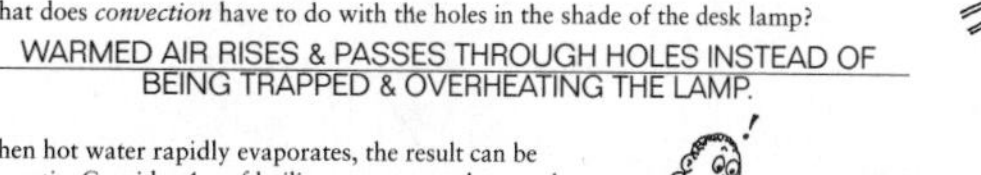

   WARMED AIR RISES & PASSES THROUGH HOLES INSTEAD OF BEING TRAPPED & OVERHEATING THE LAMP.

5. When hot water rapidly evaporates, the result can be dramatic. Consider 4 g of boiling water spread over a large surface so that 1 g rapidly evaporates. Suppose further that the surface and surroundings are very cold so that all 540 calories for evaporation come from the remaining 3 g of water.

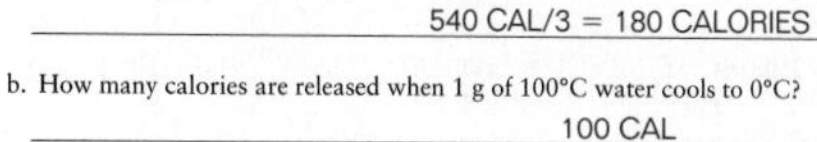

   a. How many calories are taken from each gram of water?

   540 CAL/3 = 180 CALORIES

   b. How many calories are released when 1 g of 100°C water cools to 0°C?

   100 CAL

   c. How many calories are released when 1 g of 0°C water changes to 0°C ice?

   80 CAL

   d. What happens in this case to the remaining 3 g of boiling water when 1 g rapidly evaporates?

   THE REMAINING WATER FREEZES! (EACH GRAM OF WATER RELEASES 180 CAL IN COOLING AND FREEZING.)

Electric Potential

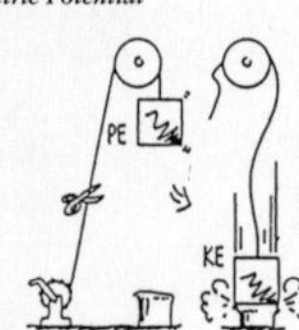

Just as PE transforms to KE for a mass lifted against the gravitation field (left), the electric PE of an electric charge transforms to other forms of energy when it changes location in an electric field (right). In both cases, how does the KE acquired compare to the decrease in PE?

SAME

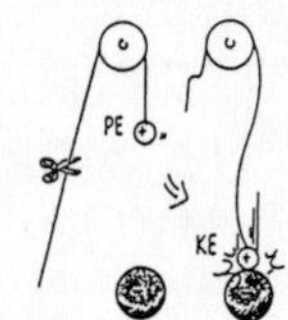

Complete the following statements:

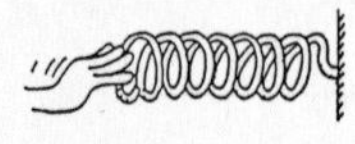

A force compresses the spring. The work done in compression is the product of the average force and the distance moved: $W = Fd$. This work increases the PE of the spring.

Similarly, a force pushes the charge (call it a *test charge*) closer to the charged sphere. The work done in moving the test charge is the product of the average FORCE and the DISTANCE moved. W = F × d. This work INCREASES the PE of the test charge.

If the test charge is released, it will be repelled and fly past the starting point. Its gain in KE at this point is EQUAL to its decrease in PE.

At any point, a greater amount of test charge means a greater amount of PE, but not a greater amount of PE *per amount* of charge. The quantities PE (measured in joules) and $\frac{PE}{charge}$ (measured in volts) are different concepts.

By definition: Electric Potential = $\frac{PE}{charge}$. 1 volt = $\frac{1 \text{ joule}}{1 \text{ coulomb}}$. So, 1 C of charge with a PE of 1 J has an electric potential of 1 V. 2 C of charge with a PE of 2 J has an electric potential of 1 V.

If a conductor connected to the terminal of a battery has an electric potential of 12 V, then each coulomb of charge on the conductor has a PE of 12 J.

You do very little work in rubbing a balloon on your hair to charge it. The PE of several thousand billion electrons (about one-millionth coulomb [10^{-6}C]) transferred may be a thousandth of a joule [10^{-3}J]. Impressively, however, the electric potential of the balloon is about 1000 V!

$$\frac{10^{-3}\text{ J}}{10^{-6}\text{ C}} = 10^{3}\text{ V}$$

Why is contact with a balloon charged to thousands of volts not as dangerous as contact with household 110 V?

HOUSEHOLD CURRENT TRANSFERS MANY COULOMBS AND MUCH ENERGY. A BALLOON TRANSFERS VERY LITTLE OF BOTH.

Series Circuits

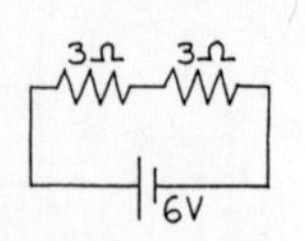

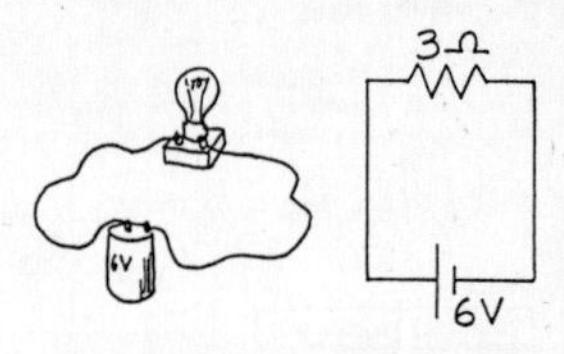

1. The simple circuit is a 6-V battery that pushes charge through a single lamp that has a resistance of 3 Ω. According to Ohm's law, the current in the lamp (and therefore the whole circuit) is 2 A.

2. If a second identical lamp is added, the 6-V battery must push charge through a total resistance of 6 Ω. The current in the circuit is then 1 A.

3. If a third identical lamp is added in series, the total resistance of the circuit (neglecting any internal resistance in the battery) is 9 Ω.

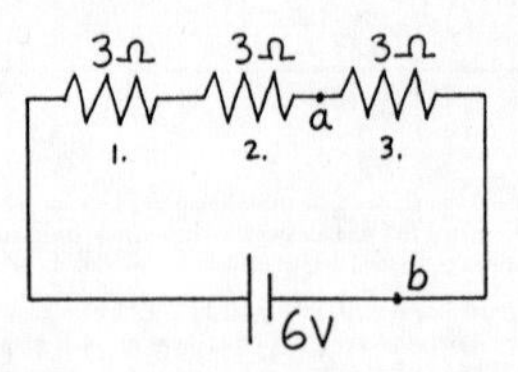

4. The current through all three lamps in series is 2/3 A. The current through each individual lamp is 2/3 A.

5. Does current in the lamps occur simultaneously, or does charge flow first through one lamp, then the other, and finally the last, in turn? SIMULTANEOUSLY (~SPEED OF LIGHT)

6. Does current flow *through* a resistor, or *across* a resistor? THROUGH Is voltage established *through* a resistor, or *across* a resistor? ACROSS

7. The voltage across all three lamps in the series is 6-V. The voltage (or commonly, *voltage drop*) across each individual lamp is 2 V.

8. Suppose a wire connects points *a* and *b* in the circuit. The voltage drop across lamp 1 is now 3 V, across lamp 2 is 3 V, and across lamp 3 is 0 V. So, the current through lamp 1 is now 1 A, through lamp 2 is 1 A, and through lamp 3 is 0 A. The current in the battery (neglecting internal battery resistance) is ______A.

9. Which circuit dissipates more power, the 3-lamp circuit or the 2-lamp circuit? (Another way of asking this is which circuit would glow brightest and be best seen on a dark night from a great distance?) Defend your answer.
FOR 3 LAMPS: P = I V = $^2/_3$ × 6 = 4 W FOR 2 LAMPS P = I V = 1 × 6 = 6 W ∴ THE 2-LAMP CIRCUIT IS BRIGHTEST (IT WOULD BE EVEN BRIGHTER, 12 W, IF THERE WERE 1 LAMP)

Parallel Circuits

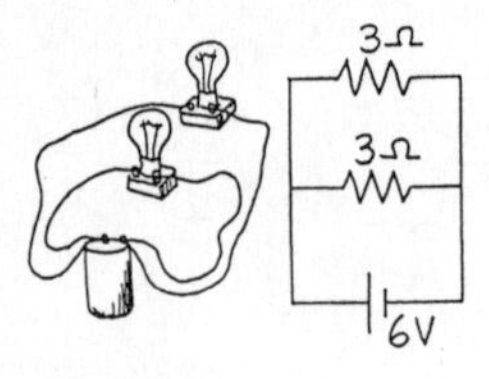

1. In the circuit shown to the left there is a voltage drop of 6-V across each 3-Ω lamp. By Ohm's law, the current in each lamp is 2 A. The current through the battery is the sum of the currents in the lamps, 4 A.

THE SUM OF THE CURRENTS IN THE TWO BRANCH PATHS EQUALS THE CURRENT BOTH BEFORE AND AFTER IT DIVIDES!

2. Fill in the current in the eight blank spaces in the view of the same circuit shown again at the right.

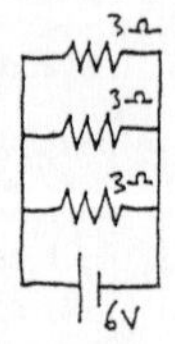

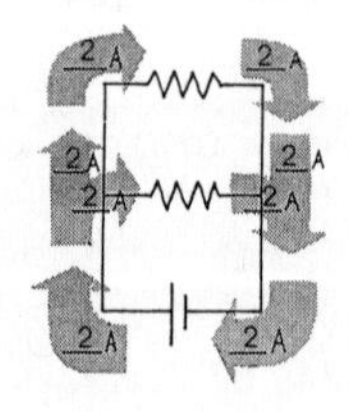

3. Suppose a third identical lamp is added in parallel to the circuit. Sketch a schematic diagram of the 3-lamp circuit in the space at the right.

4. For the three identical lamps in parallel, the voltage drop across each lamp is 6 V. The current through each lamp is 2 A. The current through the battery is now 6 A. Is the circuit resistance now greater or less than before the third lamp was added? Explain.
LESS, BECAUSE OF MORE PATHS, WHICH MEANS LESS RESISTANCE BETWEEN BATTERY TERMINALS

5. Which circuit dissipates more power, the 3-lamp circuit or the 2-lamp circuit? (Another way of asking this is which circuit would glow brightest and be best seen on a dark night from a great distance?) Defend your answer and compare this to the similar case for 2-and 3-lamp series circuits.
3 LAMPS: P = I V = 6 × 6 = 36 W 2 LAMPS: P = I V = 4 × 6 = 24 W
SO 3-LAMP CIRCUIT IS BRIGHTEST; MORE CURRENT FLOWS (BECAUSE OF REDUCED RESISTANCE) FOR THE SAME VOLTAGE. OPPOSITE FOR SERIES CIRCUIT.

Compound Circuits

The table beside circuit *a* below shows the current through each resistor, the voltage across each resistor, and the power dissipated as heat in each resistor. Find the similar correct values for circuits *b*, *c*, and *d*, and put your answers in the tables shown.

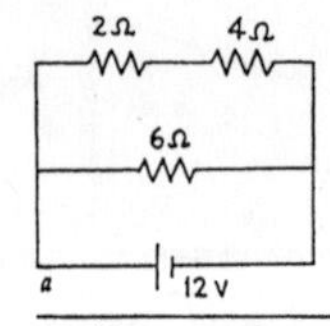

RESISTANCE	CURRENT ×	VOLTAGE =	POWER
2Ω	2 A	4 V	8 W
4Ω	2 A	8 V	16 W
6Ω	2 A	12 V	24 W

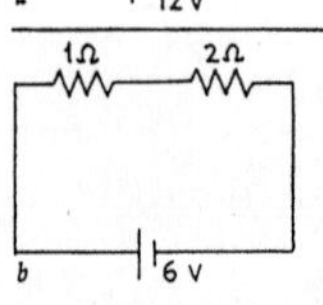

RESISTANCE	CURRENT ×	VOLTAGE =	POWER
1Ω	2 A	2 V	4 W
2Ω	2 A	4 V	8 W

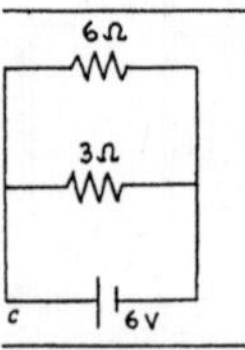

RESISTANCE	CURRENT ×	VOLTAGE =	POWER
6Ω	1 A	6 V	6 W
3Ω	2 A	6 V	12 W

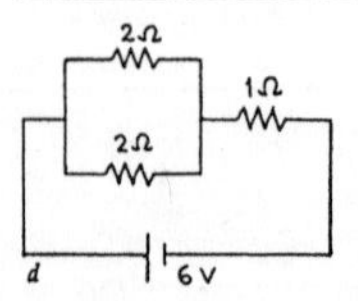

RESISTANCE	CURRENT ×	VOLTAGE =	POWER
2Ω	1.5 A	3 V	4.5 W
2Ω	1.5 A	3 V	4.5 W
1Ω	3 A	3 V	9 W

NOTE THAT TOTAL POWER DISSIPATED BY ALL RESISTORS IN A CIRCUIT EQUALS THE POWER SUPPLIED BY THE BATTERY: (VOLTAGE OF BATTERY × CURRENT THRU BATTERY)

A VOLT IS A UNIT OF POTENTIAL (OR "PRESSURE") AND AN AMPERE IS A UNIT OF CURRENT

DOES VOLTAGE CAUSE CURRENT, OR DOES CURRENT CAUSE VOLTAGE? WHICH IS THE CAUSE AND WHICH IS THE EFFECT?

Chapter 7: Electricity and Magnetism

Magnetism

Fill in each blank with the appropriate word:

1. Attraction or repulsion of charges depends on their *signs*, positives or negatives. Attraction or repulsion of magnets depends on their magnetic **POLES** : **NORTH** or **SOUTH** .

YOU HAVE A MAGNETIC PERSONALITY!

2. Opposite poles attract; like poles **REPEL** .

3. A magnetic field is produced by the **MOTION** of electric charge.

4. Clusters of magnetically aligned atoms are magnetic **DOMAINS** .

5. A magnetic **FIELD** surrounds a current-carrying wire.

6. When a current-carrying wire is made to form a coil around a piece of iron, the result is an **ELECTROMAGNET** .

7. A charged particle moving in a magnetic field experiences a deflecting **FORCE** that is maximum when the charge moves **PERPENDICULAR** to the field.

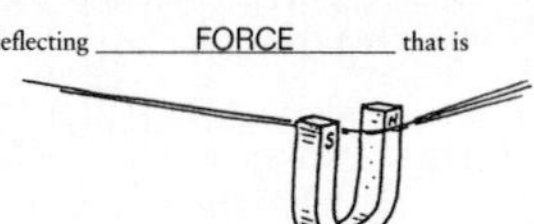

8. A current-carrying wire experiences a deflecting **FORCE** that is maximum when the wire and magnetic field are **PERPENDICULAR** to one another.

9. A simple instrument designed to detect electric current is the **GALVANOMETER**; when calibrated to measure current, it is an **AMMETER** ; when calibrated to measure voltage, it is a **VOLTMETER**.

10. The largest size magnet in the world is the **WORLD** itself.

THEN TO REALLY MAKE THINGS "SIMPLE," THERE'S THE RIGHT-HAND RULE!
F
B
I

Chapter 7: Electricity and Magnetism

Field Patterns

1. The illustration below is similar to Figure 7.26 in your textbook. Iron filings trace out patterns of magnetic field lines about a bar magnet. In the field are some magnetic compasses. The compass needle in only one compass is shown. Draw in the needles with proper orientation in the other compasses.

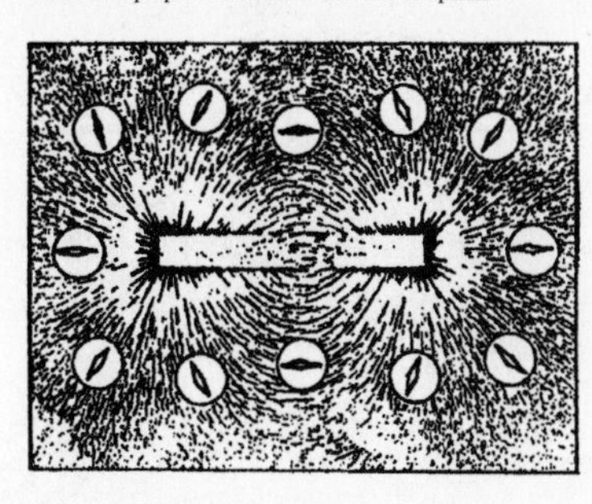

2. The illustration below is similar to Figure 7.33b in your textbook. Iron filings trace out the magnetic field pattern about the loop of current-carrying wire. Draw in the compass needle orientations for all the compasses.

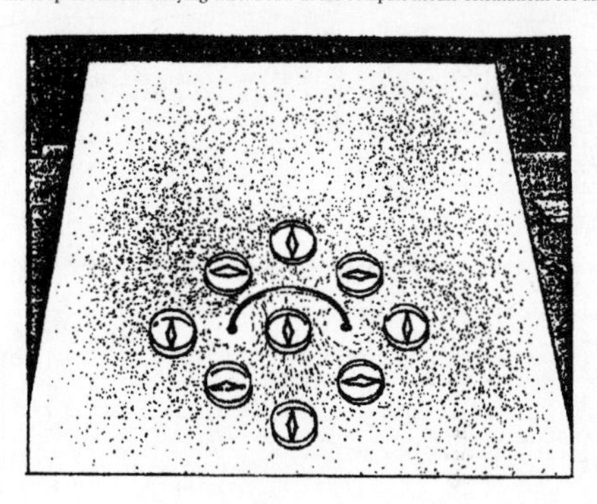

Chapter 7: Electricity and Magnetism

Electromagnetism

1. Hans Christian Oersted discovered that magnetism and electricity are
   (**related**) (independent of each other).
   Magnetism is produced by
   (batteries) (**the motion of electric charges**).

Faraday and Henry discovered that electric current can be produced by
(batteries) (**motion of a magnet**).
More specifically, voltage is induced in a loop of wire if there is a change in the
(batteries) (**magnetic field in the loop**).
This phenomenon is called
(electromagnetism) (**electromagnetic induction**).

2. When a magnet is plunged in and out of a coil of wire, voltage is induced in the coil. If the rate of the in-and-out motion of the magnet is doubled, the induced voltage
   (**doubles**) (halves) (remains the same).
   If instead, the number of loops in the coil is doubled, the induced voltage
   (**doubles**) (halves) (remains the same).

3. A rapidly changing magnetic field in any region of space induces a rapidly changing
   (**electric field**) (magnetic field) (gravitational field),
   which in turn induces a rapidly changing
   (**magnetic field**) (electric field) (baseball field).
   This generation and regeneration of electric and magnetic fields makes up
   (**electromagnetic waves**) (sound waves) (both of these).

Vibration and Wave Fundamentals

1. A sine curve that represents a transverse wave is drawn below. With a ruler, measure the wavelength and amplitude of the wave.

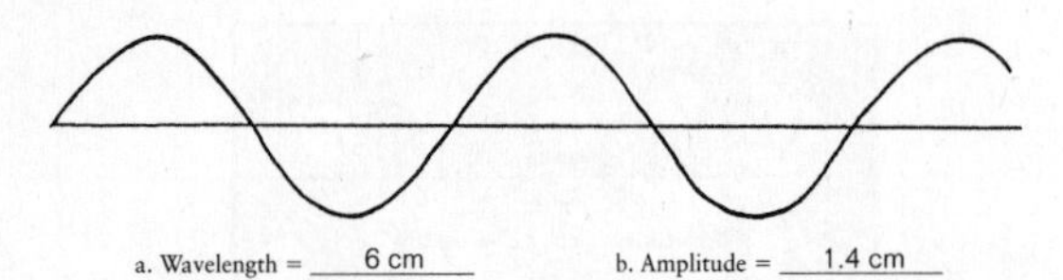

a. Wavelength = 6 cm b. Amplitude = 1.4 cm

2. A girl on a playground swing makes a complete to-and-fro swing each 2 seconds. The frequency of swing is

   (0.5 hertz) (1 hertz) (2 hertz)

   and the period is

   (0.5 second) (1 second) (2 seconds).

3. *Complete the following statements:*

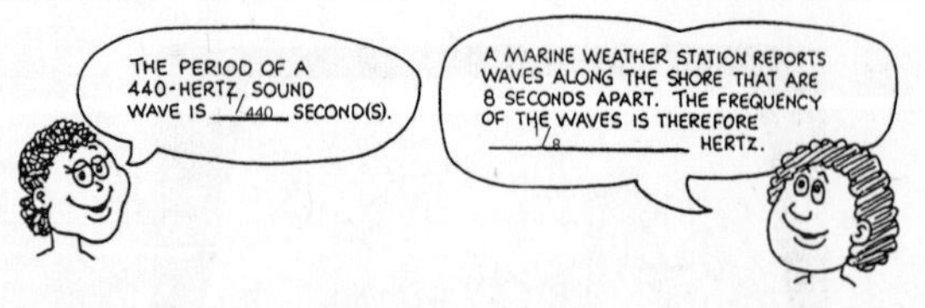

4. The annoying sound from a mosquito occurs because it beats its wings at the average rate of 600 wingbeats per second.

   a. What is the frequency of the soundwaves?

   600 Hz

   b. What is the wavelength? (Assume the speed of sound is 340 m/s.)

   0.57 m

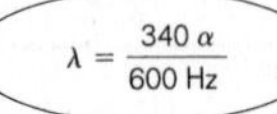

5. A machine gun fires 10 rounds per second. The speed of the bullets is 300 m/s.

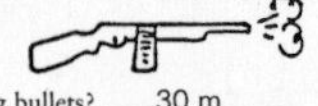

   a. What is the distance in the air between the flying bullets? 30 m

   b. What happens to the distance between the bullets if the rate of fire is increased?

   DISTANCE BETWEEN BULLETS DECREASES

6. Consider a wave generator that produces 10 pulses per second. The speed of the waves is 300 cm/s.

   a. What is the wavelength of the waves? 30 cm

   b. What happens to the wavelength if the frequency of pulses is increased?

   λ DECREASES, JUST AS DISTANCE BETWEEN BULLETS IN #5 DECREASES

7. The bird at the right watches the waves. If the portion of a wave between 2 crests passes the pole each second, what is the speed of the wave?

   $v = f\lambda = 2 \times 1\ \text{m} = 2\ \text{m/s}$

   What is its period?

   $T = \frac{1}{P} = \frac{1}{2} = 0.5\ \text{s}$

8. If the distance between crests in the above question were 1.5 meters apart, and 2 crests pass the pole each second, what would be the speed of the wave?

   $v = f\lambda = 2 \times 1.5 = 3\ \text{m/s}$

   What would be its period?

   SAME (0.5 s)

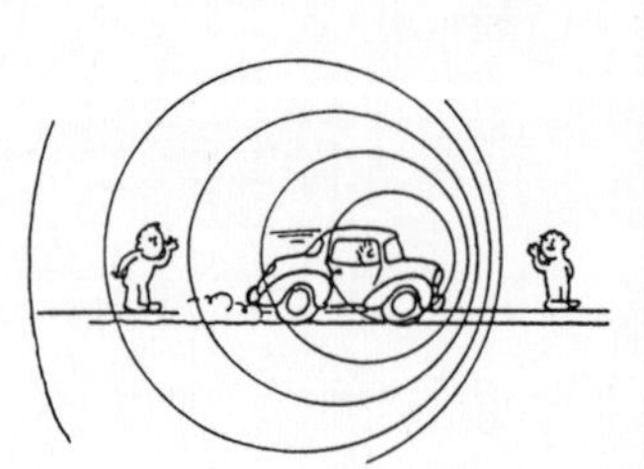

9. When an automobile moves toward a listener, the sound of its horn seems relatively

   (low pitched) (normal) (high pitched)

   When moving away from the listener, its horn seems

   (low pitched) (normal) (high pitched).

10. The changed pitch of the Doppler effect is due to changes in

    (wave speed) (wave frequency).

Color

The sketch to the right shows the shadow of an instructor in front of a white screen in a dark room. The light source is red, so the screen looks red and the shadow looks black. Color the sketch, or label the colors with a pen or pencil.

A green lamp is added and makes a second shadow. The shadow cast by the red lamp is no longer black, but is illuminated by green light, so it is green. Color or mark it green. The shadow cast by the green lamp is not black because it is illuminated by the red lamp. Indicate its color. Do the same for the background, which receives a mixture of red and green light.

A blue lamp is added and three shadows appear. Indicate the appropriate colors of the shadows and the background.

The lamps are placed closer together so the shadows overlap. Indicate the colors of all screen areas.

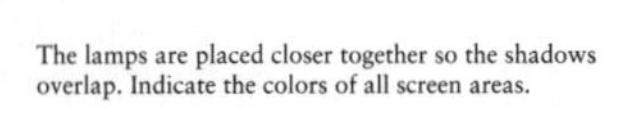

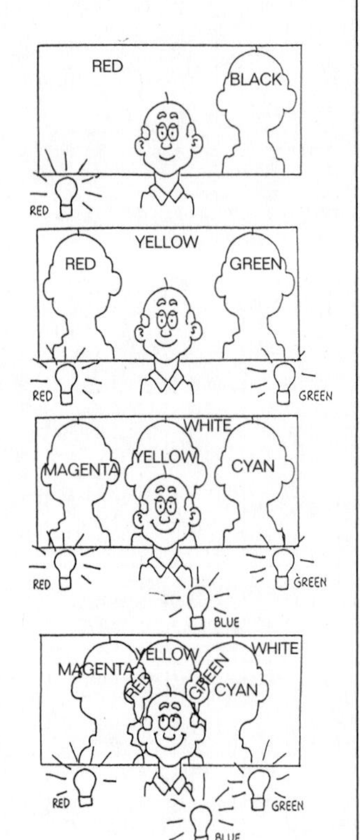

If you have colored pencils or markers, have a go at these.

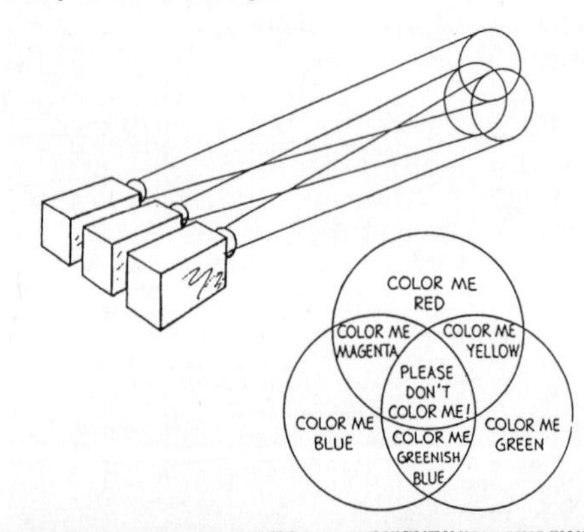

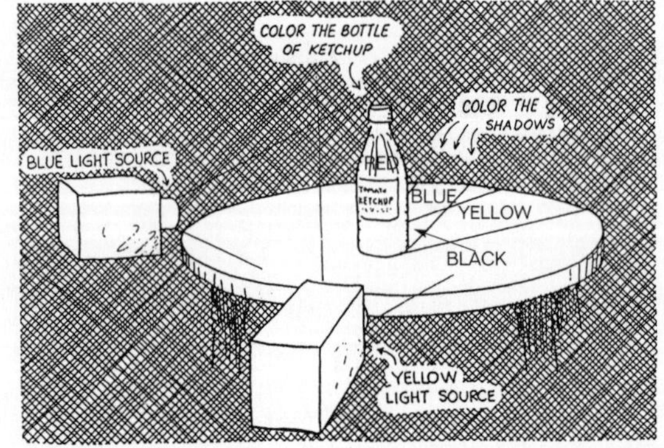

Chapter 8: Waves—Sound and Light

Diffraction and Interference

Shown below are concentric solid and dashed circles, each different in radius by 1 cm. Consider the circular pattern a top view of water waves, where the solid circles are crests and the dashed circles are troughs.

1. Draw another set of the same concentric circles with a compass. Choose any part of the paper for your center (except the present central point). Let the circles run off the edge of the paper.
2. Find where a dashed line crosses a solid line and draw a large dot at the intersection. Do this for ALL places where a solid and dashed line intersect.
3. With a wide felt marker, connect the dots with smooth lines. These *nodal lines* lie in regions where the waves have cancelled—where the crest of one wave overlaps the trough of another.

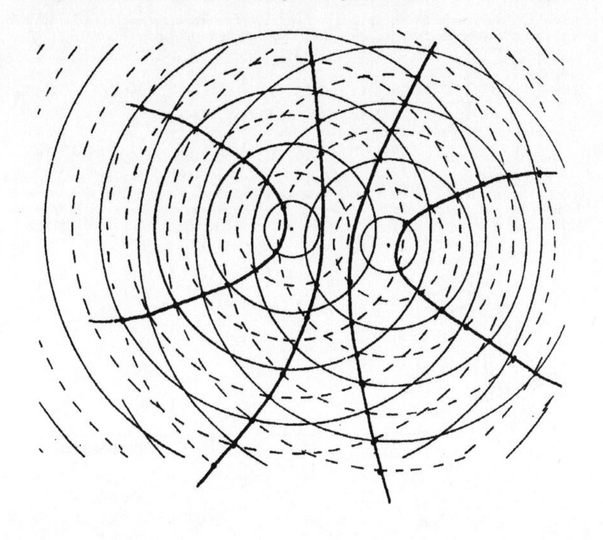

Chapter 8: Waves—Sound and Light

Reflection

1. Light from a flashlight shines on a mirror and illuminates one of the cards. Draw the reflected beam to indicate the illuminated card.

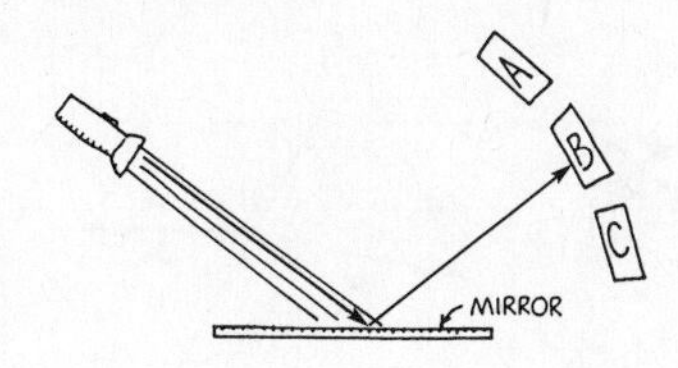

2. A periscope has a pair of mirrors in it. Draw the light path from the object "O" to the eye of the observer.

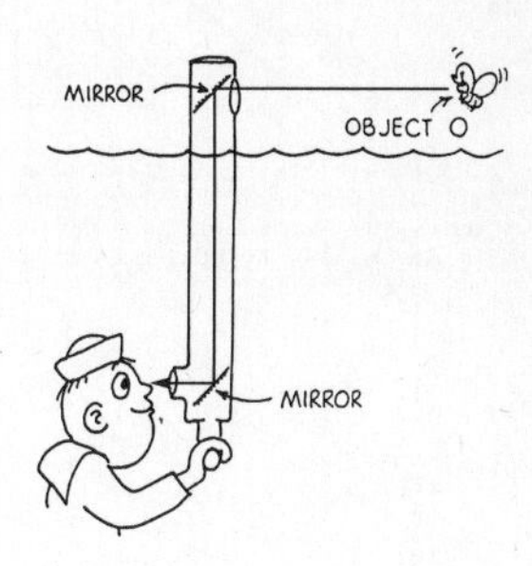

3. The ray diagram below shows the extension of one of the reflected rays from the plane mirror. Complete the diagram by (1) carefully drawing the three other reflected rays, and (2) extending them behind the mirror to locate the image of the flame. (Assume the candle and image are viewed by an observer on the left.)

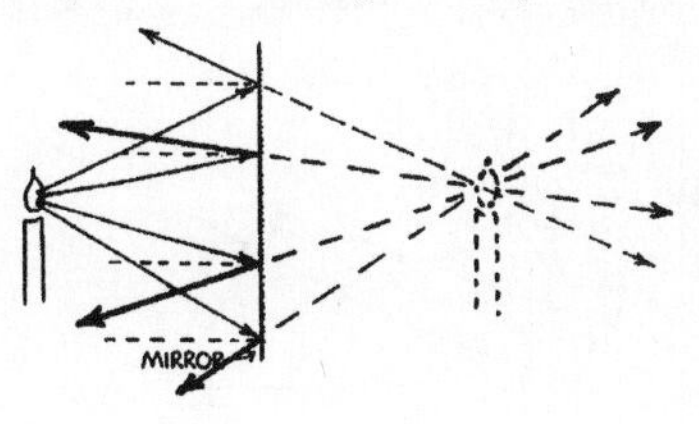

Reflection—continued

4. The ray diagram below shows the reflection of one of the rays that strikes the parabolic mirror. Notice that the law of reflection is observed, and the angle of incidence (from the normal, the dashed line) equals the angle of reflection (from the normal). Complete the diagram by drawing the reflected rays of the other three rays that are shown. (Do you see why parabolic mirrors are used in automobile headlights?)

5. A girl takes a photograph of the bridge as shown. Which of the two sketches below correctly shows the reflected view of the bridge? Defend your answer.

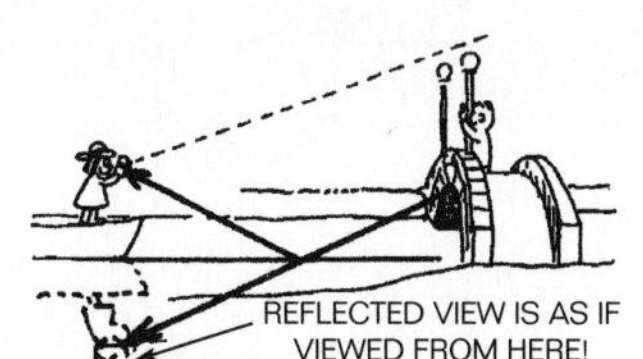

THE RIGHT VIEW IS CORRECT, SHOWING THE UNDERSIDE OF BRIDGE, OR WHAT YOUR EYE WOULD SEE IF IT WERE AS FAR BELOW THE REFLECTING SURFACE AS IT IS ABOVE! THE REFLECTION IS SEEN FROM BELOW THE EYE.

Chapter 8: Waves—Sound and Light

Refraction—Part 1

1. A pair of toy cart wheels are rolled obliquely from a smooth surface onto two plots of grass—a rectangular plot as shown at the left, and a triangular plot as shown at the right. The ground is on a slight incline so that after slowing down in the grass, the wheels speed up again when emerging on the smooth surface. Finish each sketch and show some positions of the wheels inside the plots and on the other side. Clearly indicate their paths and directions of travel.

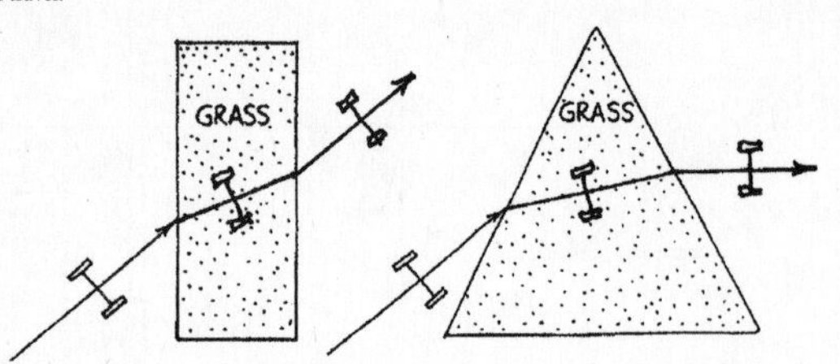

2. Red, green, and blue rays of light are incident upon a glass prism as shown. The average speed of red light in the glass is less than in air, so the red ray is refracted. When it emerges into the air it regains its original speed, and travels in the direction shown. Green light takes longer to get through the glass. Because of its slower speed, it is refracted as shown. Blue light travels even slower in glass. Complete the diagram by estimating the path of the blue ray.

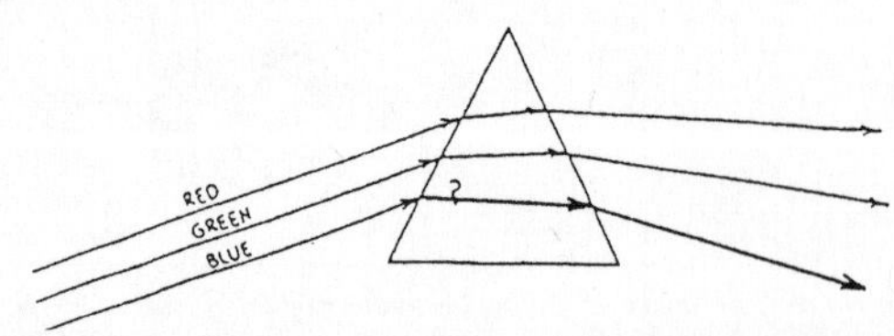

3. Below, we consider a prism-shaped hole in a piece of glass—that is, an "air prism." Complete the diagram; showing likely paths of the beams of red, green, and blue light as they pass through this "prism" and back to glass.

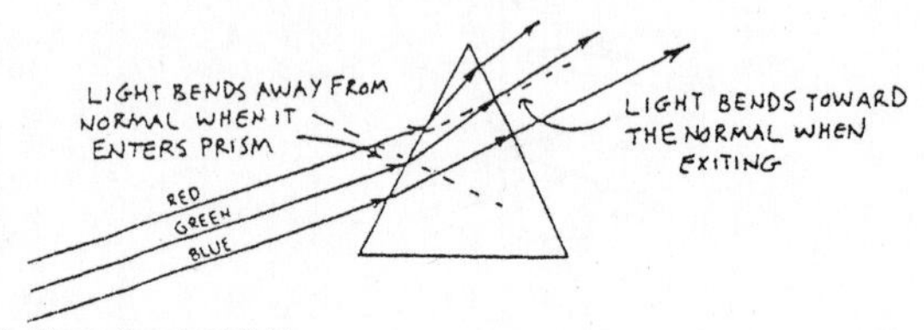

Refraction—Part 1—continued

4. Light of different colors diverges when emerging from a prism. Newton showed that with a second prism he could make the diverging beams become parallel again. Which placement of the second prism will do this?

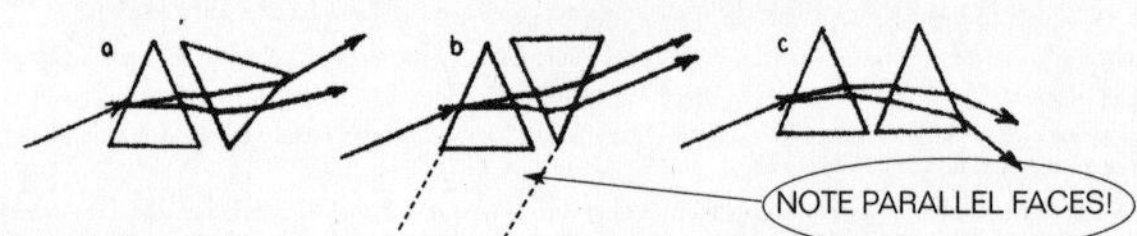

5. The sketch shows that due to refraction, the man sees the fish closer to the water surface than it actually is.
   a. Draw a ray beginning at the fish's eye to show the line of sight of the fish when it looks upward at 50° to the normal at the water surface. Draw the direction of the ray after it meets the surface of the water.
   b. At the 50° angle, does the fish see the man, or does it see the reflected view of the starfish at the bottom of the pond? Explain.
   FISH SEES REFLECTED VIEW OF STARFISH (50° > 48° CRITICAL ANGLE, SO THERE IS TOTAL INTERNAL REFLECTION)
   c. To see the man, should the fish look higher or lower than the 50° path?
   HIGHER, SO LINE OF SIGHT TO THE WATER IS LESS THAN 48° WITH NORMAL
   d. If the fish's eye were barely above the water surface, it would see the world above in a 180° view, horizon to horizon. The fish's-eye view of the world above as seen beneath the water, however, is very different. Due to the 48° critical angle of water, the fish sees a normally 180° horizon-to-horizon view compressed within an angle of 96°.

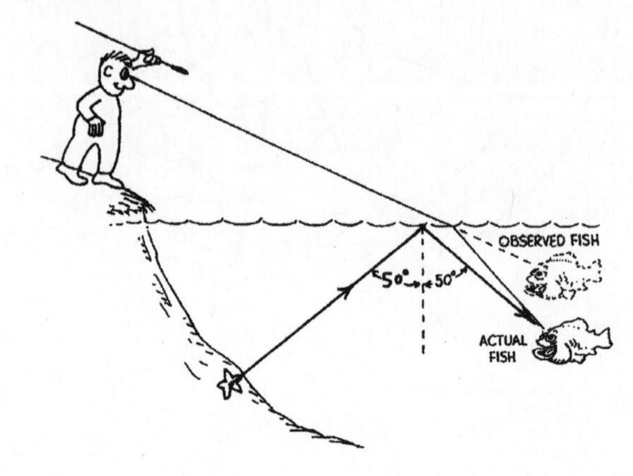

Chapter 8: Waves—Sound and Light

Refraction—Part 2

1. The sketch to the right shows a light ray moving from air into water, at 45° to the normal. Which of the three rays indicated with capital letters is most likely the light ray that continues inside the water?
   C

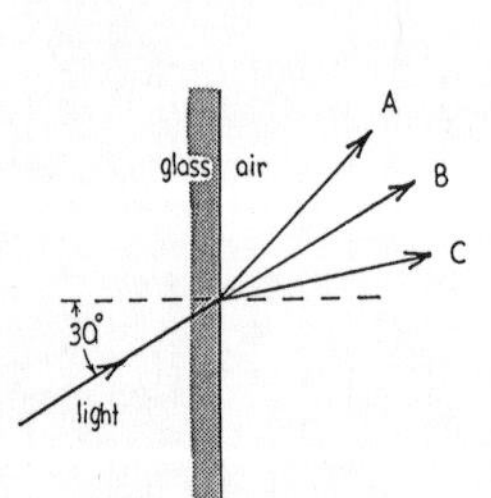

2. The sketch on the left shows a light ray moving from glass into air, at 30° to the normal. Which of the three is most likely the light ray that continues in the air?
   A

3. To the right, a light ray is shown moving from air into a glass block, at 40° to the normal. Which of the three rays is most likely the light ray that travels in the air after emerging from the opposite side of the block?
   A

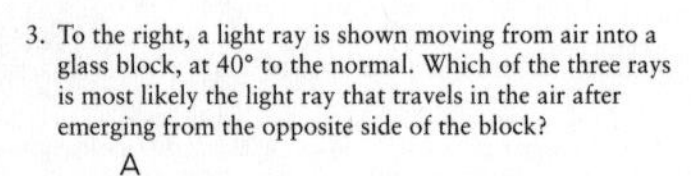

Sketch the path the light would take inside the glass.

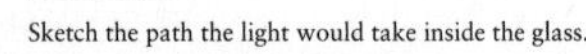

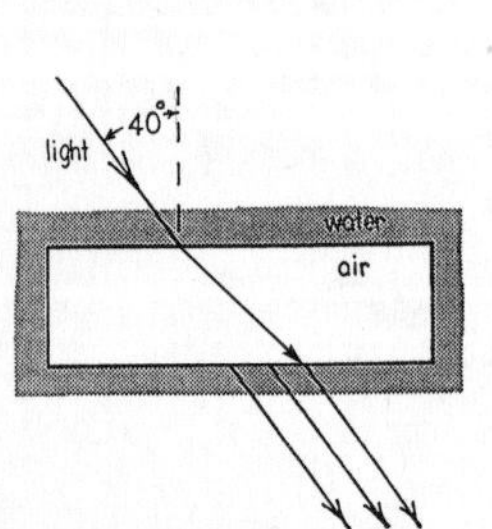

4. To the left, a light ray is shown moving from water into a rectangular block of air (inside a thin-walled plastic box), at 40° to the normal. Which of the three rays is most likely the light ray that continues into the water on the opposite side of the block?
   C

Sketch the path the light would take inside the air.

Refraction—Part 2—continued

5. The two transparent blocks (right) are made of different materials. The speed of light in the left block is greater than the speed of light in the right block. Draw an appropriate light path through and beyond the right block. Is the light that emerges displaced more or less than light emerging from the left block?
   MORE

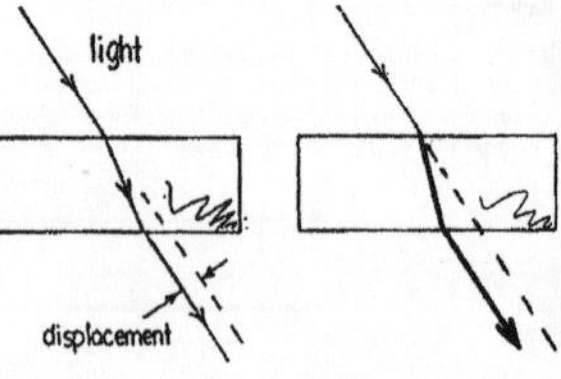

6. Light from the air passes through plates of glass and plastic below. The speeds of light in the different materials is shown to the right (these different speeds are often implied by the "index of refraction" of the material). Construct a rough sketch showing an appropriate path through the system of four plates.

   Compared to the 50° incident ray at the top, what can you say about the angles of the ray in the air between and below the block pairs?
   SAME 50°

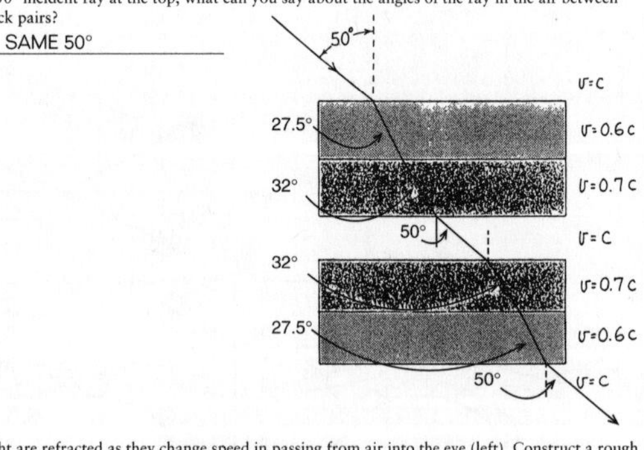

7. Parallel rays of light are refracted as they change speed in passing from air into the eye (left). Construct a rough sketch showing appropriate light paths when parallel light under water meets the same eye (right).

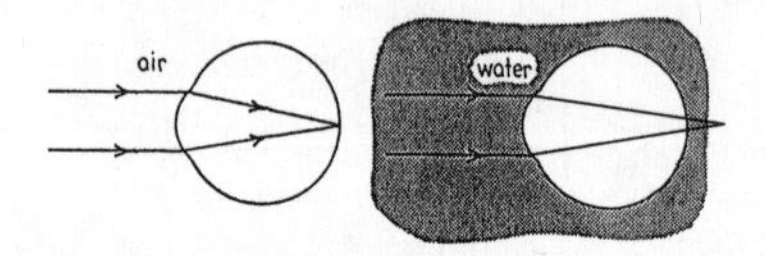

If a fish out of water wishes to clearly view objects in air, should it wear goggles filled with water or with air?

8. Why do we need to wear a face mask or goggles to see clearly when under water?
   SO THAT LIGHT GOES FROM AIR TO EYE FOR PROPER REFRACTION

Chapter 8: Waves—Sound and Light

Wave-Particle Duality

1. To say that light is quantized means that light is made up of
   (elemental units) (waves).

2. Compared to photons of low-frequency light, photons of higher-frequency light have more
   (energy) (speed) (quanta).

3. The photoelectric effect supports the
   (wave model of light) **(particle model of light).**

4. The photoelectric effect is evident when light shone on certain photosensitive materials ejects
   (photons) **(electrons).**

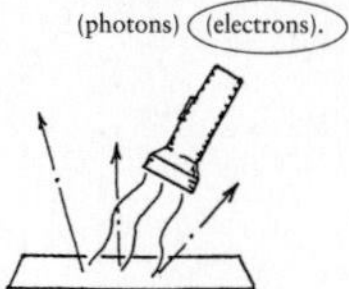

5. The photoelectric effect is more effective with violet light than with red light because the photons of violet light
   (resonate with the atoms in the material)
   (deliver more energy to the material)
   (are more numerous).

6. According to De Broglie's wave model of matter, a beam of light and a beam of electrons
   (are fundamentally different) **(are similar).**

7. According to De Broglie, the greater the speed of an electron beam, the
   (greater is its wavelength) **(shorter is its wavelength).**

8. The discreteness of the energy levels of electrons about the atomic nucleus is best understood by considering the electron to be a
   (wave) (particle).

9. Heavier atoms are not appreciably larger in size than lighter atoms. The main reason for this is the greater nuclear charge
   (pulls surrounding electrons into tighter orbits)
   (holds more electrons about the atomic nucleus)
   (produces a denser atomic structure).

10. Whereas in the everyday macroworld the study of motion is called *mechanics*, in the microworld the study of quanta is called
    (Newton mechanics) **(quantum mechanics).**

Chapter 9: The Atom

Subatomic Particles

Three fundamental particles of the atom are the PROTON, NEUTRON, and ELECTRON. At the center of each atom lies the atomic NUCLEUS, which consists of PROTONS and NEUTRONS The **atomic number** refers to the number of PROTONS in the nucleus. All atoms of the same element have the same number of PROTONS, hence, the same atomic number.

Isotopes are atoms that have the same number of PROTONS, but a different number of NEUTRONS An isotope is identified by its **atomic mass number,** which is the total number of NEUTRONS and PROTONS in the nucleus. A carbon isotope that has 6 NEUTRONS and 6 PROTONS is identified as carbon-12, where 12 is the atomic mass number. A carbon isotope having 6 PROTONS and 8 NEUTRONS on the other hand, is carbon-14.

1. Complete the following table:

Isotope	Electrons	Number of... Protons	Neutrons
Hydrogen-1	1	1	0
Chlorine-36	17	17	19
Nitrogen-14	7	7	7
Potassium-40	19	19	21
Arsenic-75	33	33	42
Gold-197	79	79	118

2. Which results in a more valuable product—*adding* or *subtracting* protons from gold nuclei?
   SUBTRACT FOR PLATINUM (MORE VALUABLE)

3. Which has more mass, a helium atom or a neon atom?
   NEON

4. Which has a greater number of atoms, a gram of helium or a gram of neon?
   HELIUM!

Chapter 10: Nuclear Physics

Radioactivity

1. Complete the following statements.
   a. A lone neutron spontaneously decays into a proton plus an ELECTRON.
   b. Alpha and beta rays are made of streams of particles, whereas gamma rays are streams of PHOTONS.
   c. An electrically charged atom is called an ION.
   d. Different ISOTOPES of an element are chemically identical but differ in the number of neutrons in the nucleus.
   e. Transuranic elements are those beyond atomic number 92.
   f. If the amount of a certain radioactive sample decreases by half in four weeks, in four more weeks the amount remaining should be 1/4 the original amount.
   g. Water from a natural hot spring is warmed by RADIOACTIVITY inside Earth.

2. The gas in the little girl's balloon is made up of former alpha and beta particles produced by radioactive decay.
   a. If the mixture is electrically neutral, how many more beta particles than alpha particles are in the balloon?
      TWICE AS MANY BETA PARTICLES AS ALPHA PARTICLES
   b. Why is your answer not "same"?
      ALPHA HAS DOUBLE CHARGE; THE CHARGE OF 2 BETAS = MAGNITUDE OF CHARGE OF 1 ALPHA
   c. Why are the alpha and beta particles no longer harmful to the child?
      THEY HAVE LOST THEIR HIGH KE, WHICH IS NOW REDUCED TO THERMAL ENERGY OF RANDOM MOLECULAR MOTION.
   d. What element does this mixture make?
      HELIUM

Radioactivity—continued

Draw in a decay-scheme diagram below, similar to Figure 10.15 in your text. In this case, you begin at the upper right with U-235 and end up with a different isotope of lead. Use the table at the left and identify each element in the series by its chemical symbol.

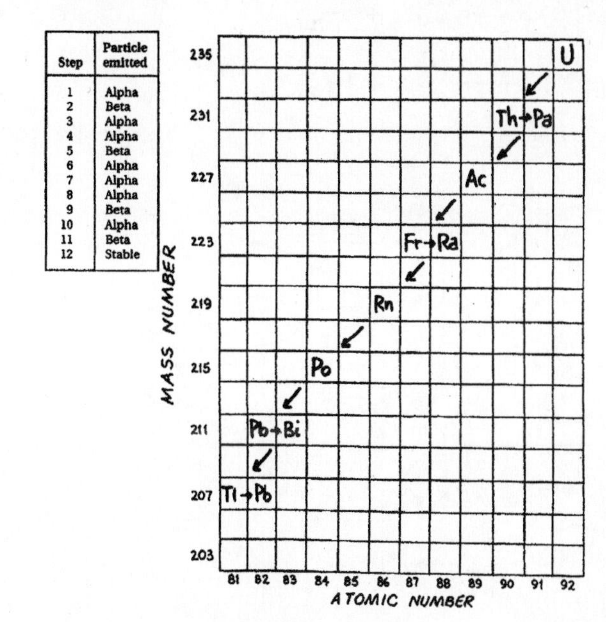

Step	Particle emitted
1	Alpha
2	Beta
3	Alpha
4	Alpha
5	Beta
6	Alpha
7	Alpha
8	Alpha
9	Beta
10	Alpha
11	Beta
12	Stable

What isotope is the final product? $^{207}_{82}$Pb (LEAD-207)

Chapter 10: Nuclear Physics

Radioactive Half-Life

You and your classmates will now play the "half-life game." Each of you should have a coin to shake inside cupped hands. After it has been shaken for a few seconds, the coin is tossed on the table or on the floor. Students with tails up fall out of the game. Only those who consistently show heads remain in the game. Finally, everybody has tossed a tail and the game is over.

1kg
1/2 kg
1/4 kg
1/8 kg
1620 3240 4860 Years

1. The graph to the left shows the decay of Radium-226 with time. Note that each 1620 years, half remains (the rest changes to other elements). In the grid below, plot the number of students left in the game after each toss. Draw a smooth curve that passes close to the points on your plot. What is the similarity of your curve with that of the curve of Radium-226?

   SHOULD BOTH DECREASE RAPIDLY

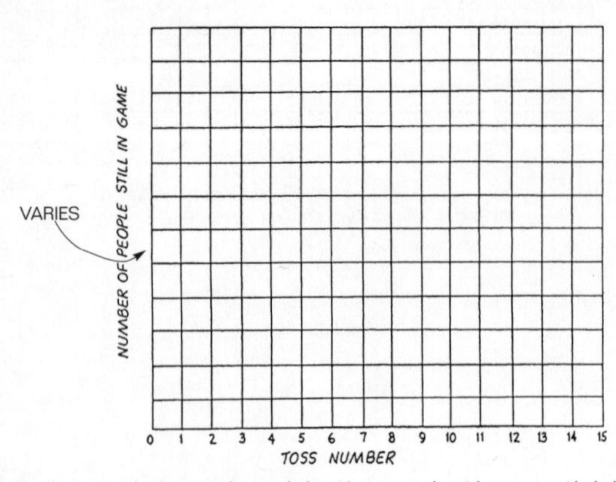

2. Was the person to last longest in the game *lucky*, with some sort of special powers to guide the long survival? What test could you make to decide the answer to this question?

   TEST! REPEAT TO SEE IF "LUCKY" PERSON REMAINS LUCKY!

Chapter 10: Nuclear Physics

Nuclear Fission and Fusion

1. Complete the table for a chain reaction in which two neutrons from each step individually cause a new reaction.

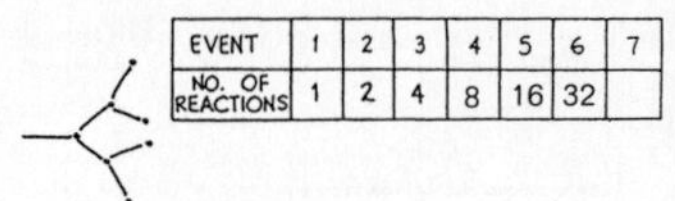

EVENT	1	2	3	4	5	6	7
NO. OF REACTIONS	1	2	4	8	16	32	

2. Complete the table for a chain reaction where three neutrons from each reaction cause a new reaction.

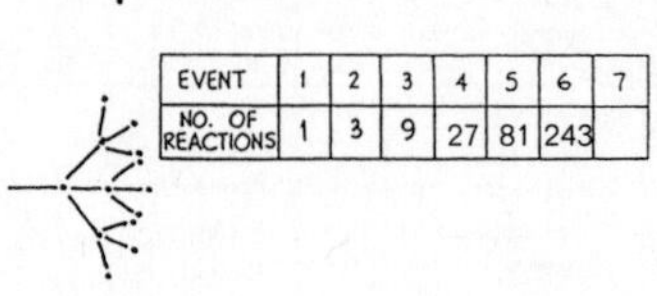

EVENT	1	2	3	4	5	6	7
NO. OF REACTIONS	1	3	9	27	81	243	

3. Complete these beta reactions, which occur in a fission breeder reactor.

$${}^{239}_{92}U \rightarrow {}^{239}_{93}Np + {}^{0}_{-1}e$$

$${}^{239}_{93}Np \rightarrow {}^{239}_{94}Pu + {}^{0}_{-1}e$$

4. Complete the following fission reactions.

$${}^{1}_{0}n + {}^{235}_{92}U \rightarrow {}^{143}_{54}Xe + {}^{90}_{38}Sr + 3\,({}^{1}_{0}n)$$

$${}^{1}_{0}n + {}^{235}_{92}U \rightarrow {}^{152}_{60}Nd + {}^{80}_{32}Ge + 4\,({}^{1}_{0}n)$$

$${}^{1}_{0}n + {}^{239}_{94}Pu \rightarrow {}^{141}_{54}Xe + {}^{97}_{40}Zr + 2\,({}^{1}_{0}n)$$

5. Complete the following fusion reactions.

$${}^{2}_{1}H + {}^{2}_{1}H \rightarrow {}^{3}_{2}He + {}^{1}_{0}n$$

$${}^{2}_{1}H + {}^{3}_{1}H \rightarrow {}^{4}_{2}He + {}^{1}_{0}n$$

Chapter 10: Nuclear Physics

Nuclear Reactions

Complete these nuclear reactions:

1. $${}^{230}_{90}Th \rightarrow {}^{226}_{88}Ra + {}^{4}_{2}He$$
2. $${}^{218}_{85}At \rightarrow {}^{214}_{83}Bi + {}^{4}_{2}He$$
3. $${}^{14}_{6}C \rightarrow {}^{0}_{-1}e + {}^{14}_{7}N$$
4. $${}^{80}_{35}Br \rightarrow {}^{80}_{36}Kr + {}^{0}_{-1}e$$
5. $${}^{214}_{83}Bi \rightarrow {}^{4}_{2}He + {}^{210}_{81}Tl$$

NUCLEAR PHYSICS··· IT'S THE SAME TO ME WITH THE FIRST TWO LETTERS INTERCHANGED!

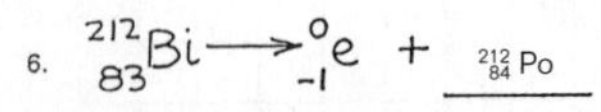

6. $${}^{212}_{83}Bi \rightarrow {}^{0}_{-1}e + {}^{212}_{84}Po$$

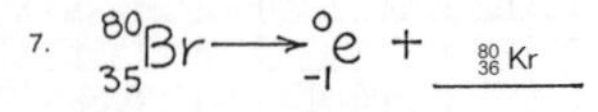

7. $${}^{80}_{35}Br \rightarrow {}^{0}_{-1}e + {}^{80}_{36}Kr$$
8. $${}^{80}_{35}Br \rightarrow {}^{0}_{+1}e + {}^{80}_{34}Se$$
9. $${}^{1}_{1}H + {}^{7}_{3}Li \rightarrow {}^{4}_{2}He + {}^{4}_{2}He$$
10. $${}^{2}_{1}H + {}^{3}_{1}H \rightarrow {}^{4}_{2}He + {}^{1}_{0}n$$

Chapter 11: Investigating Matter

Melting Points of the Elements

There is a remarkable degree of organization in the periodic table. As discussed in your textbook, elements within the same atomic group (vertical column) share similar properties. Also, the chemical reactivity of an element can be deduced from its position in the periodic table. Two additional examples of the periodic table's organization are the melting points and densities of the elements.

The periodic table below shows the melting points of nearly all the elements. Note the melting points are not randomly oriented, but, with only a few exceptions, either gradually increase or decrease as you move in any particular direction. This can be clearly illustrated by color coding each element according to its melting point.

Use colored pencils to color in each element according to its melting point. Use the suggested color legend. Color lightly so that symbols and numbers are still visible.

Color	Temperature Range, °C	Color	Temperature Range, °C
Violet	-273 — -50	Yellow	1400 — 1900
Blue	-50 — 300	Orange	1900 — 2900
Cyan	300 — 700	Red	2900 — 3500
Green	700 — 1400		

Melting Points of the Elements (°C)

1	2	3	4	5	6	7	8	9	10	11	12	13	14	15	16	17	18
H -259																	He -272
Li 180	Be 1278											B 2079	C 3550	N -210	O -218	F -219	Ne -248
Na 97	Mg 648											Al 660	Si 1410	P 44	S 113	Cl -100	Ar -189
K 63	Ca 839	Sc 1541	Ti 1660	V 1890	Cr 1857	Mn 1244	Fe 1535	Co 1495	Ni 1453	Cu 1083	Zn 419	Ga 30	Ge 937	As 817	Se 217	Br -7	Kr -156
Rb 39	Sr 769	Y 1522	Zr 1852	Nb 2468	Mo 2617	Tc 2172	Ru 2310	Rh 1966	Pd 1554	Ag 961	Cd 320	In 156	Sn 231	Sb 630	Te 449	I 113	Xe -111
Cs 28	Ba 725	La 921	Hf 2227	Ta 2996	W 3410	Re 3180	Os 3045	Ir 2410	Pt 1772	Au 1064	Hg -38	Tl 303	Pb 327	Bi 271	Po 254	At 302	Rn -71
Fr 27	Ra 700	Ac 1050	--	--	--	--	--	--									

TUNGSTEN

Lanthanides:	Ce 799	Pr 931	Nd 1021	Pm 1168	Sm 1077	Eu 822	Gd 1313	Tb 1356	Dy 1412	Ho 1474	Er 1159	Tm 1545	Yb 819	Lu 1663
Actinides:	Th 1750	Pa 1600	U 1132	Np 640	Pu 641	Am 994	Cm 1340	Bk --	Cf --	Es --	Fm --	Md --	No --	Lr --

1. Which elements have the highest melting points?

   THE ONES CLOSER TO TUNGSTEN

2. Which elements have the lowest melting points?

   ELEMENTS TOWARD UPPER RIGHT

3. Which atomic groups tend to go from higher to lower melting points reading from top to bottom? (Identify each group by its group number.)

   1, 2, 3, 12, 13, 14

4. Which atomic groups tend to go from lower to higher melting points reading from top to bottom?

   4 THROUGH 10 AND 15 THROUGH 18

Chapter 11: Investigating Matter

Densities of the Elements

The periodic table below shows the densities of nearly all the elements. As with the melting points, the densities of the elements either gradually increase or decrease as you move in any particular direction. Use colored pencils to color in each element according to its density. Shown below is a suggested color legend. Color lightly so that symbols and numbers are still visible. (Note: All gaseous elements are marked with an asterisk and should be the same color. Their densities, which are given in units of g/L, are much less than the densities nongaseous elements, which are given in units of g/mL.)

Color	Density (g/mL)	Color	Density (g/mL)
Violet	gaseous elements	Yellow	16 — 12
Blue	5 — 0	Orange	20 — 16
Cyan	8 — 5	Red.	23 — 20
Green	12 — 8		

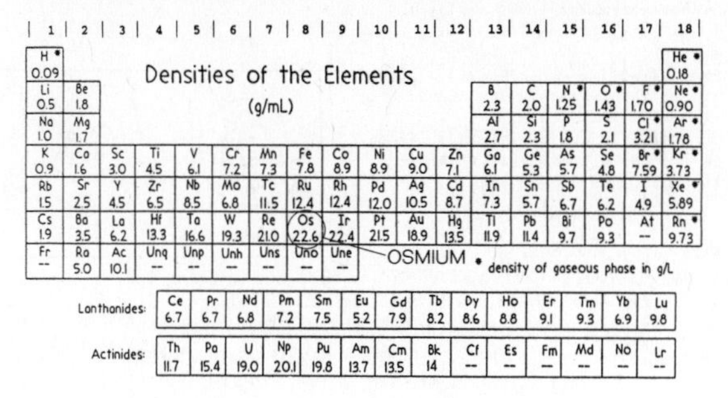
Densities of the Elements (g/mL)

1	2	3	4	5	6	7	8	9	10	11	12	13	14	15	16	17	18
H • 0.09																	He • 0.18
Li 0.5	Be 1.8											B 2.3	C 2.0	N • 1.25	O • 1.43	F • 1.70	Ne • 0.90
Na 1.0	Mg 1.7											Al 2.7	Si 2.3	P 1.8	S 2.1	Cl • 3.21	Ar • 1.78
K 0.9	Ca 1.6	Sc 3.0	Ti 4.5	V 6.1	Cr 7.2	Mn 7.3	Fe 7.8	Co 8.9	Ni 8.9	Cu 9.0	Zn 7.1	Ga 6.1	Ge 5.3	As 5.7	Se 4.8	Br • 7.59	Kr • 3.73
Rb 1.5	Sr 2.5	Y 4.5	Zr 6.5	Nb 8.5	Mo 6.8	Tc 11.5	Ru 12.4	Rh 12.4	Pd 12.0	Ag 10.5	Cd 8.7	In 7.3	Sn 5.7	Sb 6.7	Te 6.2	I 4.9	Xe • 5.89
Cs 1.9	Ba 3.5	La 6.2	Hf 13.3	Ta 16.6	W 19.3	Re 21.0	Os 22.6	Ir 22.4	Pt 21.5	Au 18.9	Hg 13.5	Tl 11.9	Pb 11.4	Bi 9.7	Po 9.3	At --	Rn • 9.73
Fr --	Ra 5.0	Ac 10.1	Unq --	Unp --	Unh --	Uns --	Uno --	Une --									

OSMIUM (Os circled)

• density of gaseous phase in g/L

Lanthanides:	Ce 6.7	Pr 6.7	Nd 6.8	Pm 7.2	Sm 7.5	Eu 5.2	Gd 7.9	Tb 8.2	Dy 8.6	Ho 8.8	Er 9.1	Tm 9.3	Yb 6.9	Lu 9.8
Actinides:	Th 11.7	Pa 15.4	U 19.0	Np 20.1	Pu 19.8	Am 13.7	Cm 13.5	Bk 14	Cf --	Es --	Fm --	Md --	No --	Lr --

1. Which elements are the most dense?
   THE ONES CLOSER TO OSMIUM, OS
2. How variable are the densities of the lanthanides compared to the densities of the actinides?
   THE ACTINIDES ARE MUCH MORE VARIABLE
3. Which atomic groups tend to go from higher to lower densities reading from top to bottom? (Identify each group by its group number).
   NONE
4. Which atomic groups tend to go from lower to higher densities reading from top to bottom?
   ALL

Chapter 11: Investigating Matter

The Submicroscopic

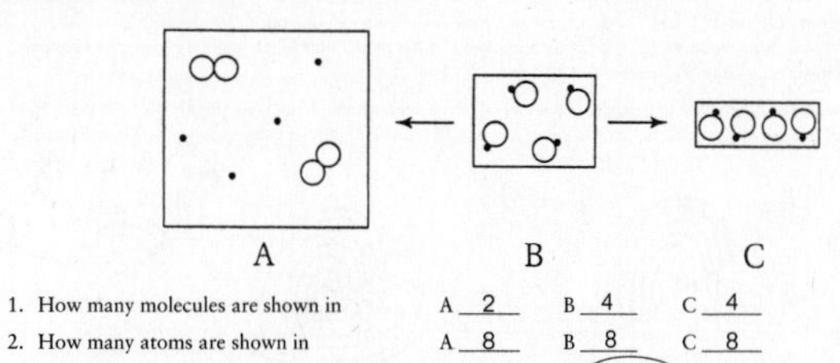

1. How many molecules are shown in A 2 B 4 C 4
2. How many atoms are shown in A 8 B 8 C 8
3. Which represents a physical change? B → A (B → C) *(circle one)*
4. Which represents a chemical change? (B → A) B → C *(circle one)*
5. Which box(es) represent(s) a mixture? A ✓ B ___ C ___
6. Which box contains the most mass? A ✓ B ✓ C ✓ ALL WITH SAME MASS
7. Which box is coldest? A ↗ MAY BE WARMER OR B ___ C ✓
8. Which box contains the most air between molecules? A NONE B ___ C ___
   THERE IS NO AIR BETWEEN THE MOLECULES.

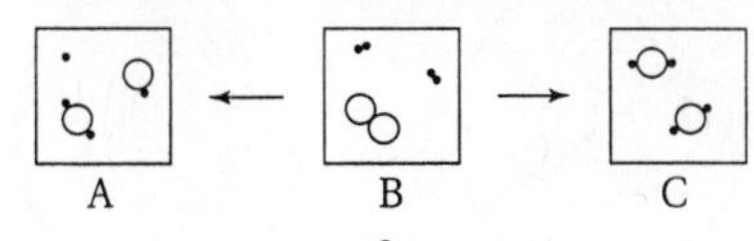

9. How many molecules are shown in A 2 B 3 C 2
10. How many atoms are shown in A 6 B 6 C 6
11. Which represents a physical change? B → A B → C *(circle one)* NEITHER
12. Which represents a chemical change? (B → A) (B → C) *(circle one)* BOTH
13. Which box(es) represent(s) a mixture? A ✓ B ✓ C ___
14. Which box contains the most mass? A ✓ B ✓ C ✓ ALL WITH SAME MASS
15. Which should take longer? B → A (B → C) *(circle one)*
    ONE LESS STEP IS REQUIRED TO GO FROM B → A
16. Which box most likely contains ions? A ✓ B ___ C ___

Chapter 11: Investigating Matter

Physical and Chemical Changes

1. What distinguishes a chemical change from a physical change?
   DURING A CHEMICAL CHANGE ATOM CHANGE PARTNERS
2. Based upon observations alone, why is distinguishing a chemical change from a physical change not always so straight-forward?
   BOTH INVOLVE A CHANGE IN PHYSICAL APPEARANCE

Try your hand at categorizing the following processes as either chemical or physical changes. Some of these examples are debatable! Be sure to discuss your reasoning with fellow classmates or your instructor.

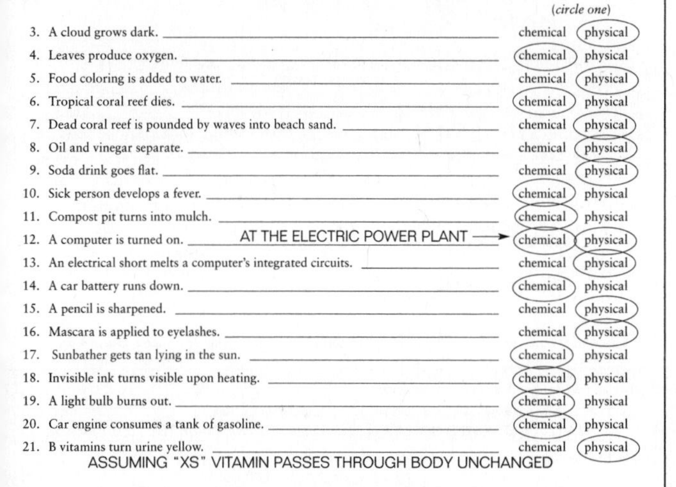

(circle one)

3. A cloud grows dark. — chemical (physical)
4. Leaves produce oxygen. — (chemical) physical
5. Food coloring is added to water. — chemical (physical)
6. Tropical coral reef dies. — (chemical) physical
7. Dead coral reef is pounded by waves into beach sand. — chemical (physical)
8. Oil and vinegar separate. — chemical (physical)
9. Soda drink goes flat. — chemical (physical)
10. Sick person develops a fever. — (chemical) physical
11. Compost pit turns into mulch. — (chemical) physical
12. A computer is turned on. AT THE ELECTRIC POWER PLANT → (chemical) (physical)
13. An electrical short melts a computer's integrated circuits. — chemical (physical)
14. A car battery runs down. — (chemical) physical
15. A pencil is sharpened. — chemical (physical)
16. Mascara is applied to eyelashes. — chemical (physical)
17. Sunbather gets tan lying in the sun. — (chemical) physical
18. Invisible ink turns visible upon heating. — (chemical) physical
19. A light bulb burns out. — (chemical) physical
20. Car engine consumes a tank of gasoline. — (chemical) physical
21. B vitamins turn urine yellow. — chemical (physical)
    ASSUMING "XS" VITAMIN PASSES THROUGH BODY UNCHANGED

Chapter 12: The Nature of Chemical Bonds

Losing Valence Electrons

The shell model described in Section 12.1 can be used to explain a wide variety of properties of atoms. Using the shell model, for example, we can explain how atoms within the same group tend to lose (or gain) the same number of electrons. Let's consider the case of three group 1 elements: lithium, sodium, and potassium. Look to a periodic table and find the nuclear charge of each of these atoms:

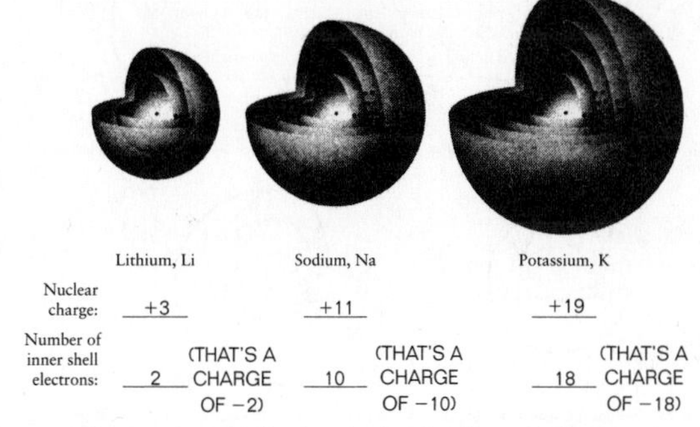

	Lithium, Li	Sodium, Na	Potassium, K
Nuclear charge:	+3	+11	+19
Number of inner shell electrons:	2 (THAT'S A CHARGE OF −2)	10 (THAT'S A CHARGE OF −10)	18 (THAT'S A CHARGE OF −18)

How strongly the valence electron is held to the nucleus depends on the strength of the nuclear charge—the stronger the charge, the stronger the valence electron is held. There's more to it, however, because inner-shell electrons weaken the attraction outer-shell electrons have for the nucleus. The valence shell in lithium, for example, doesn't experience the full effect of three protons. Instead, it experiences a diminished nuclear charge of about +1. We get this by subtracting the number of inner-shell electrons from the actual nuclear charge. What do the valence electrons for sodium and potassium experience?

Diminished nuclear charge:	(+3 − 2 = +1) ABOUT +1	(+11 − 10 = +1) ABOUT +1	(+19 − 18 = +1) ABOUT +1

Question: Potassium has a nuclear charge many times greater than that of lithium. Why is it actually *easier* for a potassium atom to lose its valence electron than it is for a lithium atom to lose its valence electron?

Hint: Remember from Chapter 7 what happens to the electric force as distance is increased!

POTASSIUM'S VALENCE ELECTRON IS MUCH FARTHER FROM THE NUCLEUS. BECAUSE THE ELECTRIC FORCE DECREASES WITH DISTANCE, THE +1 CHARGE FOR POTASSIUM'S VALENCE ELECTRON IS NOT SO EFFECTIVE AT HOLDING TO THE ATOM. HENCE, IT IS EASILY LOST.

Chapter 12: The Nature of Chemical Bonds

Drawing Shells

Atomic shells can be represented by a series of concentric circles as shown in your textbook. With a little effort, however, it's possible to show these shells in three dimensions. Grab a pencil and blank sheet of paper and follow the steps shown below. Practice makes perfect.

1. Lightly draw a diagonal guideline. Then, draw a series of seven semicircles. Note how the ends of the semicircles are not perpendicular to the guideline. Instead, they are parallel to the length of the page, as shown in Figure 1.

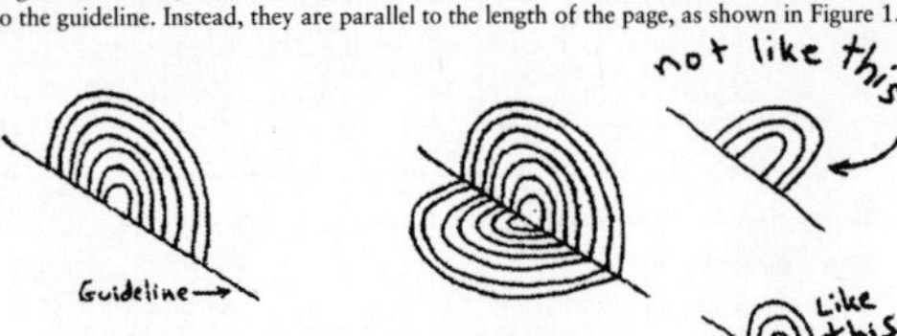

Figure 1 Figure 2

2. Connect the ends of each semicircle with another semicircle such that a series of concentric hearts is drawn. The ends of these new semicircles should be drawn perpendicular to the ends of the previously drawn semicircles, as shown in Figure 2.
3. Now the hard part. Draw a portion of a circle that connects the apex of the largest vertical and horizontal semicircles, as in Figure 3.

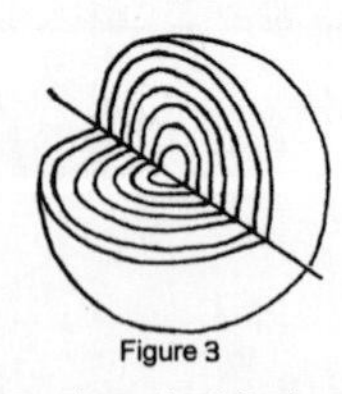

Figure 3

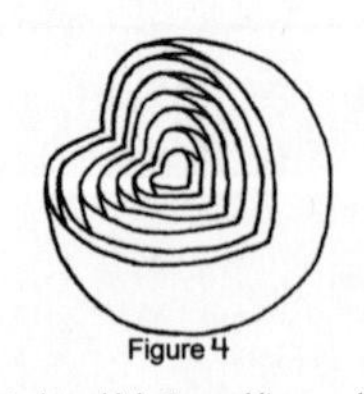

Figure 4

4. Now the fun part. Erase the pencil guideline drawn in Step 1, then add the internal lines, as shown in Figure 4, that create a series of concentric shells.

You need not draw all the shells for each atom. Oxygen, for example, is nicely represented drawing only the first two inner shells, which are the only ones that contain electrons. Remember that these shells are not to be taken literally. Rather, they are a highly simplified view of how electrons tend to organize themselves with an atom. You should know that each shell represents a set of atomic orbitals of similar energy levels as shown in your textbook.

Chapter 12: The Nature of Chemical Bonds

Atomic Size

1. Complete the shells for the following atoms using arrows to represent electrons.

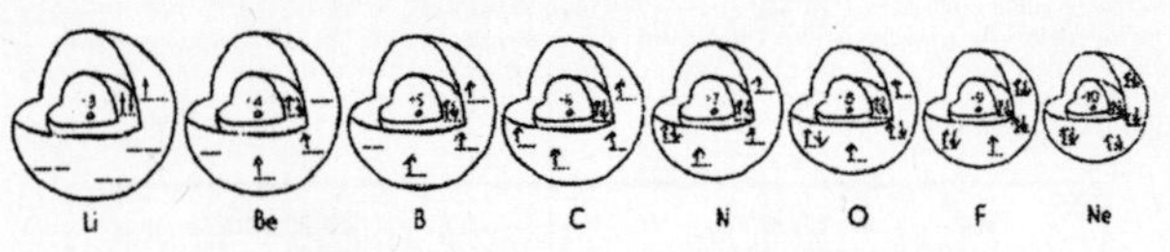

2. Neon, Ne, has many more electrons than lithium, Li, yet it is a much smaller atom. Why?
NEON HAS A STRONGER NUCLEAR CHARGE (+10) THAT PULLS THE ELECTRON IN CLOSER TO IT.
3. Draw the shell model for a sodium atom, Na (atomic number 11), adjacent to the neon atom in the box shown below. Use a pencil because you may need to erase.

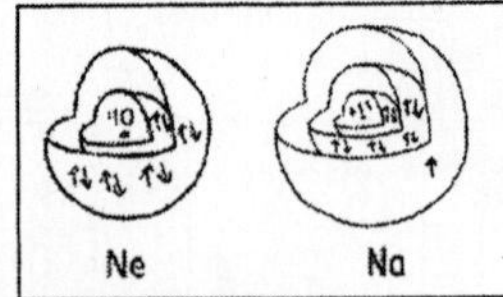

a. Which should be larger: neon's first shell or sodium's first shell. Why? Did you represent this accurately within your drawing?
NEON'S FIRST SHELL IS LARGER BECAUSE OF THE WEAKER NUCLEAR CHARGE.

b. Which has a greater nuclear charge, Ne or Na?
SODIUM, Na

c. Which is a larger atom, Ne or Na?
SODIUM, BUT NOT BECAUSE OF A GREATER NUCLEAR CHARGE, BUT BECAUSE OF THE EXTRA SHELL OF ELECTRONS.

4. Moving from left to right across the periodic table, what happens to the nuclear charge within atoms? What happens to atomic size?
THE NUCLEAR CHARGE INCREASES FROM LEFT TO RIGHT ACROSS THE PERIODIC TABLE, WHICH IS WHY THE ATOMIC SIZE DECREASES.
5. Moving from top to bottom down the periodic table, what happens to the number of occupied shells? What happens to atomic size?
MOVING DOWN A GROUP, THE NUMBER OF OCCUPIED SHELLS INCREASES, WHICH IS WHY THE ATOMIC SIZE ALSO INCREASES.
6. Where in the periodic table are the smallest atoms found? Where are the largest atoms found?
THE SMALLEST ATOMS ARE FOUND TO THE UPPER RIGHT WHILE THE LARGEST ATOMS ARE FOUND TO THE LOWER LEFT.

Chapter 12: The Nature of Chemical Bonds

Effective Nuclear Charge

The magnitude of the nuclear charge sensed by an orbiting electron depends upon several factors, including the number of positively-charged protons in the nucleus, the number of inner shell electrons shielding it from the nucleus, and its distance from the nucleus.

1. Place the proper number of electrons in each shell for carbon and silicon (use arrows to represent electrons).

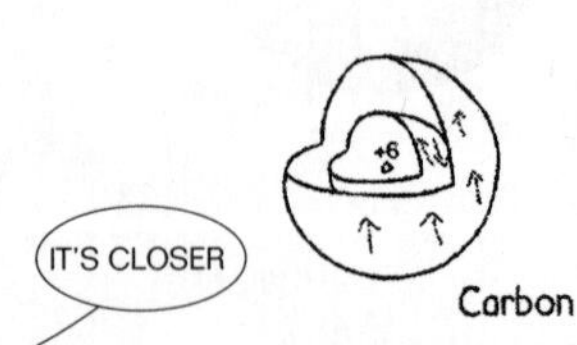

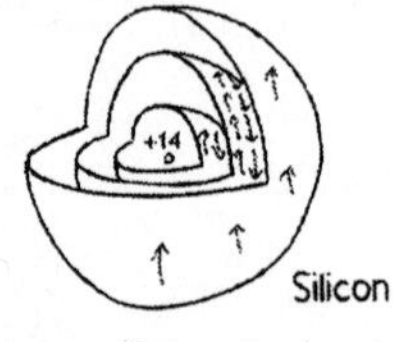

IT'S CLOSER

2. According to the shell model, which should experience the greater effective nuclear charge: an electron in
   a. carbon's 1st shell or (silicon's 1st shell?) (circle one)
   b. carbon's 2nd shell or (silicon's 2nd shell?) (circle one)
   c. (carbon's 2nd shell) or silicon's 3rd shell? (circle one)
3. List the shells of carbon and silicon in order of decreasing effective nuclear charge.
SILICON'S 1ST (~+14) > SILICON'S 2ND (~+12) > CARBON'S 1ST (~+6) > CARBON'S 2ND (~+4) > SILICON'S 3RD (≤+4)
4. Which should have the greater ionization energy, the carbon atom or the silicon atom? Defend your answer. THE CARBON ATOM BECAUSE ITS OUTERMOST ELECTRON IS EXPERIENCING A GREATER EFFECTIVE NUCLEAR CHARGE.
5. How many additional electrons are able to fit in the outermost shell of carbon? 4 silicon? 4
6. Which should be stronger, a (C-H bond) or an Si-H bond? Defend your answer. C-H IS STRONGER. THEIR VALENCE ELECTRONS EXPERIENCE A GREATER EFFECTIVE NUCLEAR CHARGE, HENCE, THEY ARE HELD TIGHTER.
7. Which should be larger in size, the (ion C^{4+}) or the ion Si^{4+}? Why?
THE ARRANGEMENTS OF ELECTRONS ARE THE SAME. SILICON, HOWEVER, HAS A GREATER NUCLEAR CHARGE, WHICH PULLS ELECTRONS INWARD MAKING S^{4+} SMALLER.

Chapter 12: The Nature of Chemical Bonds

Solutions

1. Use these terms to complete the following sentences. Some terms may be used more than once.

solution	**solvent**	**solute**
dissolve	**concentrated**	**dilute**
saturated	**concentration**	**mole**
molarity	**solubility**	**soluble**
insoluble	**precipitate**	

Sugar is SOLUBLE in water for the two can be mixed homogeneously to form a SOLUTION. The SOLUBILITY of sugar in water is so great that CONCENTRATED homogeneous mixtures are easily prepared. Sugar, however, is not infinitely SOLUBLE in water for when too much of this SOLUTE is added to water, which behaves as the SOLVENT, the solution becomes SATURATED. At this point any additional sugar is INSOLUBLE for it will not DISSOLVE. If the temperature of a saturated sugar solution is lowered, the SOLUBILITY of the sugar in water is also lowered. If some of the sugar comes out of solution, it is said to form a PRECIPITATE. If, however, the sugar remains in solution despite the decrease in solubility, then the solution is said to be supersaturated. Adding only a small amount of sugar to water results in a DILUTE solution. The CONCENTRATION of this solution or any solution can be measured in terms of MOLARITY, which tells us the number of solute molecules per liter of solution. If there are 6.022×10^{23} molecules in 1 liter of solution, then the CONCENTRATION of the solution is 1 MOLE per liter.

2. Temperature has a variety of effects on the solubilities of various solutes. With some solutes, such as sugar, solubility increases with increasing temperature. With other solutes, such as sodium chloride (table salt), changing temperature has no significant effect. With some solutes, such as lithium sulfate, Li_2SO_4, the solubility actually decreases with increasing temperature.
   a. Describe how you would prepare a supersaturated solution of lithium sulfate.
   FORM A SATURATED SOLUTION AND THEN SLOWLY RAISE THE TEMPERATURE
   b. How might you cause a saturated solution of lithium sulfate to form a precipitate?
   INCREASE ITS TEMERATURE

Pure Mathematics

Using a scientist's definition of *pure*, identify whether each of the following is 100% pure:

	100% pure?
Freshly squeezed orange juice....	Yes (No)
Country air	Yes (No)
Ocean water................	Yes (No)
Fresh drinking water	Yes (No)
Skim milk..................	Yes (No)
Stainless steel	Yes (No)
A single water molecule	(Yes) No

A glass of water contains on the order of a trillion trillion (1×10^{24}) molecules. If the water in this were 99.9999% pure, you could calculate the percent of impurities by subtracting from 100.0000%.

$$\begin{array}{r} 100.0000\% \text{ water + impurity molecules} \\ -\ 99.9999\% \text{ water molecules} \\ \hline 0.0001\% \text{ impurity molecules} \end{array}$$

Pull out your calculator and calculate the number of impurity molecules in the glass of water. Do this by finding 0.0001% of 1×10^{24}, which is the same as multiplying 1×10^{24} by 0.000001.

$$(1 \times 10^{24})(0.000001) = \underline{1 \times 10^{18}}$$

1. How many impurity molecules are there in a glass of water that's 99.9999% pure?
   a. 1000 (one thousand: 10^3)
   b. 1,000,000 (one million: 10^6)
   c. 1,000,000,000 (one billion: 10^9)
   (d.) 1,000,000,000,000,000,000 (one million trillion: 10^{18}).
2. How does your answer make you feel about drinking water that is 99.9999 percent free of some poison, such as pesticide?
   THAT THERE ARE A MILLION TRILLION POISON MOLECULES IN A GLASS OF WATER MIGHT MAKE ONE HESITATE . . . BUT READ ON!
3. For every one impurity molecule, how many water molecules are there? (Divide the number of water molecules by the number of impurity molecules.)
   $10^{24}/10^{18} = 10^6 = 1{,}000{,}000$ = one million
4. Would you describe these impurity molecules within water that's 99.9999% pure as "rare" or "common"?
   FOR EVERY ONE IMPURITY MOLECULE THERE ARE ONE MILLION WATER MOLECULES. ONE IN A MILLION IS RARE!
5. A friend argues that he or she doesn't drink tap water because it contains thousands of molecules of some impurity in each glass. How would you respond in defense of the water's purity, if it indeed does contain thousands of molecules of some impurity per glass?
   ONLY 1,000 IMPURITY MOLECULES IN THIS GLASS OF WATER WOULD MAKE THIS WATER INCREDIBLY PURE . . . ABOUT 99.9999999999999999999% PURE!

Chemical Bonds

1. Based upon their positions in the periodic table, predict whether each pair of elements will form an ionic bond, covalent bond, or neither (atomic number in parenthesis).
   a. Gold (79) and platinum (78) N
   b. Rubidium (37) and iodine (53) C
   c. Sulfur (16) and chlorine (17) I
   d. Sulfur (16) and magnesium (12) N
   e. Calcium (20) and chlorine (17) C
   f. Germanium (32) and arsenic (33) I
   g. Iron (26) and chromium (24) I
   h. Chlorine (17) and iodine (53) C
   i. Carbon (6) and bromine (35) C
   j. Barium (56) and astatine (85) I
2. The most common ions of lithium, magnesium, aluminum, chlorine, oxygen, and nitrogen and their respective charges are as follows:

Positively Charged Ions	Negatively Charged Ions
Lithium ion: Li^{1+}	Chloride ion: Cl^{1-}
Barium ion: Ba^{2+}	Oxide ion: O^{2-}
Aluminum ion: Al^{3+}	Nitride ion: N^{3-}

Use this information to predict the chemical formulas for the following ionic compounds:

a. Lithium chloride: $LiCl$
b. Barium chloride: Li_2O
c. Aluminum chloride: Li_3O
d. Lithium oxide: $BaCl_2$
e. Barium oxide: BaO
f. Aluminum oxide: Ba_3N_2
g. Lithium nitride: $AlCl_2$
h. Barium nitride: Al_2O_3
i. Aluminum nitride: AlN
j. How are elements that form positive ions grouped in the periodic table relative to elements that form negative ions? POSITIVE ION ELEMENTS TOWARD THE LEFT AND NEGATIVE IONS TOWARD THE RIGHT.

3. Specify whether the following chemical structures are polar or nonpolar:

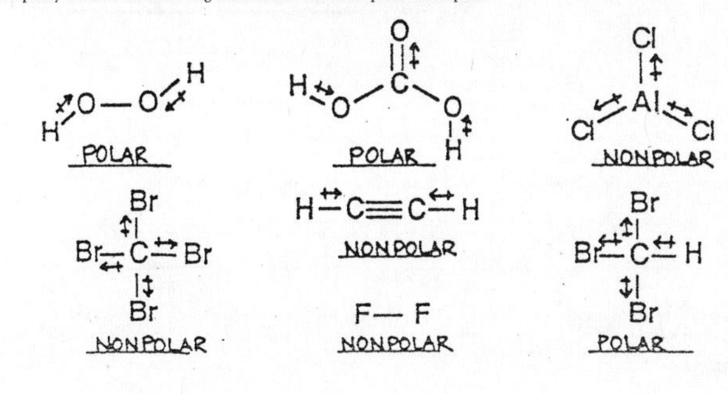

Shells and the Covalent Bond

When atoms bond covalently, their atomic shells overlap so that shared electrons can occupy both shells at the same time.

Nonbonded hydrogen atoms

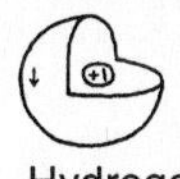
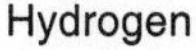

Covalently bonded hydrogen atoms

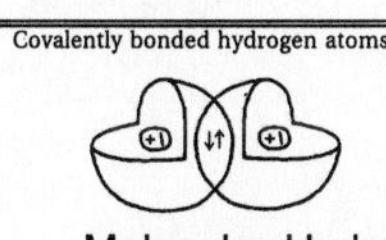

Formula: H_2

Fill each shell model shown below with enough electrons to make each atom electrically neutral. Use arrows to represent electrons. Within the box draw a sketch showing how the two atoms bond covalently. Draw hydrogen shells more than once when necessary so that no electrons remain unpaired. Write the name and chemical formula for each compound.

A.

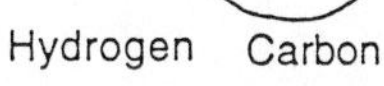

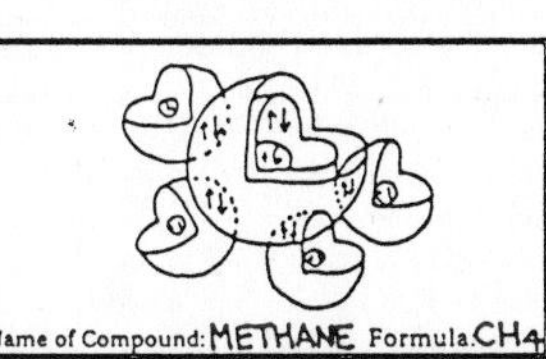

Name of Compound: METHANE Formula: CH_4

B.

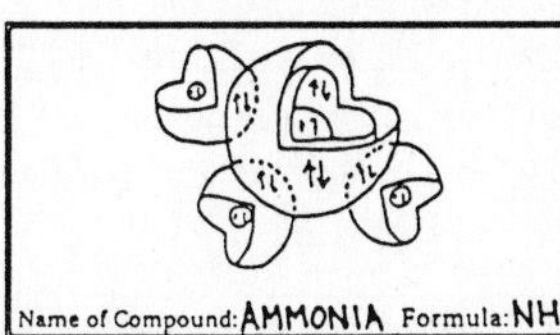

Name of Compound: AMMONIA Formula: NH_3

C.

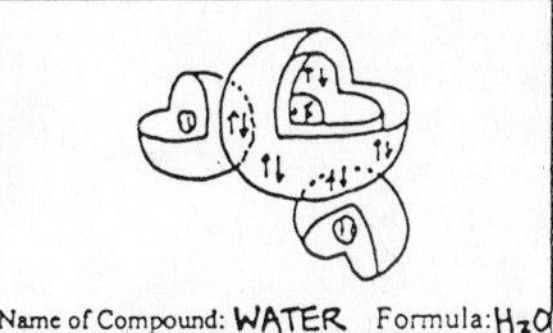

Hydrogen Oxygen

Name of Compound: WATER Formula: H_2O

D.

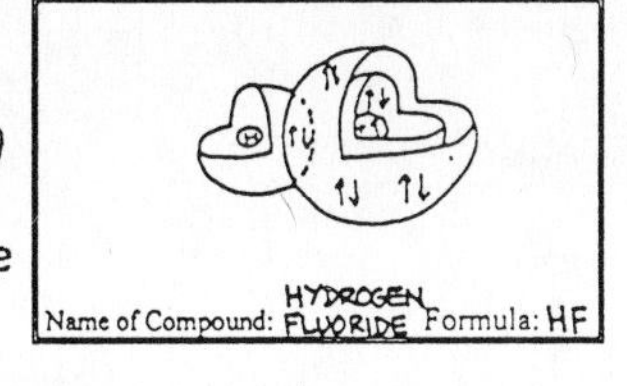

Name of Compound: HYDROGEN FLUORIDE Formula: HF

E.

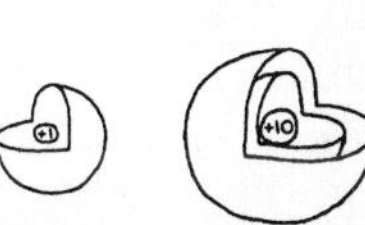
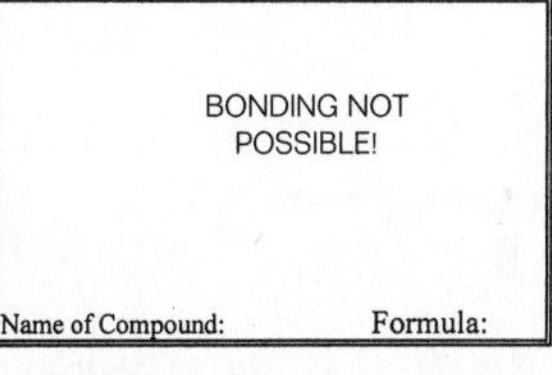

Hydrogen Neon

BONDING NOT POSSIBLE!

Name of Compound: Formula:

1. Note the relative positions of carbon, nitrogen, oxygen, fluorine, and neon in the periodic table. How does this relate to the number of times each of these elements is able to bond with hydrogen?
   IT'S IN A DESCENDING ORDER FROM LEFT TO RIGHT.
2. How many times is the element boron (atomic number 5) able to bond with hydrogen? Use the shell model to help you with your answer.
   ONLY 3 VALENCE ELECTRONS, THEREFORE, ONLY 3 BONDS

Chapter 12: The Nature of Chemical Bonds

Bond Polarity

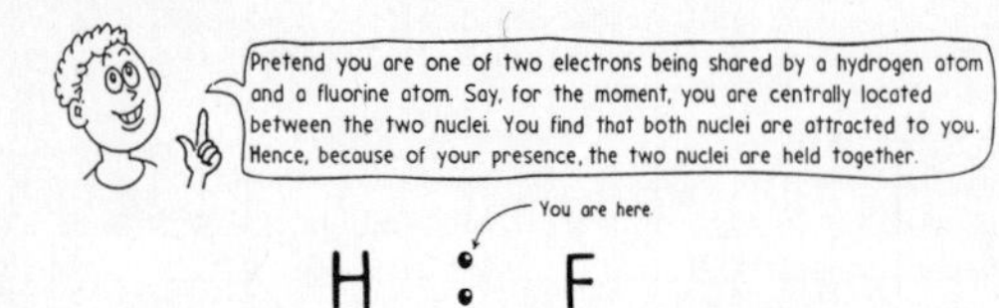

You are here

H : F

1. Why are the nuclei of these atoms attracted to you? BECAUSE OF YOUR NEGATIVE CHARGE
2. What type of chemical bonding is this? COVALENT

You are held within hydrogen's 1st shell and at the same time within fluorine's 2nd shell. Draw a sketch using the shell models below to show how this is possible. Represent yourself and all other electrons using arrows. Note your particular location with a circle.

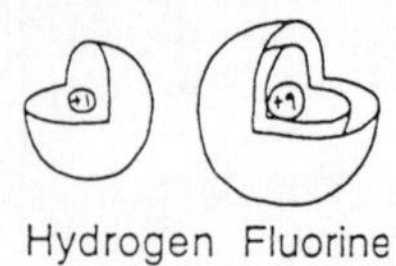

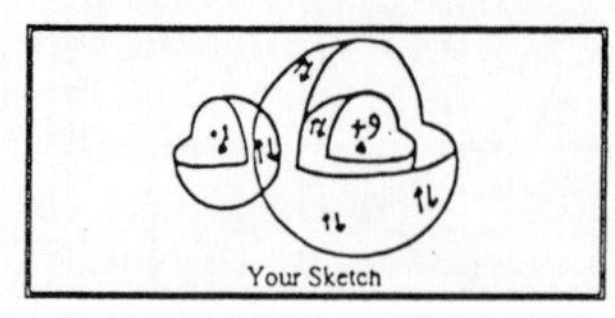

According to the laws of physics, if the nuclei are both attracted to you, then you are attracted to both of the nuclei.

3. You are pulled toward the hydrogen nucleus, which has a positive charge. How strong is this charge from your point of view—what is its *electronegativity*? ~+1
4. You are also attracted to the fluorine nucleus. What is its electronegativity? ~+7

You are being shared by the hydrogen and fluorine nuclei. But as a moving electron you have some choice as to your location.

5. Consider the electronegativities you experience from both nuclei. Which nucleus would you tend to be closest to? FLUORINE

Stop pretending you are an electron and observe the hydrogen-fluorine bond from outside the hydrogen fluoride molecule. Bonding electrons tend to congregate to one side because of the differences in effective nuclear charges. This makes one side slightly negative in character and the opposite side slightly positive. Indicate this on the following structure for hydrogen fluoride using the symbols δ- and δ+

H : F

By convention, bonding electrons are not shown. Instead, a line is simply drawn connecting the two bonded atoms. Indicate the slightly negative and positive ends.

H — F

6. Would you describe hydrogen fluoride as a polar or nonpolar molecule? POLAR
7. If two hydrogen fluoride molecules were thrown together, would they stick or repel? (Hint: What happens when you throw two small magnets together?) STICK
8. Place bonds between the hydrogen and fluorine atoms to show many hydrogen fluoride molecules grouped together. Each element should be bonded only once. Circle each molecule and indicate the slightly negative and slightly positive ends.

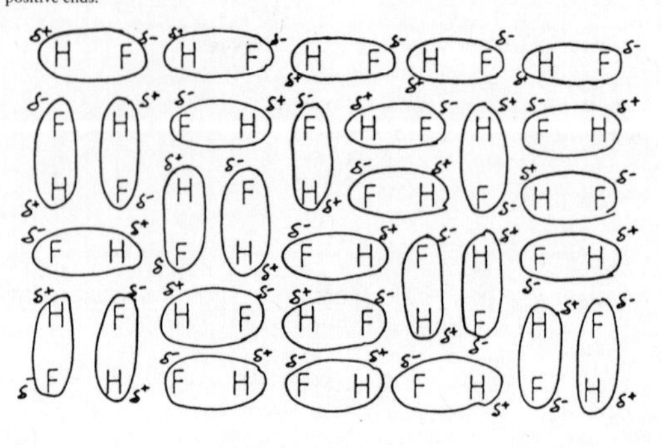

Chapter 12: The Nature of Chemical Bonds

Atoms to Molecules

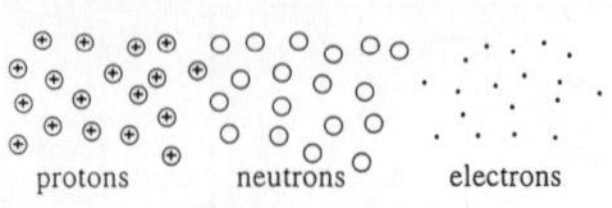

Subatomic particles are the fundamental building blocks of all ATOMS.

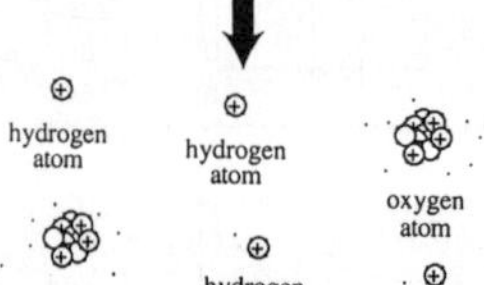

An atom is a group of SUBATOMIC PARTICLES held tightly together. An oxygen atom is a group of 8 PROTONS, 8 NEUTRONS, and 8 ELECTRONS. A hydrogen atom is a group of only 1 PROTON and 1 ELECTRON.

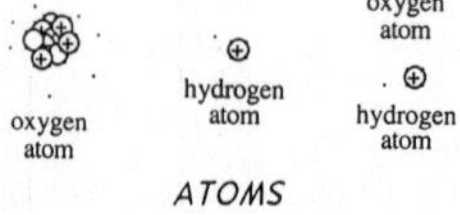

A MOLECULE is a group of atoms held tightly together. A water MOLECULE consists of 2 OXYGEN atoms and 1 HYDROGEN atom.

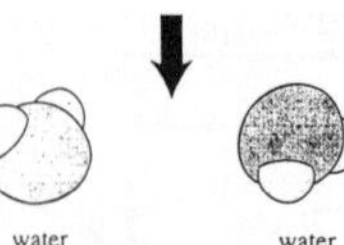

Water is a material made up of billions upon billions of water MOLECULES. The physical properties of water are based upon how these water MOLECULES interact with one another. The electronic attractions between MOLECULES is one of the major topics of Chapter 12.

Chapter 13: Chemical Reactions

Balancing Chemical Equations

In a balanced chemical equation the number of times each element appears as a reactant is equal to the number of times it appears as a product. For example,

$$2\ H_2 + O_2 \rightarrow 2\ H_2O$$

Recall that *coefficients* (the integer appearing before the chemical formula) indicate the number of times each chemical formula is to be counted and *subscripts* indicate when a particular element occurs more than once within the formula.

Check whether or not the following chemical equations are balanced.

$3\ NO \rightarrow N_2O + NO_2$	☑ balanced ☐ unbalanced
$SiO_2 + 4\ HF \rightarrow SiF_4 + 2\ H_2O$	☑ balanced ☐ unbalanced
$4\ NH_3 + 5\ O_2 \rightarrow 4\ NO + 6\ H_2O$	☑ balanced ☐ unbalanced

Unbalanced equations are balanced by changing the coefficients. Subscripts, however, should never be changed because this changes the chemical's identity—H_2O is water, but H_2O_2 is hydrogen peroxide! The following steps may help guide you:

1. Focus on balancing only one element at a time. Start with the left-most element and modify the coefficients such that this element appears on both sides of the arrow the same number of times.
2. Move to the next element and modify the coefficients so as to balance this element. Do not worry if you incidentally unbalance the previous element. You will come back to it in subsequent steps.
3. Continue from left to right balancing each element individually.
4. Repeat steps 1–3 until all elements are balanced.

Use the above methodology to balance the following chemical equations.

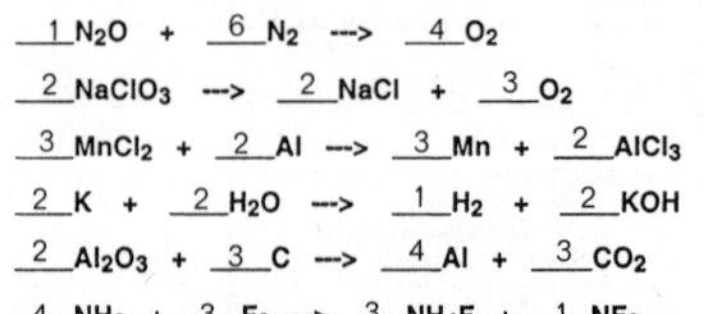

1 N_2O + 6 N_2 → 4 O_2

2 $NaClO_3$ → 2 $NaCl$ + 3 O_2

3 $MnCl_2$ + 2 Al → 3 Mn + 2 $AlCl_3$

2 K + 2 H_2O → 1 H_2 + 2 KOH

2 Al_2O_3 + 3 C → 4 Al + 3 CO_2

4 NH_3 + 3 F_2 → 3 NH_4F + 1 NF_3

This is just one of the many methods that chemists have developed to balance chemical equations.

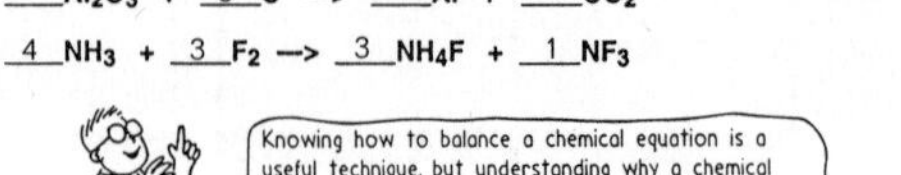

Chapter 13: Chemical Reactions

Exothermic and Endothermic Reactions

During a chemical reaction atoms are neither created nor destroyed. Instead, atoms rearrange—they change partners. This rearrangement of atoms necessarily involves the input and output of energy. First, energy must be supplied to break chemical bonds that hold atoms together. Separated atoms then form new chemical bonds, which involves the release of energy. In an **exothermic** reaction more energy is released than is consumed. Conversely, in an **endothermic** reaction more energy is consumed than is released.

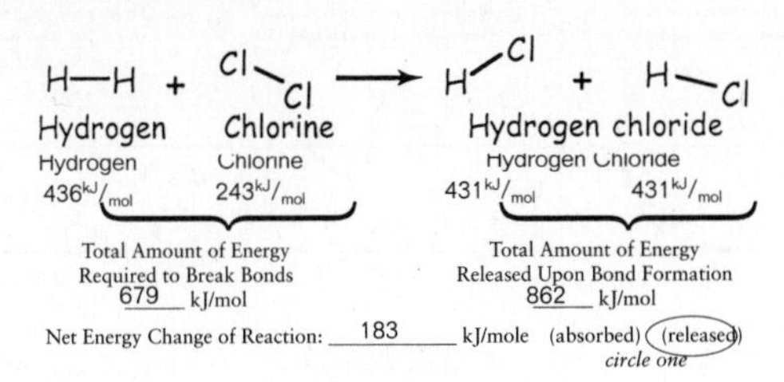

Table 1 Bond Energies

Bond	Bond Energy*	Bond	Bond Energy*
H—H	436	Cl—Cl	243
H—C	414	N—N	159
H—N	389	O=O	498
H—O	464	O=C	803
H—Cl	431	N≡N	946

*In kJ/mol

Table 1 shows bond energies—the amount of energy required to break a chemical bond, and also the amount of energy released when a bond is formed. Use these bond energies to determine whether the following chemical reactions are exothermic or endothermic.

H—H + Cl—Cl ⟶ H—Cl + H—Cl

Hydrogen, Chlorine, Hydrogen chloride

Hydrogen 436 kJ/mol; Chlorine 243 kJ/mol; Hydrogen Chloride 431 kJ/mol, 431 kJ/mol

Total Amount of Energy Required to Break Bonds: 679 kJ/mol

Total Amount of Energy Released Upon Bond Formation: 862 kJ/mol

Net Energy Change of Reaction: 183 kJ/mole (absorbed) (released) *circle one* — released circled

1. Is this reaction exothermic or endothermic? EXOTHERMIC
2. Write the balanced equation for this reaction using chemical formulas and coefficients. If it is exothermic, write "Energy" as a product. If it is endothermic, write "Energy" as a reactant.

$$H_2 + Cl_2 \longrightarrow 2\,HCl + \text{ENERGY}$$

CH_4 + O=O + O=O ⟶ O=C=O + H—O—H + H—O—H

Methane, Oxygen, Carbon Dioxide, Water

Methane	Oxygen	Carbon Dioxide	Water
414 kJ/mol	498 kJ/mol	803 kJ/mol	464 kJ/mol
× 4	× 2	× 2	× 4
1656 kJ/mol	996 kJ/mol	1606 kJ/mol	1856 kJ/mol

Total Amount of Energy Required to Break Bonds: 2652 kJ/mol

Total Amount of Energy Released Upon Bond Formation: 3462 kJ/mol

Net Energy Change of Reaction: 810 kJ/mole (absorbed/released) *circle one* — released circled

3. Is this reaction exothermic or endothermic? EXOTHERMIC
4. Write the balanced equation for this reaction using chemical formulas and coefficients. If it is exothermic write "Energy" as a product. If it is endothermic write "Energy" as a reactant.

$$CH_4 + 2\,O_2 \longrightarrow CO_2 + 2\,H_2O + \text{ENERGY}$$

N≡N + H—H + H—H ⟶ $H_2N—NH_2$

Nitrogen, Hydrogen, Hydrazine

Nitrogen	Hydrogen	Hydrazine	
946 kJ/mol	436 kJ/mol	389 kJ/mol	159 kJ/mol
	× 2	× 4	
	872 kJ/mol	1556 kJ/mol	

Total Amount of Energy Required to Break Bonds: 1818 kJ/mol

Total Amount of Energy Released Upon Bond Formation: 1715 kJ/mol

Net Energy Change of Reaction: 103 kJ/mole (absorbed/released) *circle one* — absorbed circled

5. Is this reaction exothermic or endothermic? ENDOTHERMIC
6. Write the balanced equation for this reaction using chemical formulas and coefficients. If it is exothermic write "Energy" as a product. If it is endothermic write "Energy" as a reactant.

$$\text{ENERGY} + N_2 + 2\,H_2 \longrightarrow N_2H_4$$

Chapter 13: Chemical Reactions

Donating and Accepting Hydrogen Ions

A chemical reaction that involves the transfer of a hydrogen ion from one molecule to another is classified as an acid-base reaction. The molecule that donates the hydrogen ion behaves as an acid. The molecule that accepts the hydrogen ion behaves as a base.

On paper, the acid-base process can be depicted through a series of frames:

Frame 1 — ammonium ion (NH_4^+), hydroxide ion (^-OH)

Ammonium and hydroxide ions in close proximity.

Frame 2

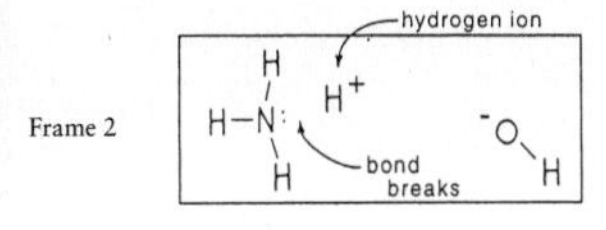

Bond is broken between the nitrogen and a hydrogen of the ammonium ion. The two electrons of the broken bond stay with the nitrogen leaving the hydrogen with a positive charge.

Frame 3

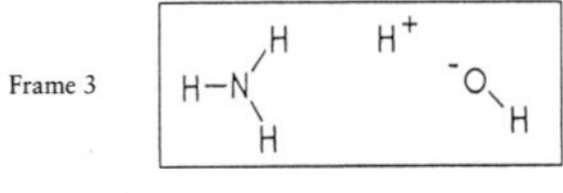

The hydrogen ion migrates to the hydroxide ion.

Frame 4

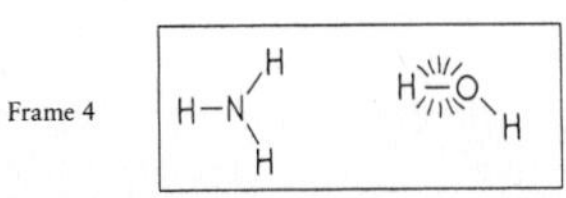

The hydrogen ion bonds with the hydroxide ion to form a water molecule.

In equation form we abbreviate this process by only showing the before and after:

NH_4^+ + ^-OH ⟶ NH_3 + H_2O

frame 1 / frame 4

We see from the previous reaction that because the ammonium ion donated a hydrogen ion, it behaved as an acid. Conversely, the hydroxide ion by accepting a hydrogen ion behaved as a base. How do the ammonia and water molecules behave during the reverse process?

NH_4^+ + ^-OH ⇌ NH_3 + H_2O

acid, base, BASE (ammonia), ACID (water)

Identify the following molecules as behaving as an acid or a base:

H_3PO_4 + H_2O ⇌ $H_2PO_4^-$ + H_3O^+	ACID	BASE	BASE	ACID
NH_3 + NaH ⇌ $NH_2^-\,Na^+$ + H-H	ACID	BASE	BASE	ACID
H-H + ^-H ⇌ H^- + H-H	ACID	BASE	BASE	ACID
HNO_3 + NH_3 ⇌ $^-NO_3$ + $^+NH_4$	ACID	BASE	BASE	ACID

Chapter 13: Chemical Reactions

Loss and Gain of Electrons

A chemical reaction that involves the transfer of an electron is classified as an oxidation–reduction reaction. Oxidation is the process of losing electrons, while reduction is the process of gaining them. Any chemical that causes another chemical to lose electrons (become oxidized) is called an *oxidizing agent*. Conversely, any chemical that causes another chemical to gain electrons is called a *reducing agent*.

1. What is the relationship between an atom's ability to behave as an oxidizing agent and its electron affinity?
   THE GREATER THE ELECTRON AFFINITY, THE GREATER ITS ABILITY TO BEHAVE AS AN OXIDIZING AGENT.
2. Relative to the periodic table, which elements tend to behave as strong oxidizing agents?
   THOSE TO THE UPPER RIGHT WITH THE EXCEPTION OF THE NOBLE GASES
3. Why don't the noble gases behave as oxidizing agents?
   THEY HAVE NO SPACE IN THEIR SHELLS TO ACCOMODATE ADDITIONAL ELECTRONS.
4. How is it that an oxidizing agent is itself reduced?
   REDUCTION IS THE GAINING OF ELECTRONS. IN PULLING AN ELECTRON AWAY FROM ANOTHER ATOM AN OXIDIZING AGENT NECESSARILY GAINS AN ELECTRON.
5. Specify whether each reactant is about to be oxidized or reduced.

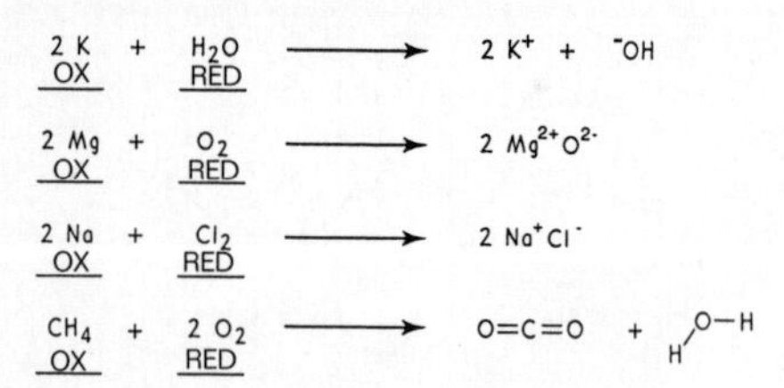

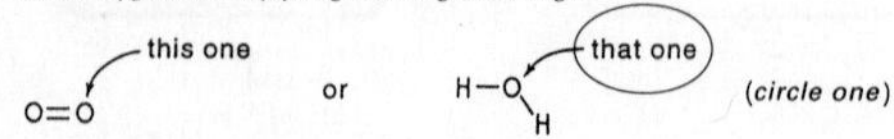

6. Which oxygen atom enjoys a greater negative charge?

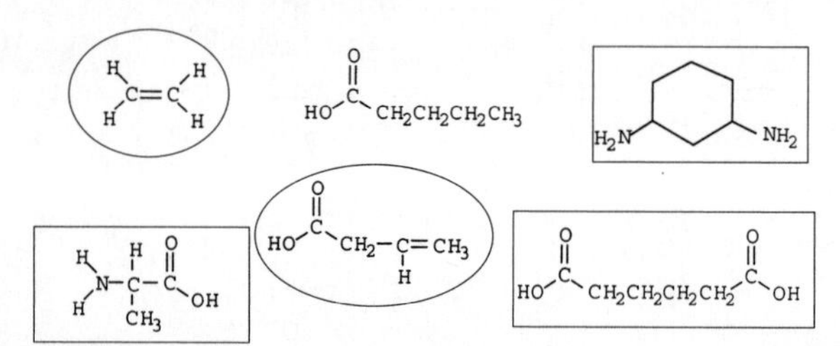

7. Relate your answer to Question 6 to how it is that O_2 is reduced upon reacting with CH_4 to form carbon dioxide and water.
   IN TRANSFORMING FROM O_2 TO H_2O, AN OXYGEN ATOM IS GAINING ELECTRONS AS BEST AS IT CAN. WITH ITS GREATER NEGATIVE CHARGE IT CAN BE THOUGHT OF AS "REDUCED."

Chapter 14: Organic Chemistry

Structures of Organic Compounds

1. What are the chemical formulas for the following structures?

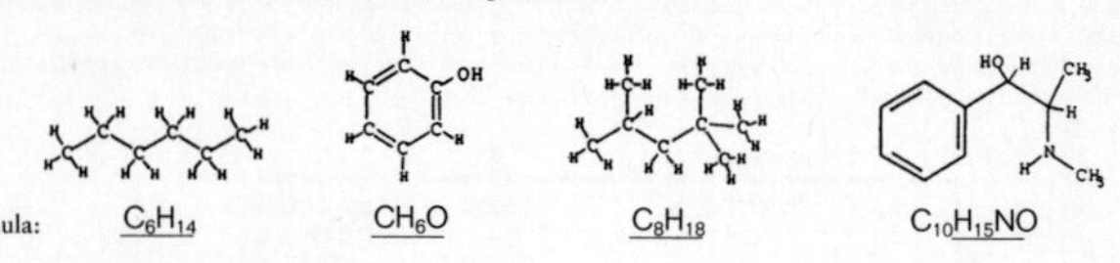

Formula: C_6H_{14} CH_6O C_8H_{18} $C_{10}H_{15}NO$

2. How many covalent bonds is carbon able to form? 4
3. What is wrong with the structure shown in the box at right?
   THE CARBON OF THE CARBONYL IS BONDED 5 TIMES
4. a. Draw a hydrocarbon that contains 4 carbon atoms.
   b. Redraw your structure and transform it into an amine.
   c. Transform your amine into an amide. You may need to relocate the nitrogen.

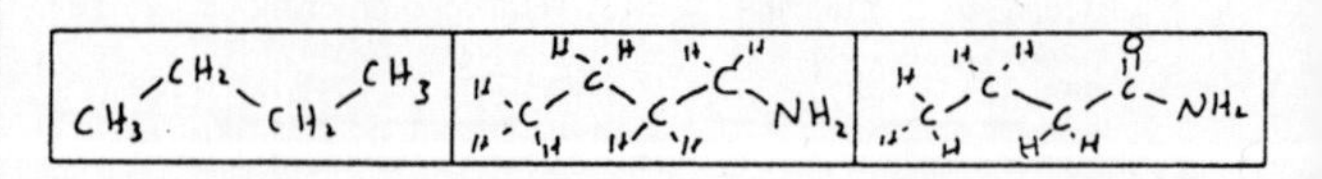

   d. Redraw your amide, transforming it into a carboxylic acid.
   e. Redraw your carboxylic acid, transforming it into an alcohol.
   f. Rearrange the carbons of your alcohol to make an ether.

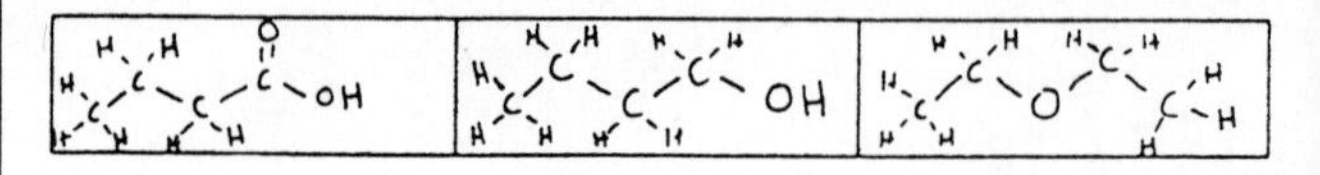

Chapter 14: Organic Chemistry

Polymers

1. Circle the monomers that may be useful for forming an addition polymer and draw a box around the ones that may be useful for forming a condensation polymer.

$H_2C=CH_2$ (circled); $HO-C(=O)-CH_2CH_2CH_2CH_3$; H_2N–cyclohexane–NH_2 (boxed); $H_2N-CH(CH_3)-C(=O)OH$ (boxed); $HO-C(=O)-CH_2-CH=CH_3$ (circled); $HO-C(=O)-CH_2CH_2CH_2CH_2-C(=O)-OH$ (boxed)

2. Which type of polymer always weighs less than the sum of its parts? Why?
   THE CONDENSATION POLYMERS LOSE SMALL MOLECULES SUCH AS WATER WHEN THEY FORM AND THUS THE POLYMER THAT FORMS WEIGHS LESS THAN THE SUM OF ITS MONOMERS.
3. Would a material with the following arrangement of polymer molecules have a relatively high or low melting point? Why?

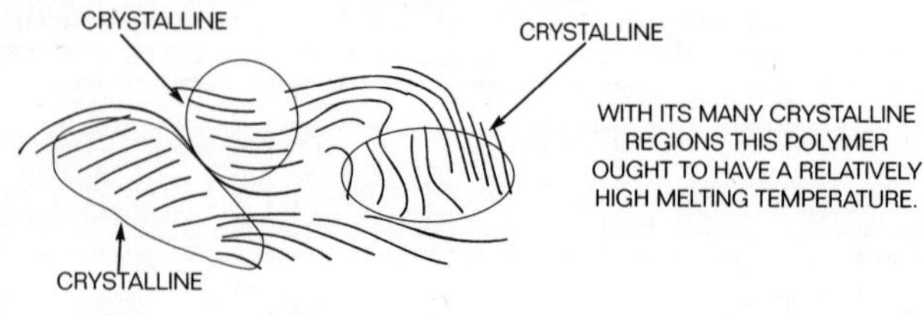

WITH ITS MANY CRYSTALLINE REGIONS THIS POLYMER OUGHT TO HAVE A RELATIVELY HIGH MELTING TEMPERATURE.

Chapter 15: The Basic Unit of Life—The Cell

Features of Prokaryotic and Eukaryotic Cells

1. Are the following associated with prokaryotic cells, eukaryotic cells, or both?
   - a. nucleic acids —BOTH
   - b. cell membrane —BOTH
   - c. nucleus —EUKARYOTIC
   - d. organelles —EUKARYOTIC
   - e. mitochondria —EUKARYOTIC
   - f. chloroplasts —EUKARYOTIC
   - g. bacteria —PROKARYOTIC
   - h. circular chromosome —PROKARYOTIC
   - i. cytoplasm —BOTH
   - j. human cells —EUKARYOTIC

2. Match the following organelles with their functions:
   - ribosome —G
   - rough endoplasmic reticulum —A
   - smooth endoplasmic reticulum —F
   - Golgi apparatus —D
   - lysosome —H
   - mitochondrion —B
   - chloroplast —C
   - cytoskeleton —E

   a. assembles proteins destined to go either to the cell membrane or to leave the cell
   b. obtains energy for the cell to use
   c. in plant cells, captures energy from sunlight to build organic molecules
   d. receives products from the endoplasmic reticulum and packages them for transport
   e. helps cell hold its shape
   f. assembles membranes and performs other specialized functions in certain cells
   g. assembles proteins for the cell
   h. breaks down organic materials

3. What are the three components of a cell membrane? Draw a portion of a cell membrane showing each of these components.

THE THREE PRIMARY MEMBRANE COMPONENTS OF THE CELL MEMBRANE ARE PHOSPHOLIPID MEMBRANE, PROTEINS, AND SHORT CARBOHYDRATES.

4. What are some functions carried out by membrane proteins?

MEMBRANE PROTEINS SERVE A VARIETY OF FUNCTIONS—THEY HELP CELLS COMMUNICATE WITH OTHER CELLS, CONTROL TRANSPORT INTO AND OUT OF CELLS, CONTROL THE CHEMICAL REACTIONS THAT OCCUR IN CELLS, AND JOIN CELLS TO ONE ANOTHER.

5. What are the functions of short carbohydrates?

THE SHORT CARBOHYDRATES PLAY AN IMPORTANT ROLE IN CELL RECOGNITION—THE ABILITY TO DISTINGUISH ONE TYPE OF CELL FROM ANOTHER. FOR EXAMPLE, IMMUNE SYSTEM CELLS USE THESE CARBOHYDRATES TO IDENTIFY FOREIGN MATERIALS SUCH AS DISEASE-CAUSING BACTERIA AND VIRUSES.

Chapter 15: The Basic Unit of Life—The Cell

Transport In and Out of Cells

1. Assume that the square-shaped molecules shown below can pass freely across the cell membrane. What is the name of the process by which they move across the cell membrane?

DIFFUSION

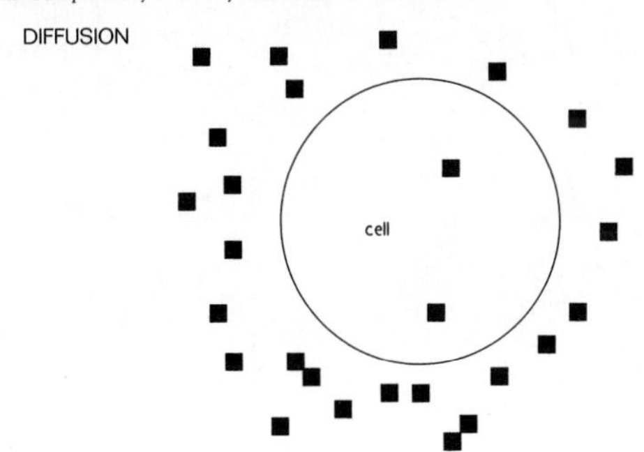

Will the square-shaped molecules tend to move out of the cell or into the cell? Why?

THE SQUARE-SHAPED MOLECULES WILL TEND TO MOVE INTO THE CELL BECAUSE MOLECULES TEND TO DIFFUSE FROM AN AREA OF HIGHER CONCENTRATION TO AN AREA OF LOWER CONCENTRATION. THERE IS HIGHER CONCENTRATION OUTSIDE THE CELL AND LOWER CONCENTRATION INSIDE THE CELL. CONSEQUENTLY, THE MOLECULES TEND TO DIFFUSE INTO THE CELL.

2. The diffusion of water has a special name. It is called OSMOSIS.
   In the figure below, a membrane allows water to move freely between two compartments. The dark circles represent solute molecules, which are not free to move between the two compartments. Which way will water tend to flow? Why?

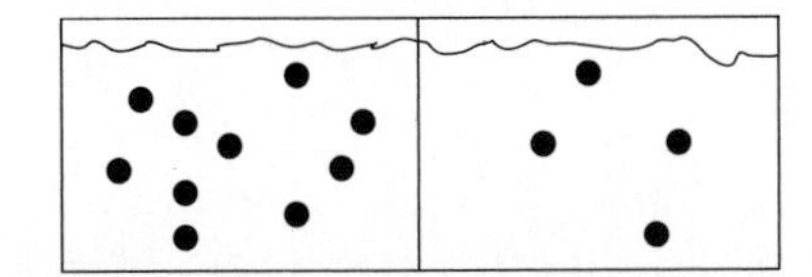

WATER WILL TEND TO MOVE TO THE LEFT, WHERE THERE IS A GREATER CONCENTRATION OF SOLUTE MOLECULES. THIS IS BECAUSE WATER, LIKE OTHER MOLECULES, MOVES FROM AN AREA OF HIGHER WATER CONCENTRATION TO AN AREA OF LOWER WATER CONCENTRATION, AND WHERE THERE IS HIGHER SOLUTE CONCENTRATION, THERE IS LOWER CONCENTRATION OF WATER MOLECULES.

3. a. If a molecule needs a carrier protein, but no energy input in order to cross a cell membrane, it provides an example of FACILITATED DIFFUSION.

   b. If a molecule requires energy input in order to cross a cell membrane, it provides an example of ACTIVE TRANSPORT.

   c. Which process is illustrated in (a)? Which is illustrated in (b)?

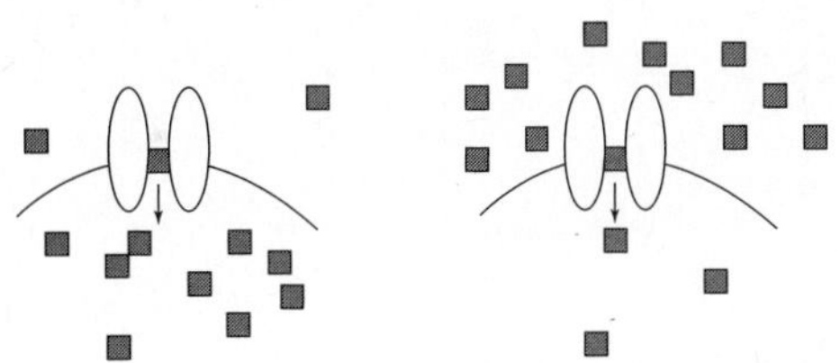

ACTIVE TRANSPORT IS ON THE LEFT; FACILITATED DIFFUSION IS ON THE RIGHT. ENERGY IS REQUIRED (ACTIVE TRANSPORT) WHEN MOLECULES ARE MOVED FROM AN AREA OF LOW CONCENTRATION TO AN AREA OF HIGH CONCENTRATION. ENERGY IS NOT REQUIRED (FACILITATED DIFFUSION) IN THE OPPOSITE SITUATION.

4. The cell below is going to take in the triangular molecule below via endocytosis. Draw what happens.

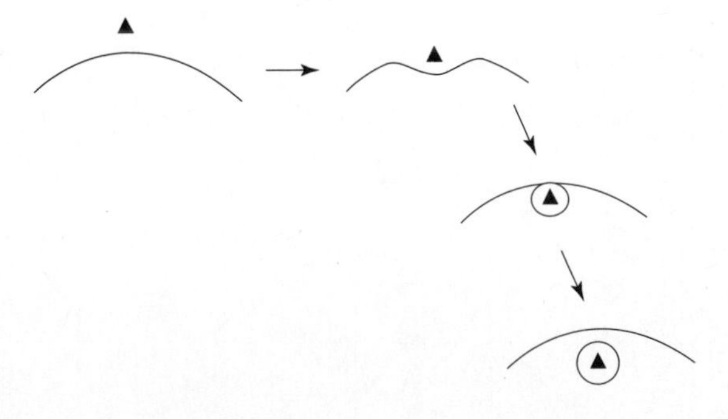

Chapter 15: The Basic Unit of Life—The Cell

Photosynthesis and Cellular Respiration

1. The chemical reaction for photosynthesis is (use words or chemical formulae):

CARBON DIOXIDE + WATER + SUNLIGHT ⟶ GLUCOSE + OXYGEN

$6CO_2 + 6H_2O$ + SUNLIGHT ⟶ $C_6H_{12}O_6 + 6O_2$.

2. a. Where in plant cells does photosynthesis take place?
   b. Where do plants get carbon dioxide from?
   c. Where do plants get water from?
   d. What parts of plants capture sunlight?

   A. CHLOROPLASTS
   B. THE ATMOSPHERE
   C. THE ENVIRONMENT—TYPICALLY, THEIR ROOTS ABSORB WATER FROM SOIL
   D. THE STEMS AND LEAVES

3. For each of the following events that occur during photosynthesis, indicate whether it occurs during the light-dependent or light-independent reactions:

a. sunlight strikes a chlorophyll molecule	LIGHT-DEPENDENT
b. Calvin cycle	LIGHT-INDEPENDENT
c. energy is generated in the form of molecules of ATP and NADPH	LIGHT-DEPENDENT
d. free oxygen is produced	LIGHT-DEPENDENT
e. carbon is fixed	LIGHT-INDEPENDENT
f. glucose is produced	LIGHT-INDEPENDENT

4. The chemical reaction for cellular respiration is

GLUCOSE + OXYGEN + ADP ⟶ CARBON DIOXIDE + WATER + ATP

$C_6H_{12}O_6 + 6O_2$ + ABOUT 38 MOLECULES OF ADP ⟶ $6CO_2 + 6H_2O$ + ABOUT 38 MOLECULES OF ATP

5. Cellular respiration allows cells to produce ATP. How do cells later obtain energy from ATP?

ENERGY IS OBTAINED FROM ATP WHEN ONE OF ITS THREE PHOSPHATE GROUPS IS REMOVED, LEAVING ADP. (CELLS EVENTUALLY TURN ADP BACK INTO ATP BY ADDING A PHOSPHATE GROUP DURING CELLULAR RESPIRATION.)

6. Which of the following processes requires oxygen?

glycolysis
Krebs cycle and electron transport
alcoholic fermentation
lactic acid fermentation

KREBS CYCLE AND ELECTRON TRANSPORT

Chapter 16: Genetics

DNA Replication, Transcription, and Translation

1. Let's start with the following strand of DNA:

AT
GC
CG
TA
TA
AT
CG
CG
GC
TA
AT
CG
GC

The strand is unwound, so that the DNA can be replicated. Fill in the nucleotides on the new strands.

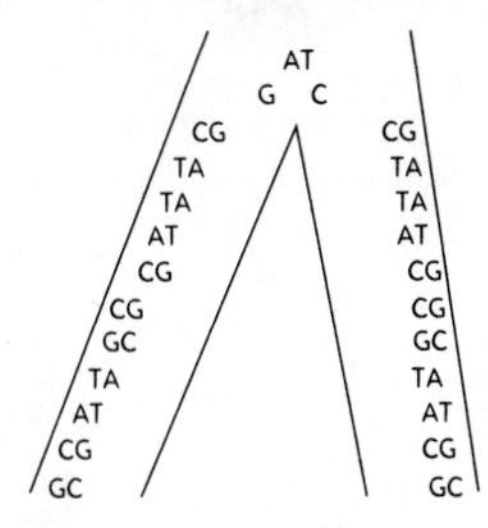

DNA Replication, Transcription, and Translation—continued

2. Transcription takes place in the NUCLEUS.
During transcription, DNA is used to make a molecule of MESSENGER RNA OR mRNA.
If the following length of DNA is being transcribed, what are the bases found on the transcript?

ATGGTCATACGTACAATG

ATGGTCATACGTACAATG (DNA TEMPLATE)
UACCAGUAUGCAUGUUAC (mRNA TRANSCRIPT)

DNA Replication, Transcription, and Translation—continued

3. Translation takes place in the _____ of cells in organelles called _____. CYTOPLASM, RIBOSOMES

During translation, _____ is used to build a _____. mRNA, PROTEIN

Divide the transcript from Problem 2 into codons and indicate the sequence of amino acids that is assembled in the ribosome. How many amino acids are coded for by this sequence?

For reference, the genetic code table is shown below:

First base	Second base: U	C	A	G	Third base
U	UUU Phenylalanine (Phe)	UCU Serine (Ser)	UAU Tyrosine (Tyr)	UGU Cysteine (Cys)	U
U	UUC Phenylalanine (Phe)	UCC Serine (Ser)	UAC Tyrosine (Tyr)	UGC Cysteine (Cys)	C
U	UUA Leucine (Leu)	UCA Serine (Ser)	UAA Stop	UGA Stop	A
U	UUG Leucine (Leu)	UCG Serine (Ser)	UAG Stop	UGG Tryptophan (Trp)	G
C	CUU Leucine (Leu)	CCU Proline (Pro)	CAU Histidine (His)	CGU Arginine (Arg)	U
C	CUC Leucine (Leu)	CCC Proline (Pro)	CAC Histidine (His)	CGC Arginine (Arg)	C
C	CUA Leucine (Leu)	CCA Proline (Pro)	CAA Glutamine (Gln)	CGA Arginine (Arg)	A
C	CUG Leucine (Leu)	CCG Proline (Pro)	CAG Glutamine (Gln)	CGG Arginine (Arg)	G
A	AUU Isoleucine (Ile)	ACU Threonine (Thr)	AAU Asparagine (Asn)	AGU Serine (Ser)	U
A	AUC Isoleucine (Ile)	ACC Threonine (Thr)	AAC Asparagine (Asn)	AGC Serine (Ser)	C
A	AUA Isoleucine (Ile)	ACA Threonine (Thr)	AAA Lysine (Lys)	AGA Arginine (Arg)	A
A	AUG Met or start	ACG Threonine (Thr)	AAG Lysine (Lys)	AGG Arginine (Arg)	G
G	GUU Valine (Val)	GCU Alanine (Ala)	GAU Aspartic acid(Asp)	GGU Glycine (Gly)	U
G	GUC Valine (Val)	GCC Alanine (Ala)	GAC Aspartic acid(Asp)	GGC Glycine (Gly)	C
G	GUA Valine (Val)	GCA Alanine (Ala)	GAA Glutamic acid (Glu)	GGA Glycine (Gly)	A
G	GUG Valine (Val)	GCG Alanine (Ala)	GAG Glutamic acid (Glu)	GGG Glycine (Gly)	G

UAC-CAG-UAU-GCA-UGU-UAC (mRNA TRANSCRIPT)

TYR-GLN-TYR-ALA-CYS-TYR

THERE ARE SIX AMINO ACIDS.

Chapter 16: Genetics

Meiosis

Consider the following diploid cell. The long, dot-filled chromosomes are homologous, but are shaded differently to distinguish them from each other. This is true for the shorter checkerboard chromosomes as well.

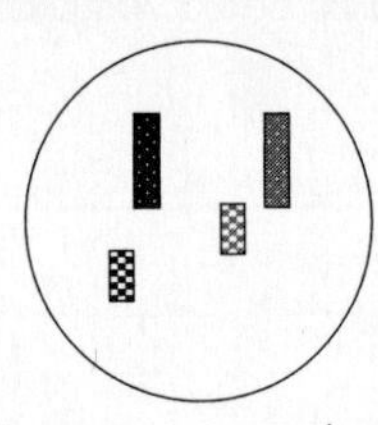

1. a. How many chromosomes are there in this diploid cell? 4

   b. If this cell were to undergo meiosis, how many chromosomes would there be in the resulting haploid cells? 2

2. Draw the cell during the following phases of meiosis:

   a. What does the cell look like when it has duplicated its genetic material in preparation for meiosis?

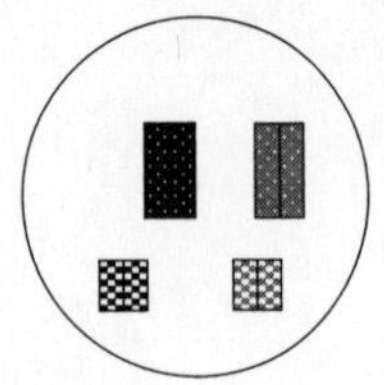

Meiosis—continued

b. What does the cell look like during metaphase I, before crossing over and recombination have occurred?

HOMOLOGOUS CHROMOSOMES ARE LINED UP AT THE EQUATORIAL PLANE OF THE CELL.

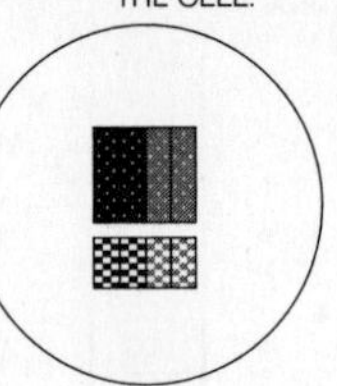

c. Suppose each chromosome experiences a single crossing over event with its homologue. Draw the cell during metaphase I, after crossing over has occurred.

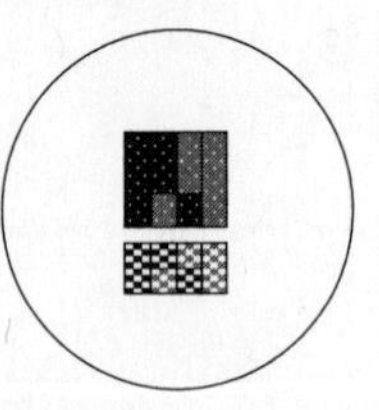

d. Draw the two daughter cells at the end of meiosis I.

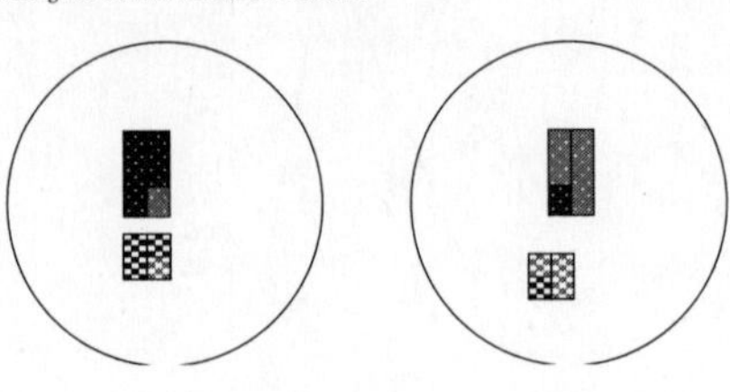

Meiosis—continued

e. Draw the four daughter cells at the end of meiosis II.

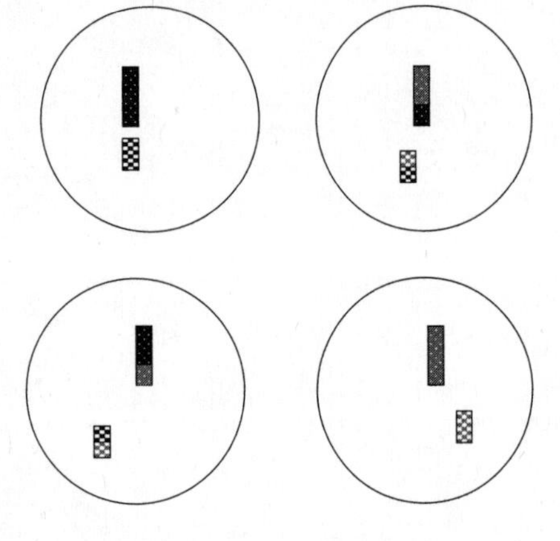

Chapter 16: Genetics

Inheritance

1. Suppose there exists a species of small woodland creature in which fur is either spotted or striped. It turns out that fur pattern is determined by a single gene, and that the striped phenotype is dominant to the spotted phenotype. Which of the following must be a homozygote?

THE SPOTTED ONE MUST BE A HOMOZYGOTE, SINCE YOU HAVE TO HAVE TWO RECESSIVE ALLELES TO HAVE SPOTS. THE STRIPED ONE COULD BE EITHER A HOMOZYGOTE OR HETEROZYGOTE.

2. Suppose, in fact, both the woodland creatures above are homozygotes. Their genotypes are aa and AA. Fill in the blanks below:

Genotype aa AA
Phenotype SPOTTED STRIPED

3. What phenotype would an Aa heterozygote have? Draw it below:

Aa HETEROZYGOTE WOULD HAVE STRIPES.

Inheritance—continued

4. Now, you breed together an aa individual with an AA individual. Draw the cross below:

Genotype aa AA
Phenotype SPOTTED STRIPED

What are the progeny like?

Genotype Aa
Phenotype STRIPED

5. Cross together two of the progeny from Problem 4 above. Fill in the boxes below.

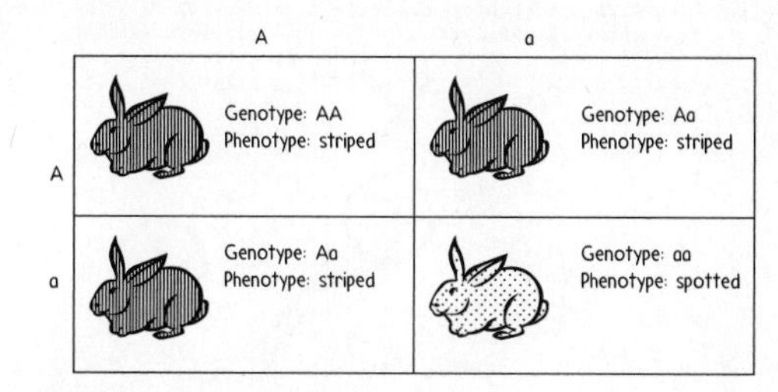

	A	a
A	Genotype: AA Phenotype: striped	Genotype: Aa Phenotype: striped
a	Genotype: Aa Phenotype: striped	Genotype: aa Phenotype: spotted

So, the progeny are spotted/striped/both, and found in a ratio _____:_____.
BOTH
3 STRIPED: 1 SPOTTED

Chapter 17: The Evolution of Life

Natural Selection

1. a. One of the best-documented instances of natural selection in the wild is the evolution of bird beak sizes during a severe drought on the Galápagos Islands in 1977. Before the drought, there was natural variation in beak size in a species of finch found on the islands. Draw a series of finches, showing this variation in beak size.

b. What happened was this: The drought made seeds scarce. Small seeds were quickly eaten up, leaving only larger, tougher seeds. Birds with larger, stronger beaks were better at cracking these larger seeds. Many birds died during the drought. Which were most likely to survive? Mark X's through some of the individuals in your drawing likely to have died, and circle individuals likely to have survived.

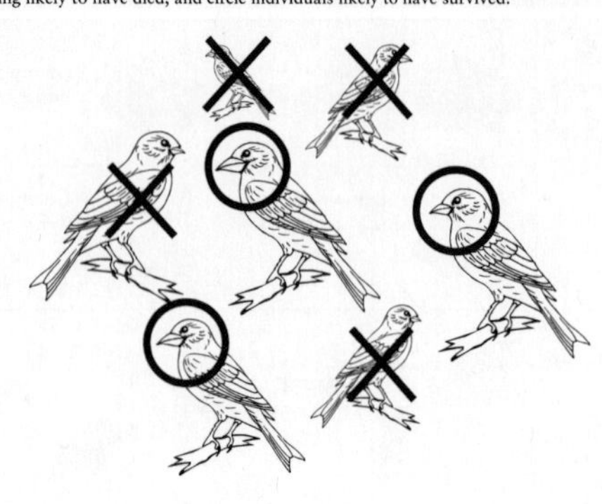

Natural Selection—continued

c. Beak size is a trait that is partly genetically determined, so parents with larger, stronger beaks tend to have offspring with larger, stronger beaks. Draw the offspring of the individuals you circled as surviving the drought.

d. How does the population you drew in part (c) compare with the population you drew in part (a)?

THEY HAVE, IN GENERAL, LARGER, STRONGER BEAKS.

Natural Selection—continued

2. Of the human traits listed below, put a V by traits that are variable and put an H by traits that are heritable. Which traits have the potential of evolving via natural selection?

a. age V
b. eye color V AND H
c. number of toes H
d. curliness or straightness of hair V AND H
e. presence or absence of dimples V AND H
f. upright posture H
g. owning versus not owning a dog V
h. height V AND H

TRAITS THAT ARE BOTH VARIABLE AND HERITABLE HAVE THE POTENTIAL OF EVOLVING BY NATURAL SELECTION—SO, OF THE TRAITS LISTED, EYE COLOR, CURLINESS OR STRAIGHTNESS OF HAIR, PRESENCE OR ABSENCE OF DIMPLES, AND HEIGHT.

Chapter 17: The Evolution of Life

Adaptation

1. The imaginary mammal below occupies temperate forests in the Eastern United States.

If a population of these mammals moved to and successfully colonized an Arctic habitat, how might you predict that it would evolve? Draw the Arctic form below.

If a population of these mammals moved to and successfully colonized a desert habitat, how might you predict that it would evolve? Draw the desert form below.

Adaptation—continued

Explain your drawings.

THE HEAT AN ANIMAL GENERATES IS PROPORTIONAL TO ITS VOLUME. THE HEAT AN ANIMAL DISSIPATES IS PROPORTIONAL TO ITS SURFACE AREA, SINCE HEAT IS LOST TO THE ENVIRONMENT THROUGH ITS BODY SURFACE. CONSEQUENTLY, ANIMALS ARE BETTER ABLE TO LOSE HEAT IF THEY HAVE A HIGH SURFACE AREA-TO-VOLUME RATIO, AND BETTER ABLE TO RETAIN HEAT IF THEY ARE HAVE A LOW SURFACE AREA-TO-VOLUME RATIO. THIS INFLUENCES BOTH THE SIZE AND SHAPE OF ANIMALS THAT OCCUPY EXTREME HABITATS. LARGER ORGANISMS, TEND TO HAVE *SMALLER* SURFACE AREA-TO-VOLUME RATIOS. THIS IS BECAUSE VOLUME INCREASES MORE QUICKLY THAN SURFACE AREA AS ORGANISMS GET BIGGER. FOR THIS REASON, ANIMALS FOUND IN COLD HABITATS ARE OFTEN LARGER THAN RELATED FORMS IN WARM HABITATS. ANIMALS ADAPTED TO HOT VERSUS COLD CLIMATES ALSO VARY IN SHAPE. DESERT SPECIES TYPICALLY HAVE LONG LEGS AND LARGE EARS THAT INCREASE THE SURFACE AREA AVAILABLE FOR HEAT DISSIPATION. THESE PARTS OF THE BODY ARE ALSO COVERED WITH EXTENSIVE BLOOD VESSELS THAT CARRY HEAT FROM THE CORE OF THE BODY TO THE SKIN, WHERE CONVECTION, THE TRANSFER OF HEAT BY MOVING AIR, COOLS THE ANIMAL. ARCTIC SPECIES TYPICALLY HAVE SHORT APPENDAGES AND SMALL EARS THAT HELP CONSERVE HEAT.

2. You are studying peppered moth populations in various locales. In order to determine whether light moths or dark moths survive better in different habitats, you mark 500 light moths and 500 dark moths and release them in different places.

If light moths survive better, you expect to recapture more LIGHT MOTHS.

Similarly, if dark moths survive better, you expect to recapture more DARK MOTHS.

Do you expect light moths or dark moths to survive better in the following habitats?

a. polluted areas DARK
b. unpolluted areas LIGHT
c. industrial centers before pollution laws were passed DARK
d. industrial centers some time after pollution laws were passed LIGHT
e. the countryside LIGHT

Chapter 17: The Evolution of Life

Speciation

1. What's the difference between a prezygotic reproductive barrier and a postzygotic reproductive barrier?
PREZYGOTIC REPRODUCTIVE BARRIERS PREVENT MEMBERS OF DIFFERENT SPECIES FROM MATING IN THE FIRST PLACE OR KEEP FERTILIZATION FROM OCCURRING IF THEY DO MATE. POSTZYGOTIC REPRODUCTIVE BARRIERS ACT AFTER FERTILIZATION HAS TAKEN PLACE. POSTZYGOTIC BARRIERS OCCUR WHEN MATING PRODUCES HYBRIDS THAT EITHER DON'T SURVIVE OR ARE STERILE—UNABLE TO BREED THEMSELVES.

2. Which of the following are prezygotic reproductive barriers and which are postzygotic reproductive barriers?
a. different courtship rituals in different bird species PREZYGOTIC
b. incompatible anatomical structures that prevent copulation in insects PREZYGOTIC
c. different-sounding calls by males in different frog species PREZYGOTIC
d. sterility in offspring produced when members of two different species mate POSTZYGOTIC
e. two species that mate at different times of year PREZYGOTIC
f. two species that use different mating sites PREZYGOTIC
g. offspring that are unable to survive when members of two species mate POSTZYGOTIC

3. What's the difference between allopatric speciation and sympatric speciation?
IN ALLOPATRIC SPECIATION NEW SPECIES ARE FORMED AFTER A GEOGRAPHIC BARRIER DIVIDES A SINGLE POPULATION INTO TWO. IN SYMPATRIC SPECIATION, SPECIATION OCCURS WITHOUT THE INTRODUCTION OF A GEOGRAPHIC BARRIER.

4. Are the following examples of allopatric or sympatric speciation?
a. speciation after developing glaciers divide a population ALLOPATRIC
b. speciation by hybridization SYMPATRIC
c. speciation after a river cuts through a population's habitat ALLOPATRIC
d. speciation after plate tectonics causes a continent to split ALLOPATRIC
e. speciation by polyploidy SYMPATRIC

Chapter 18: Biological Diversity

Classification

1. Linnaean classification groups species together based on ____.
SHARED SIMILARITIES

2. Fill in the levels of Linnaean classification from the largest group to the smallest group below.
Domain
____ KINGDOM
____ PHYLUM
____ CLASS
____ ORDER
____ FAMILY
____ GENUS
Species

3. A species' scientific name consists of its ____ name and its ____ name.
GENUS, SPECIES

4. Cladistic classification groups species together based on their ____.
EVOLUTIONARY HISTORY OR EVOLUTIONARY RELATIONSHIPS

5. The following cladogram shows evolutionary relationships between the rufous hummingbird, the honey mushroom, and the bristlecone pine.

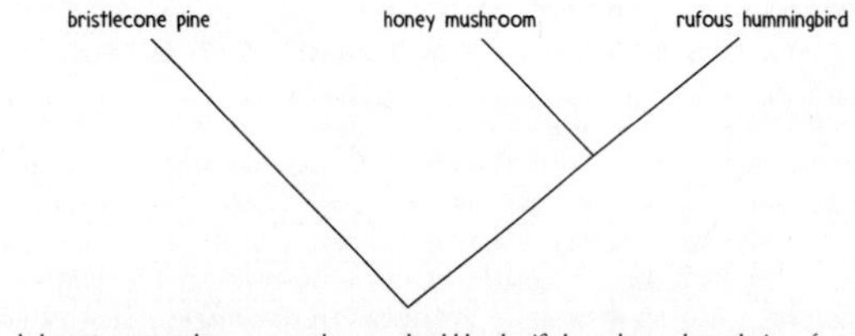

This cladogram suggests that ____ and ____ should be classified together to the exclusion of ____.
HONEY MUSHROOMS, RUFOUS HUMMINGBIRDS, BRISTLECONE PINES

Chapter 18: Biological Diversity

Biological Diversity I: Bacteria, Archaea, Protists

1. What are the three domains of life?

BACTERIA, ARCHAEA, EUKARYA

Of the three domains, BACTERIA and ARCHAEA consist of prokaryotes and EUKARYA consists of eukaryotes.

2. The four kingdoms that make up Eukarya are:

PLANTS, ANIMALS, PROTISTS, FUNGI

3. Are there any bacteria that can photosynthesize? Are there any heterotrophic bacteria?

YES, YES

4. How do bacteria typically reproduce?

BACTERIA TYPICALLY REPRODUCE ASEXUALLY BY DIVIDING.

5. Can bacteria exchange genetic material? If so, how?

YES, MOST SPECIES EXCHANGE GENETIC MATERIAL AT LEAST OCCASIONALLY, WHEN THEY TAKE UP SMALL PIECES OF NAKED DNA FROM THE ENVIRONMENT, WHEN BACTERIAL VIRUSES INADVERTENTLY TRANSFER DNA BETWEEN ORGANISMS, OR WHEN TWO BACTERIA JOIN TOGETHER AND ONE PASSES DNA TO THE OTHER.

6. Do the many bacteria that live in and on our bodies benefit us in any way?

YES, SOME PRODUCE VITAMINS AND OTHERS KEEP MORE DANGEROUS BACTERIA FROM INVADING OUR BODIES.

7. Are archaea more closely related to bacteria or to eukaryotes? What evidence supports this?

ARCHAEA ARE MORE CLOSELY RELATED TO EUKARYOTES THAN TO BACTERIA. MANY FEATURES OF ARCHAEAN GENETICS IN PARTICULAR LINK THEM TO EUKARYOTES—THEIR RIBOSOMES AND tRNA RESEMBLE THOSE OF EUKARYOTES, THEIR GENES CONTAIN INTRONS LIKE THOSE OF EUKARYOTES, AND THEIR DNA IS ASSOCIATED WITH HISTONE PROTEINS, LIKE THAT OF EUKARYOTES.

8. What's an extremophile? Are all archaea extremophiles?

EXTREMOPHILES ARE "LOVERS OF THE EXTREME," ORGANISMS THAT ARE ADAPTED TO EXTREME ENVIRONMENTS SUCH AS VERY SALTY PONDS OR THE SCALDING WATERS OF HOT SPRINGS AND HYDROTHERMAL VENTS. SOME ARCHAEA ARE EXTREMOPHILES—BUT NOT ALL. MANY OCCUR IN MORE FAMILIAR LOCALES, SUCH AS THE OPEN OCEAN OR THE DIGESTIVE TRACTS OF TERMITES AND COWS.

9. What is a chemoautotroph?

CHEMOAUTOTROPHS MAKE FOOD USING CHEMICAL ENERGY RATHER THAN ENERGY FROM SUNLIGHT. ARCHAEA IN HYDROTHERMAL VENT HABITATS, FOR EXAMPLE, OBTAIN ENERGY FROM A CHEMICAL ABUNDANT THERE, HYDROGEN SULFIDE.

10. Are each of the following groups of protists autotrophs or heterotrophs?

a. diatoms — AUTOTROPHS
b. amoebas — HETEROTROPHS
c. kelp — AUTOTROPHS
d. dinoflagellates — AUTOTROPHS OR HETEROTROPHS
e. *Plasmodium* (the protist that causes malaria) — HETEROTROPHS

Chapter 18: Biological Diversity

Biological Diversity II: Plants, Fungi, Animals

1. Match the following plant structures with their function:

Structure	Answer	Function
stomata	B	a. move water and nutrients up from the roots
roots	F	b. take in carbon dioxide
shoots	C	c. conduct photosynthesis
xylem	A	d. move sugars produced during photosynthesis
phloem	D	e. transport resources to different parts of plant
vascular system	E	f. absorb water and nutrients from soil

2. The life history of plants involves a(n) ______ in which plants move between a haploid ______ stage and a diploid ______ stage.

ALTERNATION OF GENERATIONS, GAMETOPHYTE, SPOROPHYTE

3. Mosses are unique among plants in that the ______ is much larger than the ______. When you see a moss in the forest, you are looking at a______. The sperm are released by the male ______ directly into the environment, where they use their flagella to swim through a film of water to eggs in the female gametophyte. Sperm and egg then fuse and grow into a tiny (haploid/diploid) ______ that is completely dependent on the female gametophyte for nutrients and water. Eventually, cells in the sporophyte undergo meiosis to produce (haploid/diploid) spores that scatter and grow into new ______ (moss plants).

GAMETOPHYTE; SPOROPHYTE; GAMETOPHYTE; GAMETOPHYTE; DIPLOID; SPOROPHYTE; HAPLOID; GAMETOPHYTES

4. Why are ferns less tied to a moist environment than mosses? Why are they more tied to a moist environment than seed plants?

FERNS ARE LESS TIED TO A MOIST ENVIRONMENT THAN MOSSES BECAUSE, UNLIKE MOSSES, THEY HAVE A VASCULAR SYSTEM FOR TRANSPORTING WATER AND NUTRIENTS AROUND THE PLANT. HOWEVER, THEY ARE MORE TIED TO MOIST ENVIRONMENTS THAN SEED PLANTS, BECAUSE UNLIKE SEED PLANTS, THEY HAVE SPERM THAT MUST SWIM THROUGH THE ENVIRONMENT TO FERTILIZE EGGS.

5. What is pollen? What is a seed? What is a fruit? In what groups of plants are each of these structures found?

POLLEN CONSISTS OF IMMATURE MALE GAMETOPHYTES WRAPPED IN PROTECTIVE COATINGS. POLLEN IS TRANSPORTED TO THE FEMALE GAMETOPHYTE DURING PLANT REPRODUCTION. SEEDS ARE SMALL EMBRYONIC SPOROPHYTES THAT ARE ENCASED IN A TOUGH OUTER COATING ALONG WITH A FOOD SUPPLY. SEEDS ARE ABLE TO SURVIVE IN A DORMANT STATE, DURING WHICH GROWTH AND DEVELOPMENT ARE SUSPENDED, UNTIL CONDITIONS ARE FAVORABLE FOR GROWTH. FRUITS ARE STRUCTURES SURROUNDING THE SEEDS OF FLOWERING PLANTS THAT HELP DISPERSE THE SEEDS. POLLEN AND SEEDS ARE FOUND IN SEED PLANTS. FRUITS ARE FOUND IN FLOWERING PLANTS.

6. Fungi are autotrophs/heterotrophs/both. Fungi are unicellular/multicellular/both. Fungi reproduce sexually/asexually/both.

HETEROTROPHS; BOTH; BOTH

7. Match the following animal groups with the list of features. Some groups may have more than one feature.

sponges	C, K
cnidarians	F, N
flatworms	U
roundworms	E
arthropods	I, P
mollusks	A
annelids	H, G
echinoderms	M, R
chordates	Q
catilaginous fishes	S
ray-finned fishes	B
amphibians	J, Y
reptiles	L
turtles	W
snakes	D
birds	O, X
mammals	T, V

a. muscular foot responsible for locomotion
b. swim bladder
c. the only animals that lack tissues
d. adaptations for subduing large prey and swallowing them whole
e. muscles all run longitudinally—from head to tail—down the body, resulting in a flailing whiplike motion
f. polyp stage and medusa stage alternate
g. leeches
h. segmented worms
i. segmented bodies and jointed legs
j. terrestrial vertebrates restricted to moist environments because their skins are composed of living cells that are vulnerable to drying out, and lay eggs without shells
k. maintain a constant flow of water in through numerous pores, into the central cavity, and out the top, whose purpose is for food capture
l. birds and crocodiles
m. tube feet
n. tentacles armed with barbed stinging cells
o. hollow bones, air sacs in the body, and a four-chambered heart
p. includes the most diverse group of living things on Earth, the insects
q. a notochord, gill slits, and a tail that extends beyond the anus
r. starfish
s. have a skeleton made of cartilage
t. have hair and feed their young milk
u. tapeworms
v. platypus
w. squeezes its entire body inside its ribcage
x. flying endotherms
y. frogs

Chapter 19: Human Biology I—Control and Development

The Nervous System

1. Match parts of the brain with the body functions they are responsible for. Some parts may correspond with more than one function:

brainstem	D	a. processing visual information
cerebellum	B	b. balance, posture, coordination
frontal lobes of cerebrum	F, I	c. sensory areas for temperature, touch, and pain
parietal lobes of cerebrum	C	d. involuntary activities such as heartbeat, respiration, and digestion
occipital lobes of cerebrum	A	e. comprehending language
temporal lobes of cerebrum	E	f. control of voluntary movements
thalamus	G	g. sorting and filtering of incoming information, which it then passes to the cerebral cortex
hypothalamus	H, J	h. emotions, such as pleasure and rage, and bodily drives, such as hunger and thirst
		i. speech
		j. body's internal clock

2. The two divisions of the nervous system are the _________ and _________. The central nervous system consists of the _________ and _________.

_________ carry messages from the senses to the central nervous system. _________ connect neurons to other neurons. _________ carry messages from the central nervous system to muscle cells or to other responsive organs.

Motor neurons are further divided into two groups, the _________, which controls voluntary actions and stimulates our voluntary muscles, and the _________, which controls involuntary actions and stimulates involuntary muscles and other internal organs.

The autonomic nervous system includes a _________ that promotes a "fight or flight" response and a _________ that operates in times of relaxation.

CENTRAL NERVOUS SYSTEM; PERIPHERAL NERVOUS SYSTEM; BRAIN; SPINAL CORD; SENSORY NEURONS; INTERNEURONS; MOTOR NEURONS; SOMATIC NERVOUS SYSTEM; AUTONOMIC NERVOUS SYSTEM; SYMPATHETIC DIVISION; PARASYMPATHETIC DIVISION

The Nervous System—continued

3.

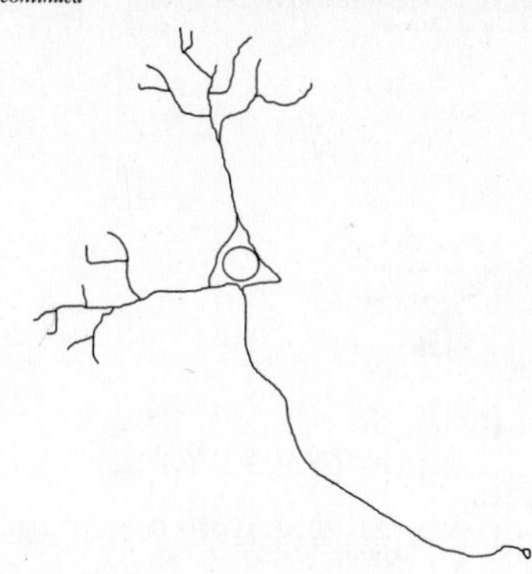

Identify the dendrites, cell body, and axon in the neuron above. What is the function of each of these parts of a neuron?

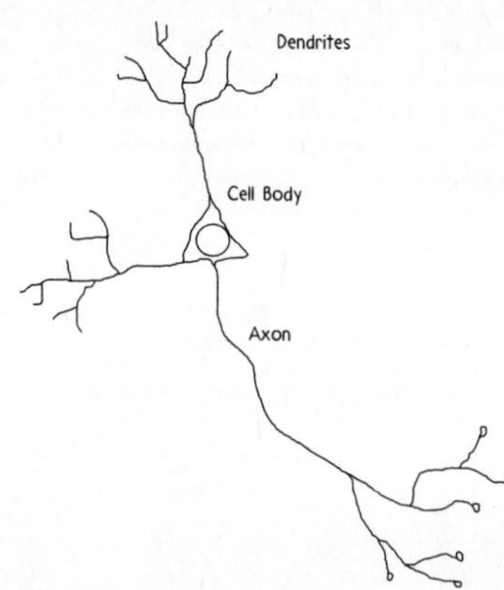

DENDRITES RECEIVE INFORMATION FROM OTHER NEURONS OR CELLS. THE CELL BODY CONTAINS THE NUCLEUS AND ORGANELLES THAT THE CELL NEEDS TO STAY ALIVE AND FUNCTION. THE AXON TRANSMITS INFORMATION TO OTHER NEURONS OR CELLS.

The Nervous System—continued

4. Place the following events that occur during an action potential in the correct order:

membrane potential decreases	1. MEMBRANE POTENTIAL INCREASES
membrane potential increases	2. MEMBRANE POTENTIAL REACHES THRESHOLD
potassium channels open	3. SODIUM CHANNELS OPEN
sodium channels open	4. SODIUM IONS FLOW INTO NEURON
membrane potential spikes	5. MEMBRANE POTENTIAL SPIKES
membrane potential reaches threshold	6. POTASSIUM CHANNELS OPEN
membrane potential returns to resting potential	7. POTASSIUM IONS FLOW OUT OF NEURON
sodium ions flow into neuron	8. MEMBRANE POTENTIAL DECREASES
potassium ions flow out of neuron	9. MEMBRANE POTENTIAL RETURNS TO RESTING POTENTIAL

5. What is the difference between an electric synapse and a chemical synapse? How does each allow a neuron to signal a target cell?

IN ELECTRICAL SYNAPSES, IONS FLOW DIRECTLY FROM A NEURON TO A TARGET CELL THROUGH GAP JUNCTIONS, TINY CHANNELS IN THE CELL MEMBRANE. IN A CHEMICAL SYNAPSE, THERE IS A NARROW SPACE BETWEEN THE NEURON AND ITS TARGET CELL. WHEN THE ACTION POTENTIAL ARRIVES AT THE END OF THE AXON, THE NEURON RELEASES CHEMICAL MESSENGERS CALLED NEUROTRANSMITTERS INTO THIS SPACE. THE NEUROTRANSMITTERS ARE RELEASED THROUGH EXOCYTOSIS; THAT IS, SMALL VESICLES CONTAINING THE NEUROTRANSMITTERS FUSE TO THE NEURON'S CELL MEMBRANE AND RELEASE THEIR CONTENTS OUTSIDE THE CELL. THE NEUROTRANSMITTERS THEN DIFFUSE ACROSS THE SPACE BETWEEN THE NEURON AND ITS TARGET CELL AND BIND TO RECEPTOR PROTEINS ON THE CELL MEMBRANE OF THE TARGET CELL. THE BINDING OF NEUROTRANSMITTERS CAUSES THE RECEPTOR PROTEINS TO CHANGE, OPENING ION CHANNELS AND ALLOWING IONS TO FLOW INTO THE TARGET CELL.

Chapter 19: Human Biology I—Control and Development

Senses

1. The light-sensitive cells in the eye are found in the _________. There are two types of light-sensitive cells, _________ and _________.

RETINA; RODS AND CONES

2. Does each of the following describe rods or cones?

responsible for vision at night or in dim light	RODS
detect color	CONES
3 types	CONES
cannot discriminate colors	RODS
very sensitive to light, responding to even a single photon	RODS
not very good at making out fine details	RODS
nonfunctional form of these causes colorblindness	CONES
cone-shaped	CONES

3. How does sound move through the ear? Place the following items in the correct order:

middle ear bones	1. PINNA
pinna	2. EARDRUM
cochlea	3. MIDDLE EAR BONES
eardrum	4. COCHLEA

4. What are the five basic tastes?

SWEET, SALTY, SOUR, BITTER, AND UMAMI

5. What is the role of prostaglandins in sensing pain?

PAIN RECEPTORS RESPOND TO STIMULI THAT CAUSE DAMAGE TO THE BODY. THEY GENERALLY REQUIRE STRONG STIMULATION BEFORE THEY WILL RESPOND. HOWEVER, DAMAGED TISSUES RELEASE CHEMICALS CALLED PROSTAGLANDINS THAT INCREASE THE SENSITIVITY OF PAIN RECEPTORS.

Chapter 19: Human Biology I—Control and Development

Skeleton and Muscles

1. Label the three layers of bone below:

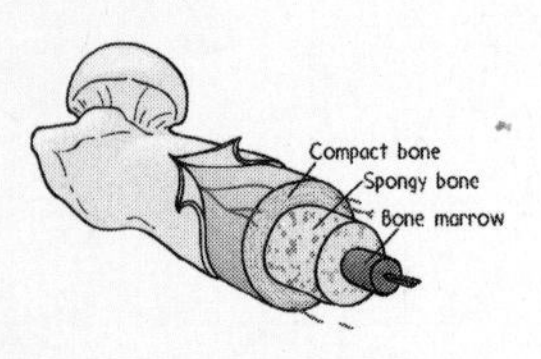

What are the two types of bone marrow?

RED AND YELLOW

What is the function of each type?

RED BONE MARROW PRODUCES RED AND WHITE BLOOD CELLS. YELLOW BONE MARROW STORES FAT.

2. Muscles are made up of a series of contractile units called ________. These contain carefully arranged fibers of two proteins, thin filaments called ________ and thick filaments called ________. When an action potential arrives at a muscle cell, ________ are released from the cell's ________. These ions allow a series of pivoting heads on the ______ fibers to attach to ________. The heads attach and pivot, ________ the length of the sarcomere a tiny bit—about 10 nanometers, to be exact—and, consequently, the length of the muscle as a whole. After pulling, the heads release, recock, reattach, and pull again. This cycle repeats until ____________________. Muscle contractions require energy, of course. ATP is required for______________, an essential step in the contraction cycle.

SARCOMERES; ACTIN; MYOSIN; CALCIUM IONS; ENDOPLASMIC RETICULUM; MYOSIN; ACTIN; REDUCING; THE SIGNAL TO CONTRACT ENDS OR UNTIL THE MUSCLE HAS FULLY CONTRACTED; THE MYOSIN HEADS TO RELEASE ACTIN

Chapter 20: Human Biology II—Care and Maintenance

Circulatory System

1. Each heartbeat begins in a part of the right atrium called the ________, or pacemaker. The pacemaker initiates an action potential that sweeps quickly through the____________, which contract simultaneously. The signal also passes to the__________, and from there to the two________, which also contract simultaneously.

SINOATRIAL NODE; RIGHT AND LEFT ATRIA; ATRIOVENTRICULAR NODE; VENTRICLES

2. Why does the heart make a "lub-dubb" sound as it beats? What is the "lub"? What is the "dubb"?

THE "LUB" IS THE SOUND OF VALVES BETWEEN THE ATRIA AND VENTRICLES SNAPPING SHUT AFTER THE ATRIA CONTRACT. THE "DUBB" IS THE SOUND OF THE VALVES BETWEEN THE VENTRICLES AND OUTGOING ARTERIES SNAPPING SHUT AFTER THE CONTRACTION OF THE VENTRICLES.

3. How does blood flow around the body? Place the following in the correct order, beginning with the right atrium:

Arteries to lungs	RIGHT ATRIUM
Right atrium	RIGHT VENTRICLE
Left atrium	ARTERIES TO LUNGS
Veins from body tissues	CAPILLARIES AROUND ALVEOLI
Veins from lungs	VEINS FROM LUNGS
Capillaries that supply body tissues	LEFT ATRIUM
Arteries to body tissues	LEFT VENTRICLE
Arterioles	ARTERIES TO BODY TISSUES
Right ventricle	ARTERIOLES
Venules	CAPILLARIES THAT SUPPLY BODY TISSUES
Left ventricle	VENULES
Capillaries around alveoli	VEINS FROM BODY TISSUES

4. The three types of cells found in blood are________, ________, and ________. ________ carry oxygen. ________ are part of the immune system and help our bodies defend against disease. ________ are involved in blood clotting.

RED BLOOD CELLS; WHITE BLOOD CELLS; PLATELETS; RED BLOOD CELLS; WHITE BLOOD CELLS; PLATELETS

5. The molecule in red blood cells that carries oxygen is ________. It can carry up to ________ oxygen molecules, each bound to a(n) ________ atom.

HEMOGLOBIN; 4; IRON

Chapter 20: Human Biology II—Care and Maintenance

Respiration, Digestion

1. Match each part of the respiratory system with its description:

nasal passages	E	a. site of gas exchange
larynx	C	b. another word for trachea
trachea	H	c. allows us to speak
alveoli	A	d. raises ribcage during breathing
diaphragm	F	e. air is warmed and moistened
muscles between ribs	D	f. dome-shaped muscle involved in breathing
bronchi	G	g. tubes leading to right and left lungs
windpipe	B	h. stiffened by cartilaginous rings

2. Where does each of the following events important in digestion occur?

a. bile is produced here	LIVER
b. a highly acidic mix of hydrochloric acid and digestive enzymes is added	STOMACH
c. absorption of nutrients	SMALL INTESTINE (BEYOND THE DUODENUM)
d. food is chewed, breaking it into smaller pieces	MOUTH
e. reabsorption of water	LARGE INTESTINE
f. food is churned by muscular action	STOMACH
g. saliva begins digestion of starches in food	MOUTH
h. bile is added to food here	SMALL INTESTINE (DUODENUM)
i. pancreatic enzymes are added to food	SMALL INTESTINE (DUODENUM)
j. home to large numbers of *Escherichia coli* and other bacteria	LARGE INTESTINE
k. bile is stored here	GALL BLADDER
l. absorption of many minerals	LARGE INTESTINE
m. synthesis of vitamin K and some of the B vitamins by bacteria	LARGE INTESTINE

Chapter 20: Human Biology II—Care and Maintenance

Excretory System, Immune System

1. We excrete nitrogen-containing wastes in the form of ____.

UREA

2. The functional unit of a kidney is the ________. Each of these units is associated with a cluster of capillaries called the ________. Blood pressure in the ________ pushes fluid out of the capillaries and into ________. This fluid is called the________ and is pretty similar to________. From Bowman's capsule, the filtrate flows into the ____________. After "good" molecules are removed from the filtrate and "bad" molecules are added to it, the filtrate moves to the ________, whose primary function is to ________. Then the filtrate moves into the ________, where additional wastes are transported into it. Finally, the filtrate moves down the ________, where ________ may be absorbed if ________ is present. Urine drips into the ________ and flows down the ________ to the ________, where it is temporarily stored. Finally, urine flows down the ________ and out the body.

NEPHRON; GLOMERULUS; GLOMERULUS; BOWMAN'S CAPSULE; FILTRATE; BLOOD PLASMA; PROXIMAL CONVOLUTED TUBULE; LOOP OF HENLE; REABSORB WATER FROM THE FILTRATE; DISTAL CONVOLUTED TUBULE; COLLECTING DUCT; WATER; ANTIDIURETIC HORMONE; RENAL PELVIS; URETER; BLADDER; URETHRA

3. Is each of the following associated with innate or acquired immunity?

skin	INNATE
T cells	ACQUIRED
B cells	ACQUIRED
acidic secretions from hair follicles in skin	INNATE
antibodies	ACQUIRED
enzymes in tears and milk	INNATE
on the order of 10 million different receptors	ACQUIRED
mucus	INNATE
large Y-shaped proteins	ACQUIRED
memory cells	ACQUIRED
immediate response	INNATE
antigen	ACQUIRED
inflammatory response	INNATE
clones	ACQUIRED
histamine	INNATE
vaccines	ACQUIRED
maximum response delayed	ACQUIRED

Chapter 21: Ecosystems and Environment

Species Interactions

Define the following terms.

1. A population is A POPULATION IS A GROUP OF INDIVIDUALS OF A SINGLE SPECIES THAT OCCUPIES A GIVEN AREA.

   A community is A COMMUNITY CONSISTS OF ALL THE ORGANISMS THAT LIVE WITHIN A GIVEN AREA.

   An ecosystem is AN ECOSYSTEM CONSISTS OF ALL THE ORGANISMS THAT LIVE WITHIN A GIVEN AREA AND ALL THE ABIOTIC FEATURES OF THEIR ENVIRONMENT.

2. Look at the food chain below:

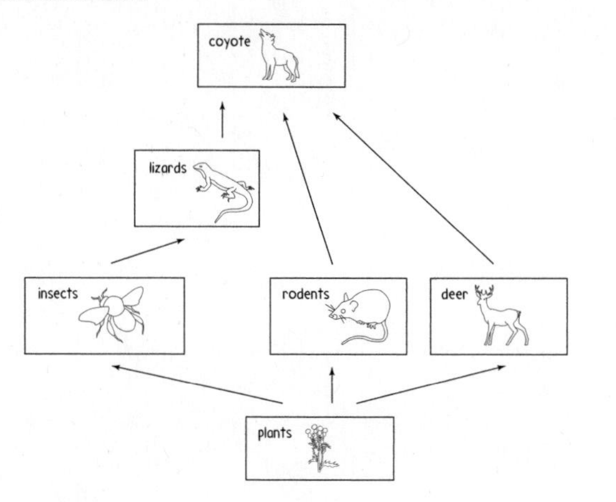

a. Who are the producers in this community?
PLANTS

b. Who are the primary consumers?
INSECTS, RODENTS, DEER

c. Who are the secondary consumers?
LIZARDS AND COYOTE

d. Who are the tertiary consumers?
COYOTE

e. Who is the top predator?
COYOTE

3. What is a species's niche?
THE TOTAL SET OF BIOTIC AND ABIOTIC RESOURCES IT USES WITHIN A COMMUNITY.

4. What is symbiosis?
A SITUATION IN WHICH INDIVIDUALS OF TWO SPECIES LIVE IN CLOSE ASSOCIATION WITH ONE ANOTHER.

Of the three types of symbiosis, ________ benefits one member of the interaction and harms the other, ________ benefits one species of the interaction while having no effect on the other, and ________ benefits both species.

PARASITISM; COMMENSALISM; MUTUALISM

Chapter 21: Ecosystems and Environment

Ecosystems

1. Match each of the following features with the appropriate biome or aquatic life zone:

Biome / zone	Answer	Feature
tropical forest	I	a. tropical grassland
temperate forest	L	b. habitats that receive very little precipitation, may be cold or hot
coniferous forest	F	c. permafrost
tundra	C	d. plants that are adapted to surviving in changing salinity conditions
savanna	A	e. in the water column
temperate grassland	J	f. evergreen trees with needlelike leaves
desert	B	g. inhabitants have adaptations that allow them to deal with exposure, temperature fluctuations, and the action of waves
littoral zone	K	h. deep water habitats in ponds and lakes
limnetic zone	M	i. more species are found in this biome than in all other biomes combined
profundal zone	H	j. grassland with fertile soil, in areas with four distinct seasons
estuary	D	k. lake habitat close to the water surface and to shore
pelagic zone	E	l. trees drop their leaves in the autumn
benthic zone	N	m. lake and pond habitats that are close to the water surface, but far from shore
intertidal zone	G	n. on the ocean bottom
neritic zone	P	o. marine habitats far from coasts
oceanic zone	O	p. underwater marine habitats near the coasts

2. Carbon is an essential component of all organic molecules. Most of the inorganic carbon on Earth exists as ________ and is found either in the ________ or dissolved in ________. Carbon moves into the biotic world when ________________________________. This carbon becomes available to other organisms as it passes up the food chain. Carbon is returned to the environment by living organisms as ________ during the process of ________. An important part of Earth's carbon supply is also found in fossil fuels such as ________. Human burning of fossil fuels has released so much carbon dioxide that atmospheric carbon dioxide levels are now higher than they have been for 420,000 years. Because atmospheric carbon dioxide traps heat on the planet, this has resulted in ________.

CARBON DIOXIDE; ATMOSPHERE; OCEAN WATERS;
PLANTS AND OTHER PRODUCERS CONVERT CARBON DIOXIDE TO GLUCOSE DURING PHOTOSYNTHESIS;
CARBON DIOXIDE; CELLULAR RESPIRATION; COAL AND OIL; GLOBAL WARMING

3. How do living things obtain water?
WATER MOVES INTO THE BIOTIC WORLD WHEN IT IS ABSORBED OR SWALLOWED BY ORGANISMS.

   How do living things return water to the abiotic world?
WATER IS RETURNED TO THE ABIOTIC ENVIRONMENT IN A VARIETY OF WAYS, INCLUDING THROUGH RESPIRATION, PERSPIRATION, EXCRETION, AND ELIMINATION.

4. The difference between primary succession and secondary succession is that PRIMARY SUCCESSION DESCRIBES THE COLONIZATION OF BARE LAND DEVOID OF SOIL, WHEREAS SECONDARY SUCCESSION OCCURS WHEN A DISTURBANCE DESTROYS EXISTING LIFE IN A HABITAT, BUT LEAVES SOIL INTACT.

   During ecological succession, the total biomass of the ecosystem typically ________, and the number of species present in the habitat typically ________. INCREASES; INCREASES

   Ecological succession ends with the ________.
CLIMAX COMMUNITY

5. What does the intermediate disturbance hypothesis state?
REGULAR DISTURBANCES, IF NOT TOO EXTREME, ACTUALLY CONTRIBUTE TO BIODIVERSITY BECAUSE DIFFERENT SPECIES MAKE USE OF DIFFERENT HABITATS, AND PERIODIC DISTURBANCES GUARANTEE THAT THERE WILL ALWAYS BE HABITAT AT VARYING STAGES OF RECOVERY.

Chapter 21: Ecosystems and Environment

Populations

1. Which of the following graphs shows exponential growth and which shows logistic growth?

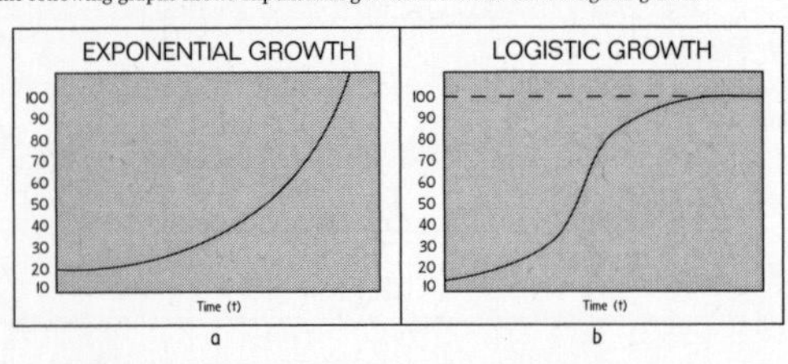

Using words, describe the difference between exponential growth and logistic growth.

EXPONENTIAL GROWTH OCCURS WHEN A POPULATION GROWS AT A RATE THAT IS PROPORTIONAL TO ITS SIZE. *LOGISTIC GROWTH* OCCURS WHEN POPULATION GROWTH SLOWS AS IT REACHES THE HABITAT'S CARRYING CAPACITY, THAT IS, THE MAXIMUM NUMBER OF INDIVIDUALS OR MAXIMUM POPULATION DENSITY THE HABITAT CAN SUPPORT.

What is the carrying capacity in the graph on the right?

100

2. Label the three survivorship curves shown below. Indicate which organisms at right correspond to which survivorship curve.

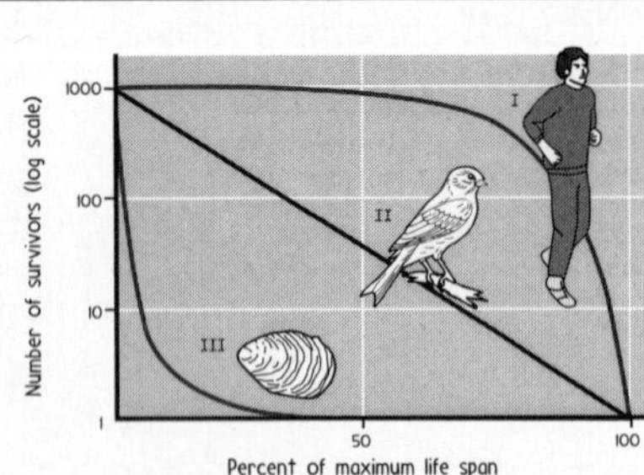

Using words, describe Type I, Type II, and Type III survivorship.

TYPE I ORGANISMS HAVE LOW DEATH RATES EARLY IN LIFE, WITH MOST INDIVIDUALS SURVIVING UNTIL FAIRLY LATE IN LIFE. TYPE II ORGANISMS HAVE A STEADY DEATH RATE THAT DOES NOT DEPEND ON AGE. INDIVIDUALS ARE AS LIKELY TO DIE EARLY IN LIFE AS LATE IN LIFE. TYPE III ORGANISMS HAVE HIGH DEATH RATES EARLY IN LIFE, WITH FEW INDIVIDUALS SURVIVING UNTIL LATE IN LIFE.

Populations—continued

3. Of the following traits, indicate which are associated with K-selected populations and which are associated with r-selected populations.

unstable environment	R-SELECTED POPULATIONS
large body size	K-SELECTED POPULATIONS
few offspring	K-SELECTED POPULATIONS
no parental care	R-SELECTED POPULATIONS
reach sexual maturity slowly	K-SELECTED POPULATIONS
long life expectancy	K-SELECTED POPULATIONS
exponential population growth	R-SELECTED POPULATIONS
Type III survivorship	R-SELECTED POPULATIONS
parental care	K-SELECTED POPULATIONS
stable environment	K-SELECTED POPULATIONS
short life expectancy	R-SELECTED POPULATIONS
small body size	R-SELECTED POPULATIONS
Type I survivorship	K-SELECTED POPULATIONS

Chapter 22: Plate Tectonics

Faults

Three block diagrams are illustrated below. Draw arrows on each diagram to show the direction of movement. Answer the questions next to each diagram.

A.

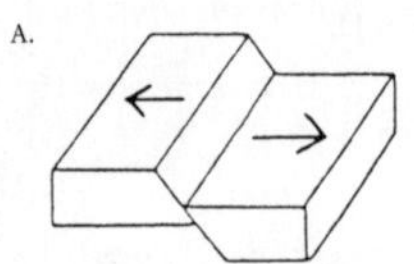

What type of force produced Fault A?

TENSIONAL

Name the fault NORMAL FAULT

Where would you expect to find this type of fault?

NEVADA, PARTS OF CA, OR, IDAHO AND UTAH

B.

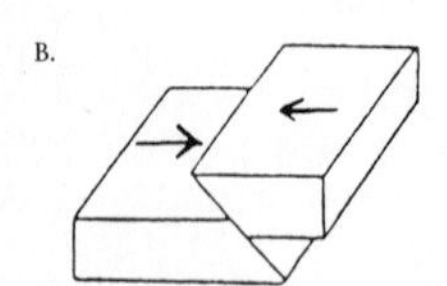

What type of force produced Fault B?

COMPRESSIONAL

Name the fault REVERSE OR THRUST FAULT

Where would you expect to find this type of fault?

ROCKY MTN FORELAND, APPALACHIAN MTNS

C.

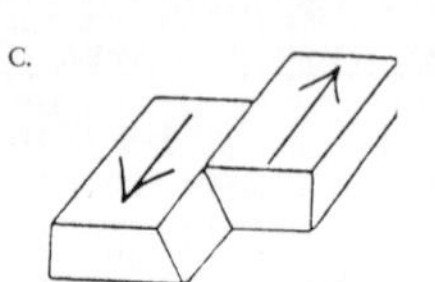

What type of force produced Fault C?

SHEAR

Name the fault HORIZONTAL MOVEMENT FAULT

Where would you expect to find this type of fault?

CALIFORNIA—THE SAN ANDREAS FAULT

Chapter 22: Plate Tectonics

Structural Geology

Much subsurface information is learned by oil companies when wells are drilled. Some of this information leads to the discovery of oil, and some reveals subsurface structures such as folds and/or faults in Earth's crust.

Four oil wells that have been drilled to the same depth are shown on the cross section below. Each well encounters contacts between different rock formations at the depths shown in the table below. (A contact is simply the boundary between two types or ages of rock). Rock formations are labeled A–F, with youngest A and oldest F.

	Depth to Contact (in meters)			
Contact	Oil Well #1	Oil Well #2	Oil Well #3	Oil Well #4
A-B	200	not encountered	200	not encountered
B-C	400	100	400	100
C-D	600	300	600	300
D-E	800	500	800	500
E-F	1000	700	1000	700

1. In the cross section below, Contacts D–E and E–F are plotted for Oil Wells 1 and 2. Plot the remainder of the data for all four wells, labeling each point you plot.
2. Draw lines to connect the contacts between the rock formations (as is done for Contacts D–E and E–F for Oil Wells 1 and 2).

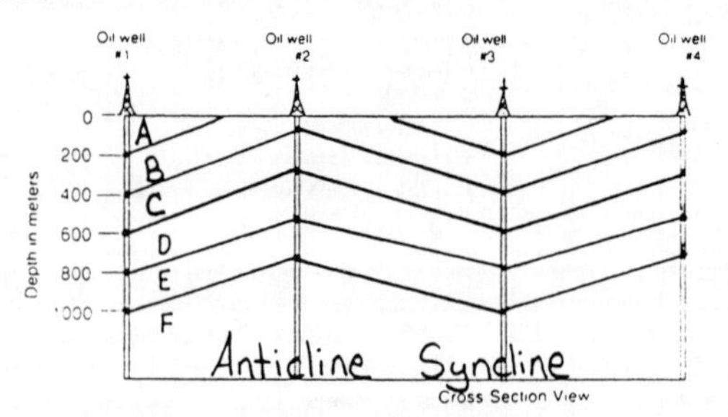

Questions

1. What explanation can you offer for no sign of Formation A in Oil Wells 2 and 4?

FORMATION A HAS BEEN ERODED AWAY.

2. What geological structures are revealed? Label them on the cross section. (Hint: Consider the shapes of the contacts.)

THE STRUCTURES ARE FOLDED. AN ANTICLINE IS EVIDENT AT WELL #2, A SYNCLINE AT WELL #3.

Chapter 22: Plate Tectonics

Plate Boundaries

Draw arrows on the plate boundaries **A, B,** and **C**, to show the relative direction of movement.

Type of plate boundary for A? CONVERGENT

What type of force generates this type of boundary?

COMPRESSIONAL

Is this a site of crustal formation, destruction, or crustal transport?

DESTRUCTION

Type of plate boundary for B? TRANSFORM FAULT

What type of force generates this type of boundary?

SHEAR

Is this a site of crustal formation, destruction, or crustal transport?

TRANSPORT

Type of plate boundary for C? DIVERGENT

What type of force generates this type of boundary?

TENSIONAL

Is this a site of crustal formation, destruction, or crustal transport?

FORMATION

Draw arrows on the transform faults below to indicate relative motion.

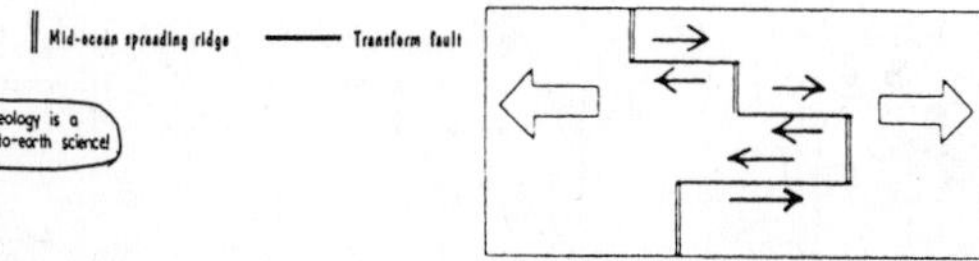

Geology is a down-to-earth science!

Chapter 22: Plate Tectonics

Seafloor Spreading

The rate of seafloor spreading is from 1 to 10 centimeters per year. If we know the distance and age between two points on the ocean floor, we can determine the rate of spreading. Diagrams A, B, and C show stages of sea floor spreading. Spreading begins at A, continues to B where rocks at location P begin to spread to the farther-apart positions we see in C. At C newer rock at the ocean crest S dated at 10 million years. Using the scale: 1 mm = 50 km, use a ruler on C to find the

1. separation rate of the two continental landmasses in the past 10 million years, in cm/yr: 2.5
2. age of the seafloor at P in Diagram C (1 cm/yr = 10 km/million years) 92 MILLION YEARS

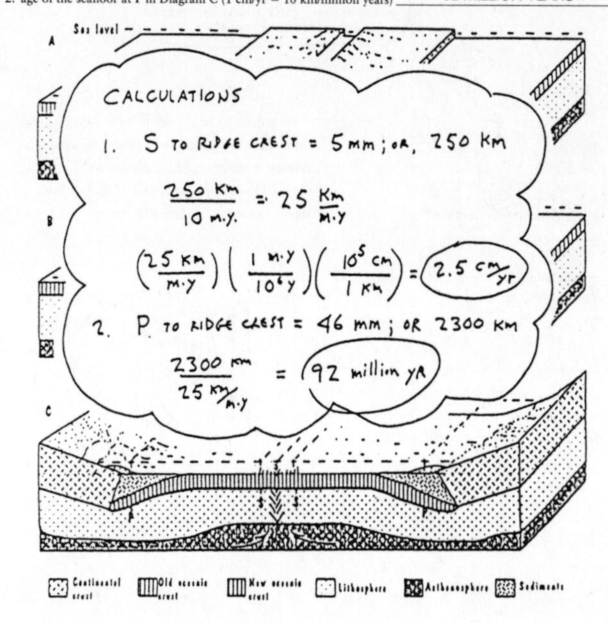

Sea level
A
B
C
CALCULATIONS
1. S TO RIDGE CREST = 5 mm; OR, 250 Km
250 Km / 10 m.y. = 25 Km/m.y
(25 Km/m.y)(1 m.y/10^6 y)(10^5 cm/1 Km) = 2.5 cm/yr
2. P TO RIDGE CREST = 46 mm; OR 2300 Km
2300 km / 25 km/m.y = 92 million yr
Continental crust
Old oceanic crust
New oceanic crust
Lithosphere
Asthenosphere
Sediments

Chapter 22: Plate Tectonics

Plate Boundaries and Magma Generation

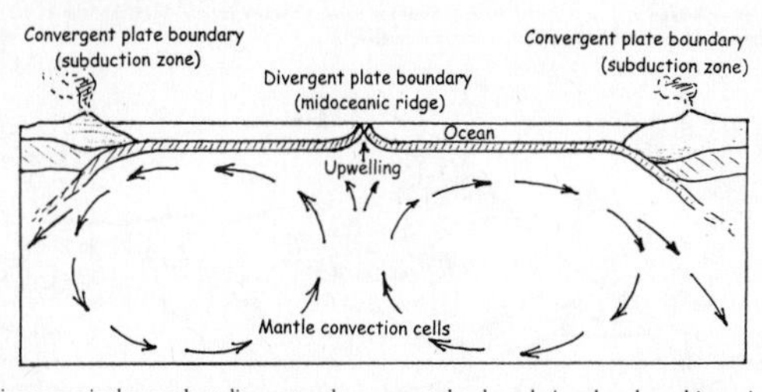

Partial melting occurs in the mantle at divergent and convergent plate boundaries when the melting point of mantle rocks is lowered.

1. What is the mechanism that lowers the melting point of mantle rock at divergent boundaries?

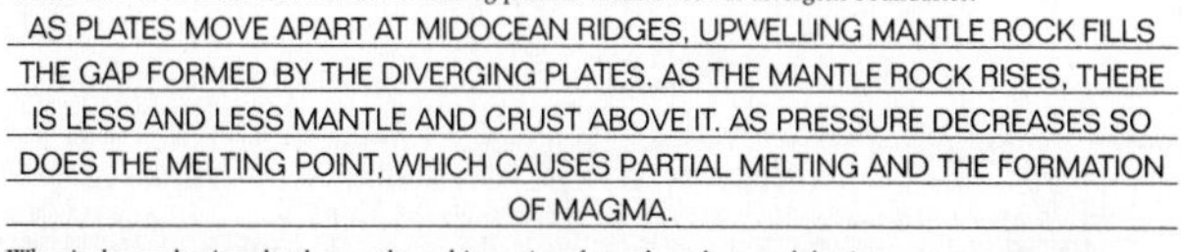

AS PLATES MOVE APART AT MIDOCEAN RIDGES, UPWELLING MANTLE ROCK FILLS THE GAP FORMED BY THE DIVERGING PLATES. AS THE MANTLE ROCK RISES, THERE IS LESS AND LESS MANTLE AND CRUST ABOVE IT. AS PRESSURE DECREASES SO DOES THE MELTING POINT, WHICH CAUSES PARTIAL MELTING AND THE FORMATION OF MAGMA.

2. What is the mechanism that lowers the melting point of mantle rock at a subduction zone at a convergent boundary?

AS THE SUBDUCTING PLATE DESCENDS FURTHER AND FURTHER INTO THE MANTLE, PRESSURE AND TEMPERATURE INCREASE, DRIVING WATER IN THE SUBDUCTING PLATE UP INTO THE MANTLE AND LOWERING ITS MELTING POINT, LIKE FLUX AT A FOUNDRY.

Chapter 23: Rocks and Minerals

Chemical Structure and Formulas of Minerals

Out of more than 3000 minerals known today, only about two dozen are abundant. Minerals are classified by their chemical composition and atomic structure and are divided into groups.

For each mineral structure diagrammed below, look for a pattern in the structure, count the number of atoms (ions) in each, and fill in the blanks.

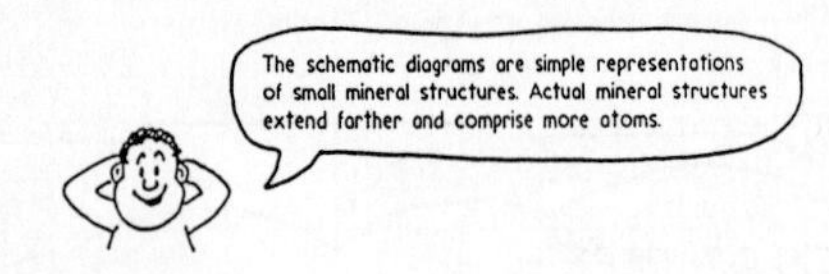

The schematic diagrams are simple representations of small mineral structures. Actual mineral structures extend farther and comprise more atoms.

Cl

Na

1. Circle pairs of Na and Cl ions in the structure and add any ion(s) needed to complete pairing. This mineral structure contains __14__ Na ions and __14__ Cl ions. The mineral's formula is __NaCl__. This mineral belongs to the __HALIDE__ group.

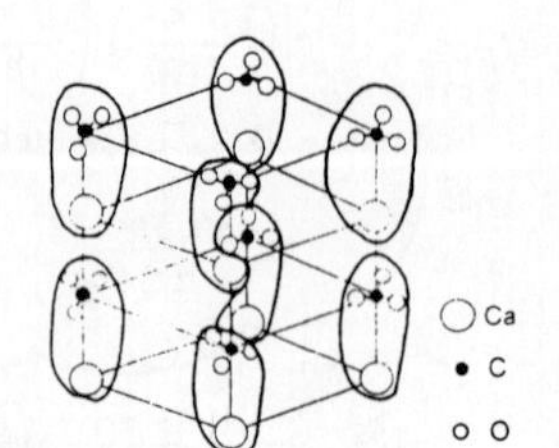

2. This mineral structure contains __8__ Ca atoms, __8__ C atoms, and __24__ O atoms. The mineral's formula is __$CaCO_3$__. This mineral belongs to the ________ group. CARBONATE

Chemical Structure and Formulas of Minerals—continued

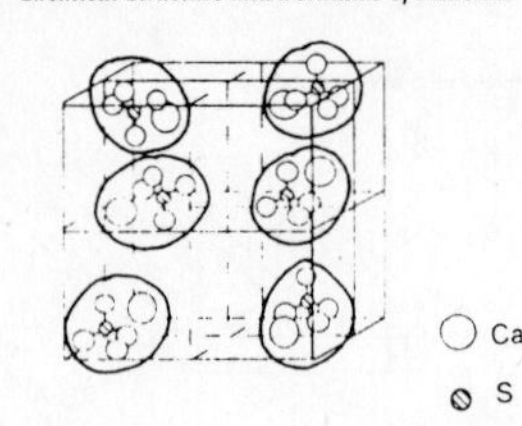

3. This mineral structure contains __6__ Ca atoms, __6__ S atoms, and __24__ O ions. The mineral's formula is __$CaSO_4$__. This mineral belongs to the __SULFATE__ group.

4. This mineral structure contains __14__ Fe atoms and __26__ S atoms. Complete the structure by adding the needed atoms(s). The mineral's formula is __FeS_2__. This mineral belongs to the __SULFITE__ group.

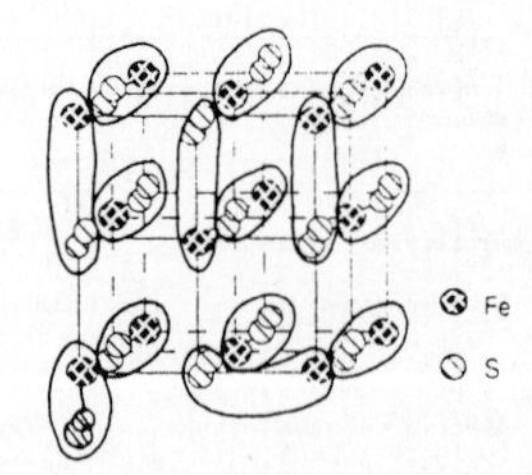

Ca

F

5. Complete the mineral structure so that each Ca atom is linked to two F atoms. Now the mineral structure contains __14__ Ca atoms and __28__ F atoms. The mineral's formula is __CaF_2__. This mineral belongs to the __HALIDE__ group.

Chapter 23: Rocks and Minerals

The Rock Cycle

Complete the illustration at right, which depicts the different paths in the rock cycle. Insert arrows to show the direction of pathways.

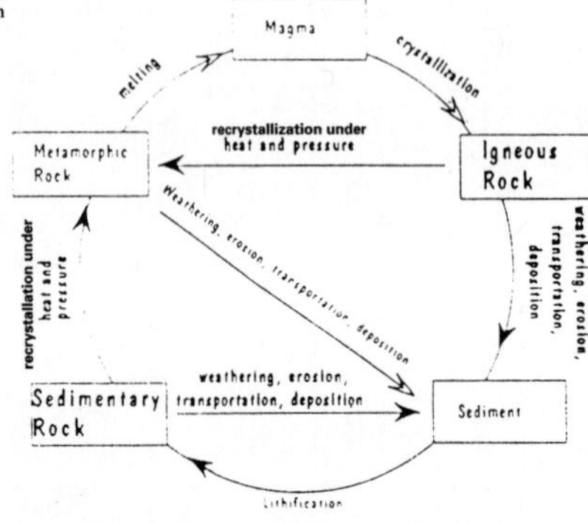

1. Can a rock that has undergone metamorphism turn into sedimentary rock? Why or why not?
   YES, METAMORPHIC ROCK SUBJECTED TO WEATHERING BREAKS DOWN INTO SEDIMENT. AS THE WEATHERED MATERIAL UNDERGOES LITHIFICATION IT BECOMES SEDIMENTARY ROCK.
2. By what process does hot molten magma become rock?
   HOT MOLTEN MAGMA BECOMES IGNEOUS ROCK AFTER IT HAS COOLED AND SOLIDIFIED (THE PROCESS OF CRYSTALLIZATION).
3. Can sedimentary rock become metamorphic rock? If so, how?
   BASALT, ANDESITE, AND GRANITE (OTHERS TOO!)
4. Can igneous rock become sedimentary rock? If so, how?
   A SHALLOW SEA ENVIRONMENT, THE REMAINS OF ANCIENT SEA FLOORS.
5. In what part(s) of the rock cycle does sandstone transform to coal?
   ALL SECTIONS. GEMS FORM FROM THE SLOW COOLING OF MAGMA—PERIDOT AND TOPAZ FORM IN IGNEOUS ENVIRONMENTS. PRESSURE AND HEAT RESULT IN THE RECRYSTALLIZATION OF MINERALS—METAMORPHISM. GARNET IS A METAMORPHIC GEM. ONLY A FEW GEMS, SUCH AS TURQUOISE AND OPAL, ARE ACTUALLY FORMED IN A SEDIMENTARY ENVIRONMENT.
6. The Big Island of Hawaii is almost 1 million years old. Yet hikers there almost never step on rock that is more than 1000 years old. Explain.
   THE ISLAND OF HAWAII IS STILL IN THE PROCESS OF FORMATION. THE ISLAND IS A LARGE SHIELD VOLCANO FORMED FROM THE ACCUMULATION OF SUCCESSIVE LAVA FLOWS. WITH EACH NEW FLOW, ROCKS FROM PREVIOUS FLOWS ARE COVERED. SO THE ROCK WE WALK ON IS NEVER MORE THAN A THOUSAND YEARS OLD.

Chapter 24: Earth's Surface—Land and Water

Groundwater Flow and Contaminant Transport

The occupants of Houses 1, 2, and 3 wish to drill wells for domestic water supply. Note that the locations of all houses are between Lakes A and B, at different elevations.

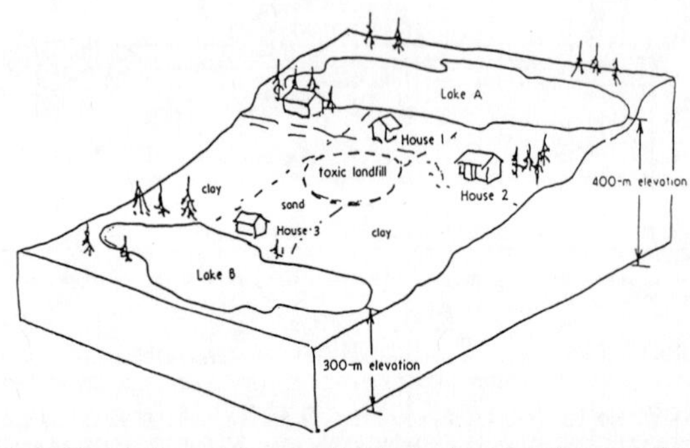

1. Show by sketching dashed lines on the drawing, the likely direction of groundwater flow beneath all the houses.
   SEE DRAWING
2. Which of the wells drilled beside Houses 1, 2, and 3 are likely to yield an abundant water supply?
   WELLS AT HOUSES 1 AND 3 SHOULD YIELD A SUFFICIENT AMOUNT OF WATER BECAUSE SAND IS QUITE PERMEABLE. THE WELL AT HOUSE 2 IS CLAY WHICH HAS A LOW PERMEABILITY AND SO WILL NOT YIELD A SUFFICIENT WATER SUPPLY.
3. Do any of the three need to worry about the toxic landfill contaminating their water supply? Explain.
   HOUSE 2 DOESN'T HAVE A DECENT WATER SUPPLY—NO CONTAMINATION WORRIES. HOUSE 1 IS UP-GRADIENT FROM THE LANDFILL SO WILL HAVE NO CONTAMINATION WORRIES UNLESS PUMPING RATE IS HIGH ENOUGH TO DRAW WATER AGAINST THE REGIONAL GRADIENT. HOUSE 3 IS IN BIG TROUBLE.
4. Why don't the homeowners simply take water directly from the lakes?
   SAND ACTS AS A GOOD FILTER FOR BACTERIA AND VIRUSES. ALSO, THE ADDITIONAL RESIDENCE TIME IN GROUNDWATER ALLOWS CHEMICAL REACTIONS TO REMOVE MANY CONTAMINANTS.
5. Suggest a potentially better location for the landfill. Defend your choice.
   A BETTER LOCATION WOULD BE IN THE CLAY. CLAY'S LOW PERMEABILITY WOULD HINDER LEACHING OF CONTAMINANTS TO THE GROUNDWATER.

Chapter 24: Earth's Surface—Land and Water

Stream Velocity

Let's explore how the average velocity of streams and rivers can change. The volume of water that flows past a given location over any given length of time depends both on the stream velocity and the cross-sectional area of the stream. We say

$$Q = A \times V$$

where Q is the volumetric flow rate (a measure of the volume passing a point per unit time). Also, A is the cross-sectional area of the stream, and V its average velocity.

Consider the stream shown below, with rectangular cross-sectional areas

$$A = \text{width} \times \text{depth}$$

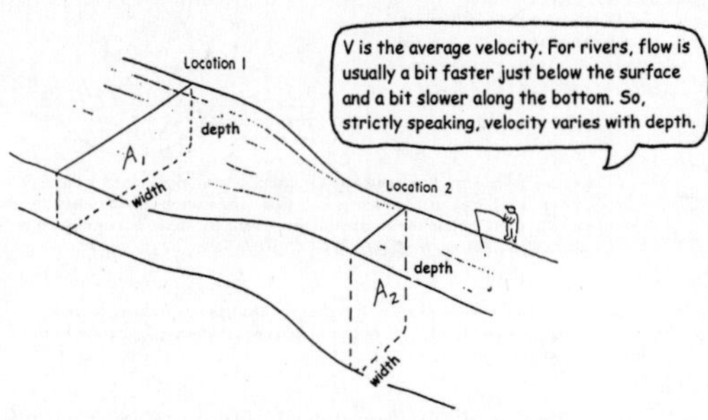

V is the average velocity. For rivers, flow is usually a bit faster just below the surface and a bit slower along the bottom. So, strictly speaking, velocity varies with depth.
Location 1
depth
width
Location 2
depth
width

1. The two locations shown have no stream inlets or outlets between them, so Q remains constant. Suppose the cross-sectional areas are also constant ($A_1 = A_2$), with Location 2 deeper but narrower than Location 1. What change, if any, occurs for the stream velocity?
   THERE IS NO CHANGE IN AVERAGE VELOCITY
2. If Q remains constant, what happens to stream velocity at Location 2 if A_2 is less than A_1?
   AVERAGE VELOCITY INCREASES AT LOCATION 2.
3. If Q remains constant, what happens to stream velocity at Location 2 if A_2 is greater than A_1?
   AVERAGE VELOCITY IDECREASES AT LOCATION 2.
4. What happens to stream velocity at Location 2 if area A_2 remains the same, but Q increases (perhaps by an inlet along the way?)
   AVERAGE VELOCITY INCREASES AT LOCATION 2.
5. What happens to stream velocity at Location 2 if both A_2 and Q increase?
   IT DEPENDS. IF Q INCREASES MORE THAN A INCREASES, AVERAGE VELOCITY INCREASES. IF A INCREASES MORE THAN Q INCREASES, AVERAGE VELOCITY DECREASES. IF THEY BOTH INCREASE AT THE SAME PROPORTION, THERE IS NO CHANGE IN AVERAGE VELOCITY.

Chapter 25: Weather

Earth's Seasons

1. The warmth of equatorial regions and coldness of polar regions on Earth can be understood by considering light from a flashlight striking a surface. If it strikes perpendicularly, light energy is more concentrated as it covers a smaller area; if it strikes at an angle, the energy spreads over a larger area. So the energy per unit area is less.

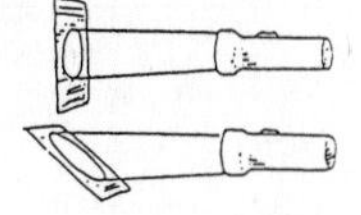

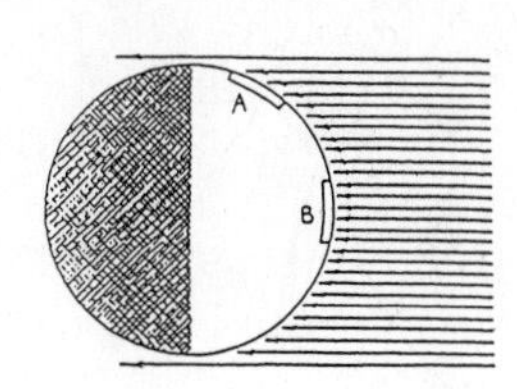

The arrows represent rays of light from the distant Sun incident upon Earth. Two areas of equal size are shown, Area A near the north pole and Area B near the equator. Count the rays that reach each region, and explain why Area B is warmer than Area A.
3 RAYS INCIDENT ON A; 6 ON B. SO REGION B GETS TWICE AS MUCH SOLAR ENERGY AND IS WARMER.

2. Earth's seasons result from the 23.5-degree tilt of Earth's daily spin axis as it orbits the Sun. When Earth is at the position shown on the right in the sketch below (not shown to scale), the Northern Hemisphere tilts toward the Sun, and sunlight striking it is strong (more rays per area). Sunlight striking the Southern Hemisphere is weak (fewer rays per area). Days in the north are warmer, and daylight lasts longer. You can see this by imagining Earth making its complete daily 24-hour spin.

   Do two things on the sketch: (1) Shade Earth in nighttime darkness for all positions, as is already done in the left position. (2) Label each position with the proper month—March, June, September, or December.

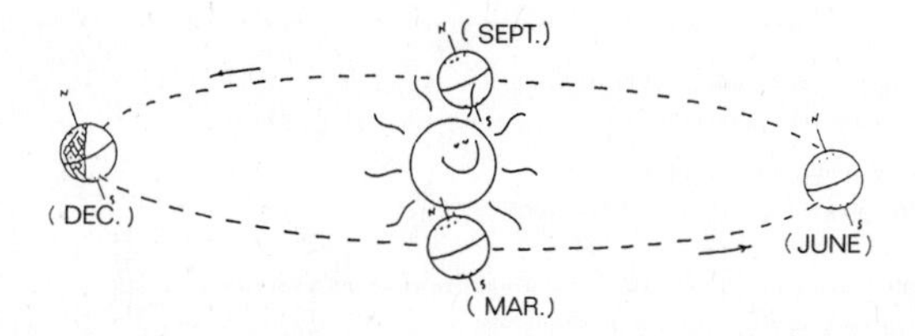

Earth's Seasons—continued

a. When Earth is in any of the four positions shown, during one 24-hour spin a location at the equator receives sunlight half the time and is in darkness the other half of the time. This means that regions at the equator always get about 12 hours of sunlight and 12 hours of darkness.

b. Can you see that in the June position regions farther north have longer days and shorter nights? Locations north of the Arctic Circle (dotted in the Northern Hemisphere) are always illuminated by the Sun as Earth spins, so they get daylight 24 hours a day.

c. How many hours of light and darkness are there in June at regions south of the Antarctic Circle (dotted line in Southern Hemisphere)?
ZERO HOURS OF LIGHT, OR 24 HOURS OF DARKNESS PER DAY.

d. Six months later, when Earth is at the December position, is the situation in the Antarctic the same or is it the reverse?
REVERSE; MORE SUNLIGHT PER AREA IN DEC. IN SOUTHERN HEMISPHERE

e. Why do South America and Australia enjoy warm weather in December instead of in June?
IN DEC. THE SOUTHERN HEMISPHERE TILTS TOWARD THE SUN AND GETS MORE SUNLIGHT PER AREA THAN IN JUNE.

3. Earth spins about its polar axis once each 24 hours, which gives us day and night. If Earth's spin was instead only one rotation per year, what difference would there be with day and night as we enjoy them now?
   ONE FACE OF THE EARTH WOULD ALWAYS BE IN SUNLIGHT, AND THE OPPOSITE SIDE WOULD ALWAYS BE IN DARKNESS.

If the spin of the Earth was the same as its revolution rate around the Sun, would we be like the Moon — one side always facing the body it orbits?

YES—BUT THE MOON HAS MONTHLY CYCLES OF DAY AND NIGHT.

Short and Long Wavelength

The sine curve is a pictorial representation of a wave—the high points being crests, and the low points troughs. The height of the wave is its *amplitude*. The wavelength is the distance between successive identical parts of the wave (like between crest to crest, or trough to trough). Wavelengths of water waves at the beach are measured in meters, wavelengths of ripples in a pond are measured in centimeters, and the wavelengths of light are measured in billionths of a meter (nanometers).

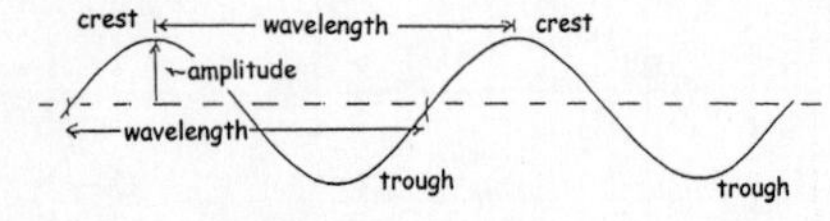

In the boxes below sketch three waves of the same amplitude—Wave A with half the wavelength of Wave B, and Wave C with wavelength twice as long as Wave B.

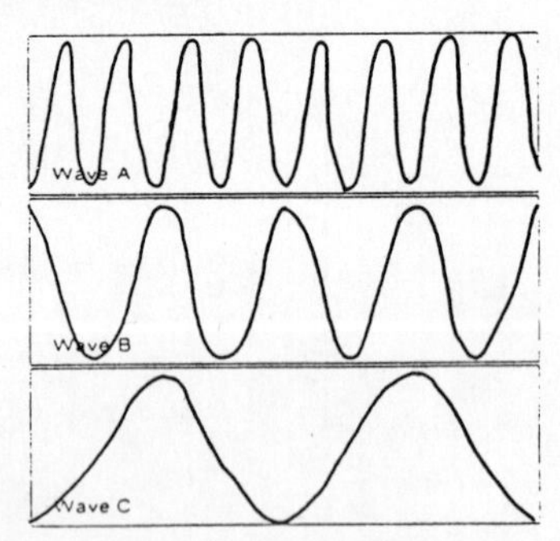

1. If all three waves have the same speed, which has the highest frequency? WAVE A
2. Compared with solar radiation, terrestrial radiation has a LONG wavelength.
3. In a florist's greenhouse, SHORT waves are able to penetrate the greenhouse glass, but LONG waves cannot.
4. Earth's atmosphere is similar to the glass in a greenhouse. If the atmosphere were to contain excess amounts of water vapor and carbon dioxide, the air would be opaque to LONG waves.

Driving Forces of Air Motion

The primary driving force of Earth's weather is SUNLIGHT. The unequal distribution of solar radiation on Earth's surface creates temperature differences which in turn result in pressure differences in the atmosphere. These pressure differences generate horizontal winds as air moves from HIGH pressure to LOW pressure. The weather patterns are not strictly horizontal though, there are other forces affecting the movement of air. Recall from Newton's Second Law that an object moves in the direction of the *net* force acting on it. The forces acting on the movement of air include: 1) pressure gradient force, 2) Coriolis force, 3) centripetal force, and 4) friction.

The greater the pressure difference, the greater the force, and the greater the wind. The "push" caused by the horizontal differences in pressure across a surface is called the *pressure gradient force*. This force is represented by isobars on a weather map. Isobars connect locations on a map that have equal atmospheric pressure. The pressure gradient force is perpendicular to the isobars and strongest where the isobars are closely spaced. So, the steeper the pressure gradient, the ______ the wind. STRONGER

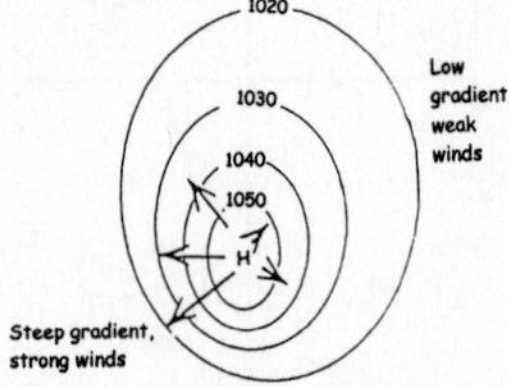

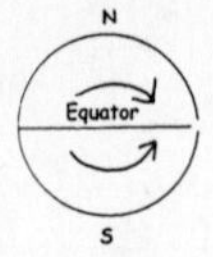

The *Coriolis effect* is a result of Earth's rotation. The Coriolis effect is the deflection of the wind from a ______ path to a ______ path. The Coriolis effect causes the wind to veer to the right of its path in the Northern Hemisphere and to the left of its path in the Southern Hemisphere. STRAIGHT, CURVED

As the wind blows around a low or high pressure center it constantly changes its direction. A change in speed or direction is acceleration. In order to keep the wind moving in a circular path the net force must be directed inward. This INWARD force is called *centripetal force*.

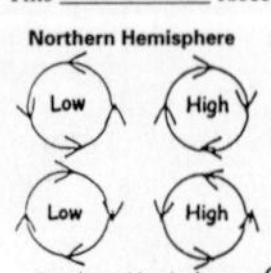

The forces described above greatly influence the flow of upper winds (winds not influenced by surface frictional forces). The interaction of these forces cause the winds in the Northern Hemisphere to rotate ______ around regions of high presssure and ______ around regions of low pressure. In the Southern Hemisphere the situation is reversed—winds rotate ______ around a high pressure zone and ______ around a low pressure zone.

CLOCKWISE, COUNTERCLOCKWISE, COUNTERCLOCKWISE, CLOCKWISE

Winds blowing near Earth's surface are slowed by *frictional forces*. In the Northern Hemisphere surface winds blow in a direction ______ into the centers of a low pressure area and ______ out of the centers of a high pressure area. The spiral direction is reversed in the Southern Hemisphere. Draw arrows to show the direction of the pressure gradient force. COUNTERCLOCKWISE, CLOCKWISE

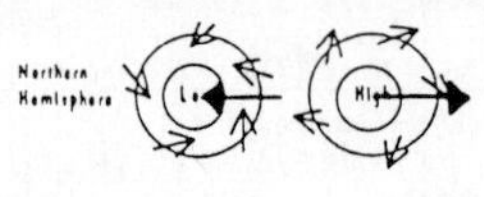

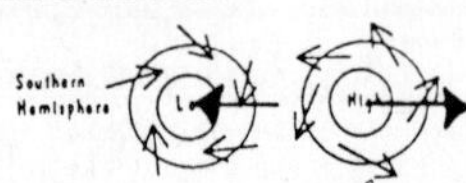

Air Temperature and Pressure Patterns

Temperature patterns on weather maps are depicted by *isotherms*—lines that connect all points having the same temperature. Each isotherm separates temperatures of higher values from temperatures of lower values.

The weather map to the right shows temperatures in degrees Fahrenheit for various locations. Using 10 degree intervals, connect same value numbers to construct isotherms. Label the temperature value at each end of the isotherm. One isotherm has been completed as an example.

Tips for drawing isotherms:

- Isotherms can never be open ended.
- Isotherms are "closed" if they reach the boundary of plotted data, or make a loop.
- Isotherms can never touch, cross, or fork.
- Isotherms must always appear in sequence; for example, there must be a 60° isotherm between a 50° and 70° isotherm.
- Isotherms should be labeled with their values.

Pressure patterns on weather maps are depicted by *isobars*—lines which connect all points having equal pressure. Each isobar separates stations of higher pressure from stations of lower pressure.

The weather map below shows air pressure in millibar (mb) units at various locations. Using an interval of 4 (for example, 1008, 1012, 1016, etc.), connect equal pressure values to construct isobars. Label the pressure value at each end of the isobar. Two isobars have been completed as an example.

Tips for drawing isobars are similar to those for drawing isotherms.

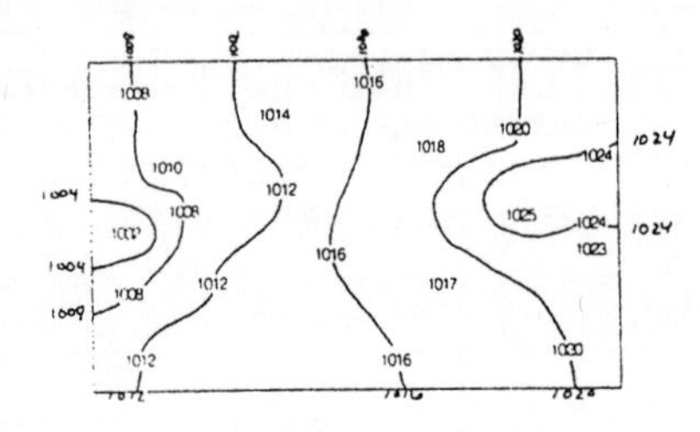

Air Temperature and Pressure Patterns—continued

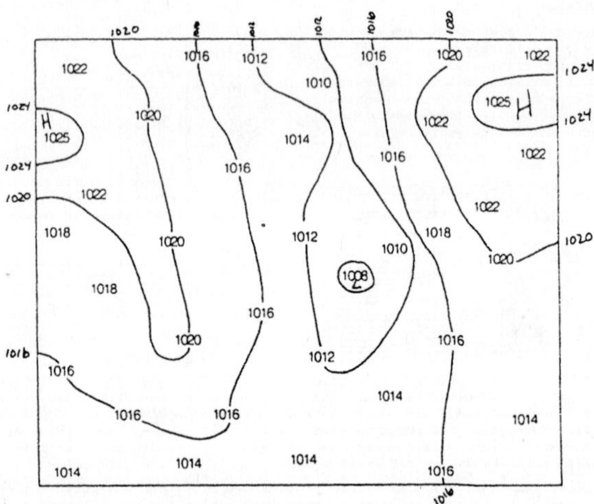

On the map above, use an interval of 4 to draw lines of equal pressure (isobars) to show the pattern of air pressure. Locate and mark regions of high pressure with an "H" and regions of low pressure with an "L".

1. On the map above, areas of high pressure are depicted by the 1024 isobar.
2. On the map above, areas of low pressure are depicted by the 1008 isobar.

Circle the correct answer.

3. Areas of high pressure are usually accompanied by (stormy weather) (fair weather).
4. In the Northern Hemisphere, surface winds surrounding a high pressure system blow in a (clockwise direction) (counterclockwise direction).
5. In the Northern Hemisphere, surface winds spiral inward into a (region of low pressure) (region of high pressure).

Chapter 25: Weather

Surface Weather Maps

Station models are used on weather maps to depict weather conditions for individual localities. Weather codes are plotted in, on, and around a central circle that describes the overall appearance of the sky. Jutting from the circle is a wind arrow, its tail in the direction from which the wind comes and its feathers indicating the wind speed. Other weather codes are in standard position around the circle.

Use the simplified station model and weather symbols to complete the statements below.

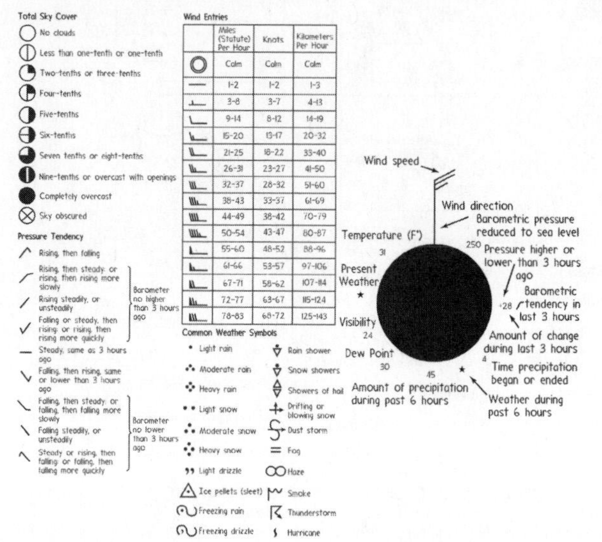

1. The overall appearance of the sky is COMPLETELY OVERCAST.
2. The wind speed is 41–50 kilometers per hour.
3. The wind direction is coming from the EAST.
4. The present weather conditions call for SNOW.
5. The barometric tendency is RISING, THEN STEADY.
6. For the past 6 hours the weather conditions have been SNOWY.

Surface Weather Maps—continued

Use the unlabeled station model to answer the questions below.

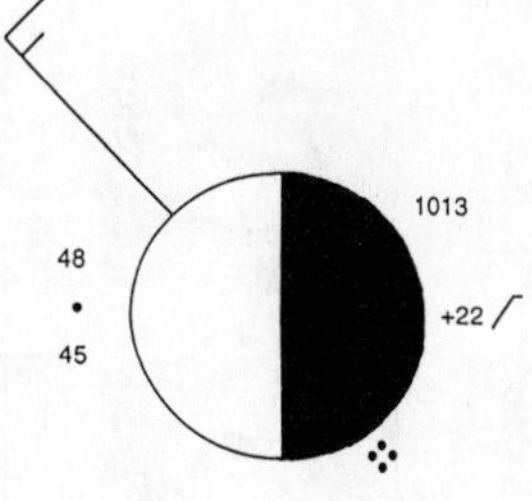

1. The overall appearance of the sky is FIVE TENTHS OVERCAST.
2. The wind speed is 20–32 kilometers per hour.
3. The wind direction is coming from the EAST.
4. The present weather conditions call for LIGHT RAIN.
5. The barometric tendency is RISING, THEN STEADY.
6. For the past 6 hours the weather conditions have been HEAVY RAIN.
7. The barometric pressure is 1013 mb.
8. The dew point is 45.
9. The current temperature is 48°F.
10. Compared to the past few hours, barometric pressure is INCREASING.

Chapter 25: Weather

Chilly Winds

Often times it feels colder outside than a thermometer indicates. The apparent difference is called *wind chill*. On November 1, 2001, the National Weather Service implemented a new Wind Chill Temperature index. The new formula uses advances in science, technology, and computer modeling to provide a more accurate, understandable, and useful formula for calculating the dangers from winter winds and freezing temperatures.

For temperatures less than 50°F and wind speeds greater than 3 mph, the new formula used to estimate the temperature we perceive when a cold wind blows is:

$$\text{Wind Chill Temperature (°F)} = 35.74 + 0.6215T - 35.75(V^{0.16}) + 0.4275T(V^{0.16})$$

where V is the wind speed in miles per hour and T is the temperature in degrees Fahrenheit.

Wind Chill Temperature Table

		Temperature (°F)			
V (mph)	$V^{0.16}$	5.00	10.00	15.00	32.00
5.00	1.2937	-4.64	1.24	7.11	27.08
10.00	1.4454	−9.74	−3.54	2.66	23.73
15.00	1.5423	-12.99	-6.59	-0.19	21.59

1. Using the formula given for Wind Chill Temperature (WCT), complete the above table. The variable $V^{0.16}$ (wind speed raised to 0.16 power) is provided in the table to simplify your calculations.
2. Which has a stronger impact on WCT, changes in wind speed or changes in temperature? Defend your answer with data from your completed table.
   FROM THE TABLE NOTE THAT CHANGES IN TEMPERATURE HAVE A STRONGER IMPACT THAN CHANGES IN WIND SPEED. EACH 5-DEGREE CHANGE IN TEMPERATURE CAUSES BIGGER CHANGES IN WCT THAN EACH 5-MPH CHANGE IN WIND SPEED.
3. How does WCT vary when wind speed remains fixed and only temperature changes? Defend your answer.
   NOTE THE WCT EQUATION IS LINEAR WITH RESPECT TO *T*. AND AS CALCULATED DIFFERENCES IN WCT ACROSS A GIVEN ROW IN THE TABLE SHOW, THE WCT VARIES AT EVEN INTERVALS AS THE TRUE TEMPERATURE CHANGES.
4. How does WCT vary when temperature stays fixed and only wind speed changes? Defend your answer.
   NOTE THE WCT EQUATION IS NON-LINEAR WITH RESPECT TO *V* (BECAUSE IT VARIES AS V^{016}). AND AS CALCULATED DIFFERENCES IN WCT DOWN A GIVEN COLUMN OF THE TABLE SHOW, FOR A FIXED TEMPERATURE, WCT DOES NOT VARY AT EVEN INTERVALS AS WIND SPEED CHANGES.

More information can be found on the National Weather Service's Web site: http://www.nws.noaa.gov/om/windchill

Chapter 26: Earth's History

Relative Time—What Came First?

The cross-section below depicts many geologic events. List to the right the sequences of geologic history starting with the oldest event to the youngest event—and where appropriate include tectonic events (such as folding, deposition of strata (beds), subsidence, uplift, erosion, and intrusion).

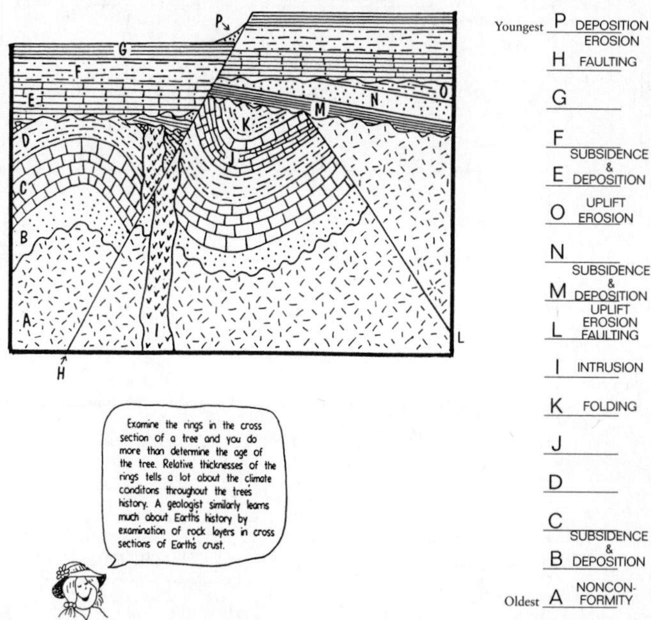

Youngest
- P DEPOSITION EROSION
- H FAULTING
- G
- F SUBSIDENCE & DEPOSITION
- E
- O UPLIFT EROSION
- N SUBSIDENCE & DEPOSITION
- M UPLIFT EROSION
- L FAULTING
- I INTRUSION
- K FOLDING
- J
- D
- C SUBSIDENCE & DEPOSITION
- B
- A NONCONFORMITY

Oldest

Examine the rings in the cross section of a tree and you do more than determine the age of the tree. Relative thicknesses of the rings tells a lot about the climate conditions throughout the trees history. A geologist similarly learns much about Earth's history by examination of rock layers in cross sections of Earth's crust.

Chapter 26: Earth's History

Age Relationships

From your investigation of the six geologic regions shown, answer the questions below. The number of each question refers to the same-numbered region.

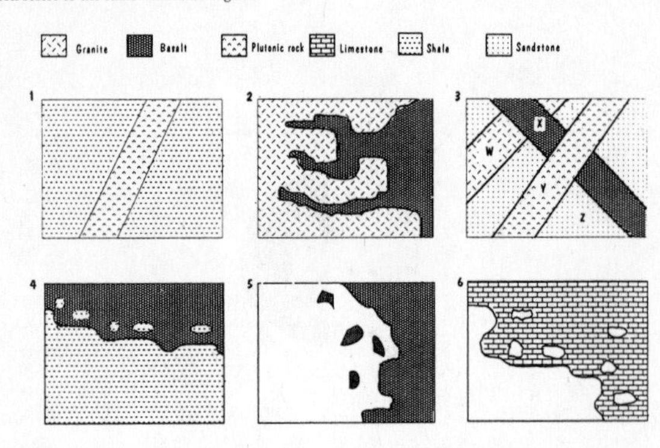

1. The shale has been cut by a dike. The radiometric age of the dike is estimated at 40 million years. Is the shale younger or older? OLDER
2. Which is older, the granite or the basalt? THE GRANITE
3. The sandstone bed, Z, has been intruded by dikes. What is the age succession of dikes, going from oldest to youngest? W, X, Y
4. Which is older, the shale or the basalt? THE SHALE
5. Which is older, the sandstone or the basalt? THE BASALT
6. Which is older, the sandstone or the limestone? THE SANDSTONE

Chapter 26: Earth's History

Unconformities and Age Relationships

The wavy lines in the four regions below represent unconformities. Investigate the regions and answer corresponding questions below.

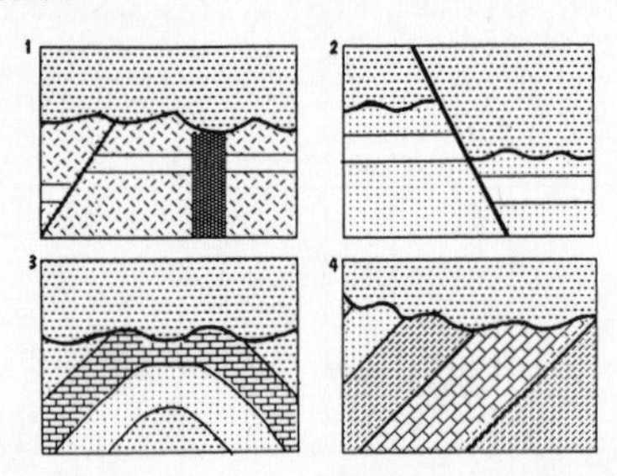

1. Did the faulting and dike occur before or after the unconformity? BEFORE

   What kind of conformity is it? NONCONFORMITY
2. Did the faulting occur before or after the unconformity? AFTER

   What kind of unconformity is represented? EROSIONAL UNCONFORMITY
3. Did the folding occur before or after the unconformity? BEFORE

   What kind of unconformity is it? ANGULAR UNCONFORMITY
4. What kind of unconformity is represented? ANGULAR UNCONFORMITY
5. Interestingly, the age of Earth is some 4.5 billion years old—yet the oldest rocks found are some 3.7 billion years old. Why do we find no 4.5-billion year old rocks?

   EARTH'S EARLIEST CRUSTAL SURFACE HAS BEEN REWORKED. ROCKS THAT FORMED 4.5 B.Y.A HAVE BEEN REMELTED INTO NEW ROCK.
6. What is the age of the innermost ring in a living redwood tree that is 2000 years old? What is the age of the outermost ring? How does this example relate to the previous question?

   THE INNERMOST RING IS 2000 YEARS OLD; THE OUTERMOST RING IS 1 YEAR OLD. NEW LAYERS COVER OLDER LAYERS.
7. What is the approximate age of the atoms that make up a 3.7-billion year old rock?

   THEY ARE OLDER THAN 3.7 BILLION YEARS.

Chapter 26: Earth's History

Radiometric Dating

Isotopes Commonly Used for Radiometric Dating

Radioactive Parent	Stable Daughter Product	Currently Accepted Half-life Value
Uranium-238	lead-206	4.5 billion years
Uranium-235	lead-207	704 million years
Potassium-40	argon-40	1.3 billion years
Carbon-14	nitrogen-14	5730 years

1. Consider a radiometric lab experiment wherein 99.98791% of a certain radioactive sample of material remains after one year. What is the decay rate of the sample?

   1.000 . . . − 0.9998791 = 0.0001209/YR.
2. What is the rate constant?
   (Assume that the decay rate is constant for the one year period.)

   $K = \frac{\text{DECAY RATE}}{\text{STARTING AMOUNT}} = \frac{0.0001209/\text{YR}}{1.000\ldots} = 0.0001209/\text{YR}$
3. What is the half-life?

   $T = \frac{0.603}{0.0001209/\text{YR}} = 5732 \text{ YR}$
4. Identify the isotope.

   CARBON 14
5. In a sample collected in the field, this isotope was found to be 1/16 of its original amount. What is the age of the sample?

   NOTE FROM GRAPH THAT $\frac{1}{16}$ IS 4 HALF LIVES 4 × 5732 = 22,928 YR

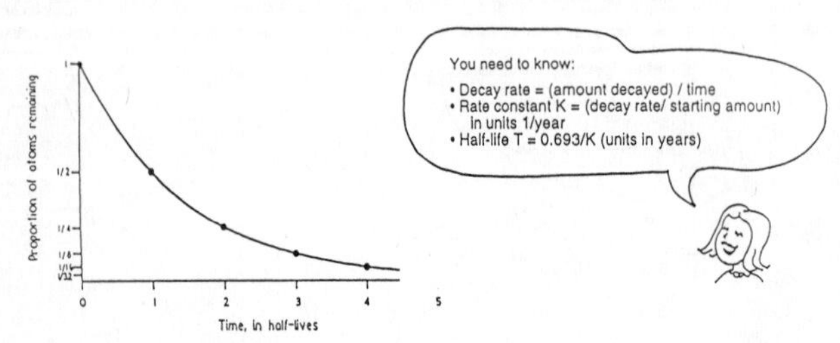

Chapter 26: Earth's History

Our Earth's Hot Interior

Scientists faced a major puzzle in the nineteenth century. Volcanoes showed that Earth's interior is semi-molten. Penetration into the crust by bore-holes and mines showed that Earth's temperature increases with depth. Scientists knew that heat flows from the interior to the surface. They assumed that the source of Earth's internal heat was primordial, the afterglow of its fiery birth. Measurements of Earth's rate of cooling indicated a relatively young Earth—some 25 to 30 million years in age, but geological evidence indicated an older Earth. This puzzle wasn't solved until the discovery of radioactivity. Then it was learned that the interior was kept hot by the energy of radioactive decay. We now know the age of Earth is some 4.5 billion years—a much older Earth.

All rock contains trace amounts of radioactive minerals. Radioactive minerals in common granite release energy at the rate 0.03 J/kg · yr. Granite at Earth's surface transfers this energy to the surroundings practically as fast as it is generated, so we don't find granite any warmer than other parts of our environment. But what if a sample of granite were thermally insulated? That is, suppose all the increase of thermal energy due to radioactive decay were contained. Then it would get hotter. How much hotter? Let's figure it out, using 790 J/kg · C° as the specific heat of granite.

Calculations to make:

1. How many joules are required to increase the temperature of 1-kg of granite by 500 C°?

   $Q = mc\Delta T = (1 \text{ kg})(\frac{790\text{J}}{\text{kgC°}})500\text{C°} = 395{,}000 \text{ J}$
2. How many years would it take radioactivity in a kilogram of granite to produce this many joules?

   $\frac{395{,}000\text{J}}{0.03\text{J/kg} \cdot \text{yr}} \times 1 \text{ kg} \approx 13$ MILLION YEARS

Let's see now... the relationship between quantity of heat Q, mass, specific heat C and temperature difference is
Q = cmΔT

Questions to answer:

1. How many years would it take a thermally insulated 1-kilogram chunk of granite to undergo a 500°C increase in temperature?

   SAME 13 MILLION YEARS
2. How many years would it take a thermally insulated one-million-kilogram chunk of granite to undergo a 500°C increase in temperature?

   SAME (CORRESPONDINGLY MORE RADIATION!)
3. Why does Earth's interior remain molten hot?

   BECAUSE OF RADIOACTIVITY
4. Rock has a higher melting temperature deep in the interior. Why?

   GREATER PRESSURE (LIKE WATER IN A PRESSURE COOKER)
5. Why doesn't Earth just keep getting hotter until it all melts?

   INTERIOR IS NOT PERFECTLY INSULATED—HEAT MIGRATES TO SURFACE.

An electric toaster stays hot while electric energy is supplied, and doesn't cool until switched off. Similarly, do you think the energy source now keeping Earth hot will one day suddenly switch off like a disconnected toaster – or gradually decrease over a long time?

Chapter 27: The Solar System

Earth-Moon-Sun Alignment

Here we see a shadow on a wall cast by an apple. Note how the rays define the darkest part of the shadow, the *umbra*, and the lighter part of the shadow, the *penumbra*. The shadows that comprise eclipses of planetary bodies are similarly formed. Below is a diagram of the Sun, Earth, and the orbital path of the Moon (dashed circle). One position of the Moon is shown. Draw the Moon in the appropriate positions on the dashed circle to represent (a) a quarter moon, (b) a half moon, (c) a solar eclipse, and (d) a lunar eclipse. Label your positions. For c and d, extend rays from the top and bottom of the Sun to show umbra and penumbra regions.

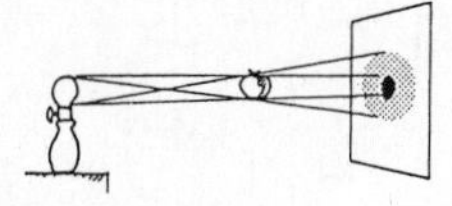

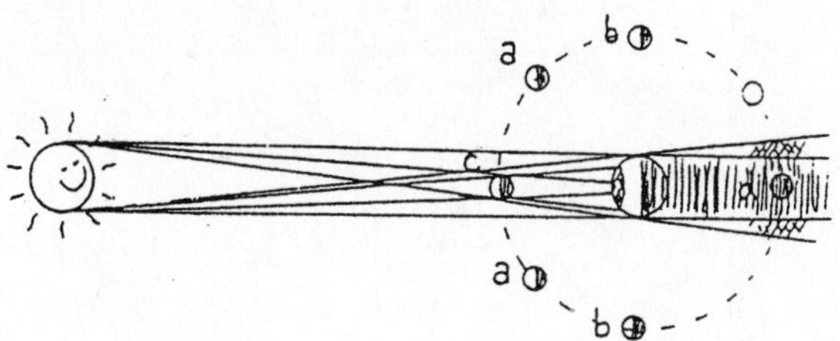

The diagram below shows three positions of the Sun, A, B, and C. Sketch the appropriate positions of the Moon in its orbit about Earth for (a) a solar eclipse, and (b) a lunar eclipse. Label your positions. Sketch solar rays similar to the above exercise.

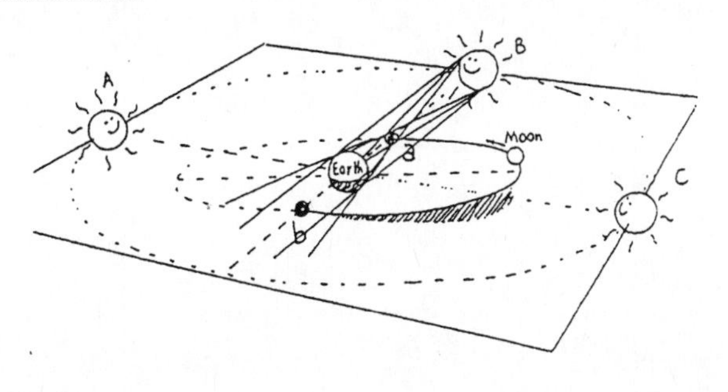

Chapter 27: The Solar System

Pinhole Image Formation

Look carefully at the round spots of light on the shady ground beneath trees. These are *sunballs*, which are images of the Sun. They are cast by openings between leaves in the trees that act as pinholes. (Did you make a pinhole "camera" back in middle school?) Large sunballs, several centimeters in diameter or so, are cast by openings that are relatively high above the ground, while small ones are produced by closer "pinholes." The interesting point is that the ratio of the diameter of the sunball to its distance from the pinhole is the same as the ratio of the Sun's diameter to its distance from the pinhole. We know the Sun is approximately 150,000,000 km from the pinhole, so careful measurements of the ratio of diameter/distance for a sunball leads you to the diameter of the Sun. That's what this page is about. Instead of measuring sunballs under the shade of trees on a sunny day, make your own easier-to-measure sunball.

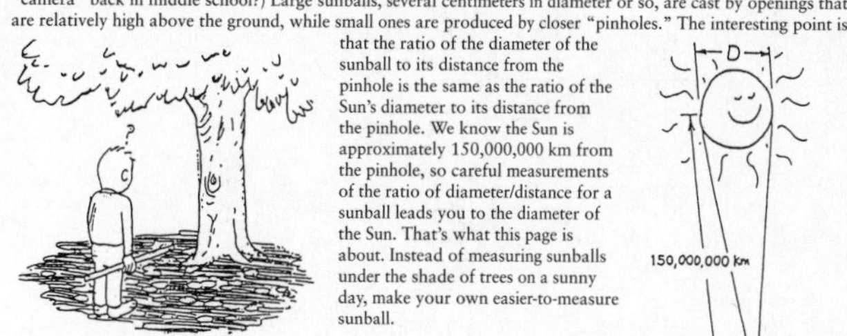

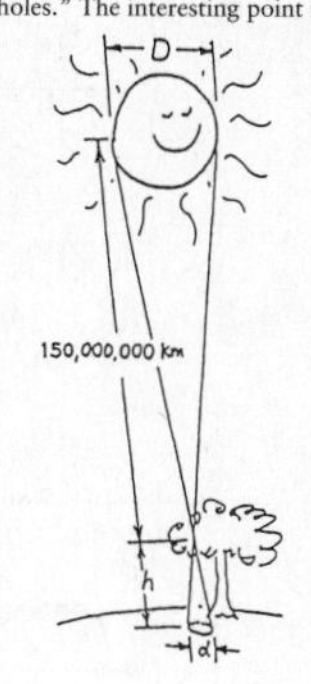

1. Poke a small hole in a piece of card. Perhaps an index card will do, and poke the hole with a sharp pencil or pen. Hold the card in the sunlight and note the circular image that is cast. This is an image of the Sun. Note that its size doesn't depend on the size of the hole in the card, but only on its distance. The image is a circle when cast on a surface perpendicular to the rays—otherwise it's "stretched out" as an ellipse.
2. Try holes of various shapes; say a square hole, or a triangular hole. What is the shape of the image when its distance from the card is large compared with the size of the hole? Does the shape of the pinhole make a difference?

   IMAGE IS ALWAYS A CIRCLE. SHAPE OF PINHOLE IS *NOT* THE SHAPE OF THE IMAGE CAST THROUGH IT.
3. Measure the diameter of a small coin. Then place the coin on a viewing area that is perpendicular to the Sun's rays. Position the card so the image of the sunball exactly covers the coin. Carefully measure the distance between the coin and the small hole in the card. Complete the following:

   $$\frac{\text{Diameter of sunball}}{\text{Distance to pinhole}} \frac{d}{h} \approx \frac{1}{110}$$ (SO SUN'S DIAM = $\frac{1}{110} \times 150{,}000{,}000$ km)

   With this ratio, estimate the diameter of the Sun. Show your work on a separate piece of paper.
4. If you did this on a day when the Sun is partially eclipsed, what shape of image would you expect to see?

   UPSIDE-DOWN CRESCENT, IMAGE OF THE PARTIALLY-ECLIPSED SUN

WHAT SHAPE DO SUNBALLS HAVE DURING A PARTIAL ECLIPSE OF THE SUN?

Chapter 28: The Universe

Stellar Parallax

Finding distances to objects beyond the solar system is based on the simple phenomenon of **parallax.** Hold a pencil at arm's length and view it against a distant background—each eye sees a different view (try it and see). The displaced view indicates distance. Likewise, when Earth travels around the Sun each year, the position of relatively nearby stars shifts slightly relative to the background stars. By carefully measuring this shift, astronomer types can determine the distance to nearby stars.

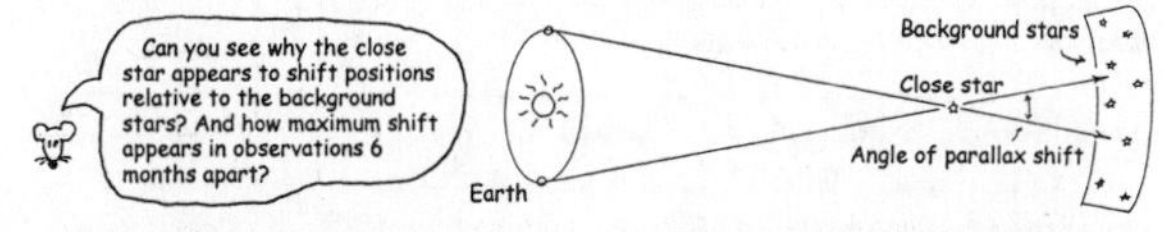

Can you see why the close star appears to shift positions relative to the background stars? And how maximum shift appears in observations 6 months apart?
Background stars
Close star
Angle of parallax shift
Earth

The photographs below show the same section of the evening sky taken at a 6-month interval. Investigate the photos carefully and determine which star appears in a different position relative to the others. Circle the star that shows a parallax shift.

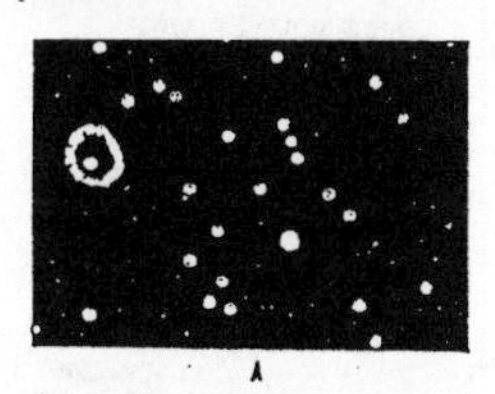

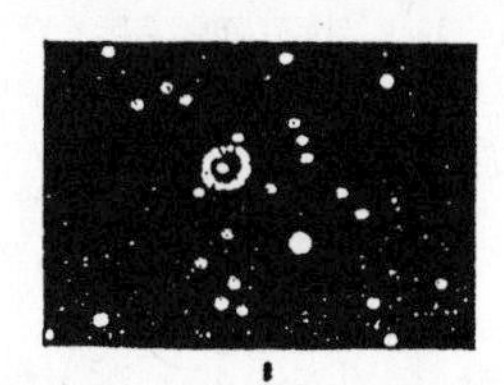

A B

Below are three sets of photographs, all taken at 6-month intervals. Circle the stars that show a parallax shift in each of the photos.

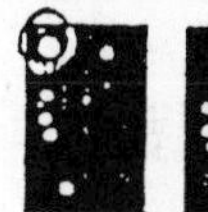

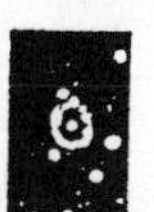

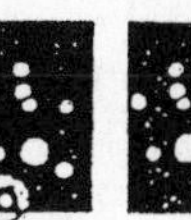

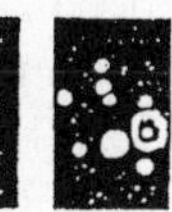

Set A Set B Set C

Use a fine ruler and measure the distance of shift in millimeters and place the values below:

Set A 19 mm Set B 6.3 mm Set C 12.6 mm

Which set of photos indicate the closest star? The most distant "parallaxed" star?

GREATER PARALLAX (ASSUMING ALL SAME SCALE); B

Appendix B: Linear and Rotational Motion

Mobile Torques

Apply what you know about torques by making a mobile. Shown below are five horizontal arms with fixed 1- and 2-kg masses attached, and four hangers with ends that fit in the loops of the arms, lettered A through R. You are to figure where the loops should be attached so that when the whole system is suspended from the spring scale at the top, it will hang as a proper mobile, with its arms suspended horizontally. This is best done by working from the bottom upward. Circle the loops where the hangers should be attached. When the mobile is complete, how many kilograms will be indicated on the scale? (Assume the horizontal struts and connecting hooks are practically massless compared to the 1- and 2-kg masses.) On a separate sheet of paper, make a sketch of your completed mobile.

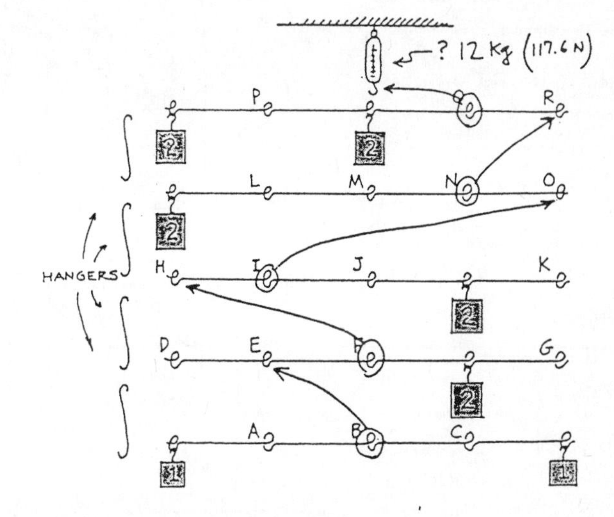

Appendix B: Linear and Rotational Motion

Torques and See-Saws

1. Complete the data for the three see-saws in equilibrium.

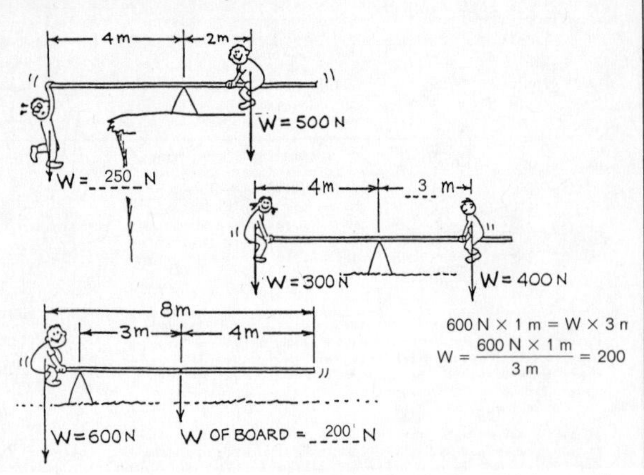
4 m
2 m
W = 500 N
W = 250 N
4 m
3 m
W = 300 N
W = 400 N

$600 \text{ N} \times 1 \text{ m} = W \times 3 \text{ m}$

$W = \frac{600 \text{ N} \times 1 \text{ m}}{3 \text{ m}} = 200$

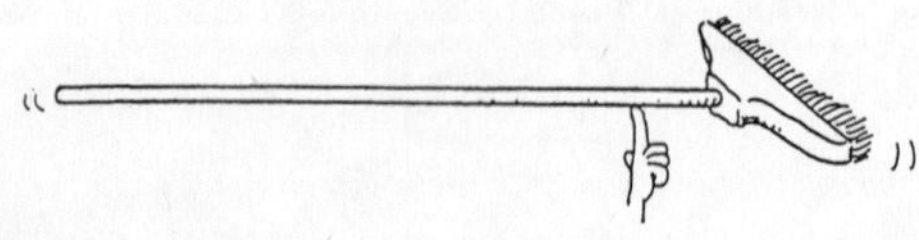
8 m
3 m
4 m
W = 600 N
W OF BOARD = 200 N

2. The broom balances at its CG. If you cut the broom in half at the CG and weigh each part of the broom, which end would weigh more?

PIECE WITH BRUSH WEIGHS MORE

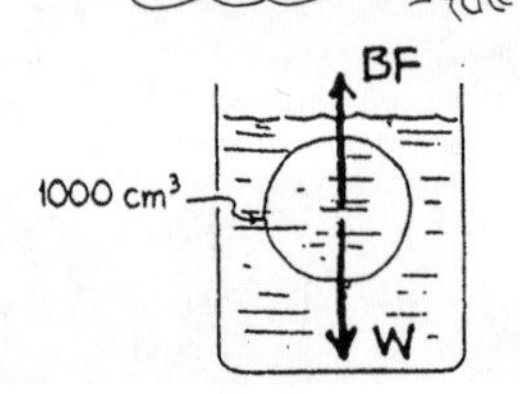

Explain why each end has or does not have the same weight? (Hint: Compare this to one of the see-saw systems above.)

WEIGHT ON EITHER SIDE ISN'T SAME, BUT TORQUE IS! LIKE SEESAWS ABOVE, SHORTER LEVER ARM HAS MORE WEIGHT

Appendix E: Physics of Fluids

Archimedes' Principle

1. Consider a balloon filled with 1 liter of water (1000 cm³) in equilibrium in a container of water, as shown in Figure 1.

a. What is the mass of the 1 liter of water?
1 kg

WATER DOES NOT SINK IN WATER!

b. What is the weight of the 1 liter of water?
9.8 N (OR 10 N)

c. What is the weight of water displaced by the balloon?
9.8 N (OR 10 N)

BF

d. What is the buoyant force on the balloon?
9.8 N (OR 10 N)

1000 cm³

W

e. Sketch a pair of vectors in Figure 1: one for the weight of the balloon and the other for the buoyant force that acts on it. How do the size and directions of your vectors compare?
SAME SIZE, OPPOSITE DIRECTIONS

2. As a thought experiment, pretend we could remove the water from the balloon but still have it remain the same size of 1 liter. The inside the balloon is a vacuum.

a. What is the mass of the liter of nothing?
0 kg

ANYTHING THAT DISPLACES 9.8 N OF WATER EXPERIENCES 9.8 N OF BUOYANT FORCE.

b. What is the weight of the liter of nothing?
0 N

c. What is the weight of water displaced by the massless balloon?
9.8 N (OR 10 N)

d. What is the buoyant force on the massless balloon?
9.8 N (OR 10 N)

IF YOU PUSH 9.8 N OF WATER ASIDE THE WATER PUSHES BACK ON YOU WITH 9.8 N!

e. In which direction would the massless balloon be accelerated?
UPWARD

Archimedes' Principle—continued

3. Assume the balloon is replaced by a 0.5-kilogram piece of wood that has exactly the same volume (1000 cm³), as shown in Figure 2. The wood is held in the same submerged position beneath the surface of the water.

1000 cm³

Figure 2

a. What volume of water is displaced by the wood?
1000 cm³ = 1 L

b. What is the mass of the water displaced by the wood?
1 kg

c. What is the weight of the water displaced by the wood?
9.8 N

d. How much buoyant force does the surrounding water exert on the wood?
9.8 N

e. When the hand is removed, what is the net force on the wood?
NET FORCE = BF − W = 9.8 N − 4.9 N = 4.9 N UPWARD

f. In which direction does the wood accelerate when released? UPWARD

THE BUOYANT FORCE ON A SUBMERGED OBJECT EQUALS THE WEIGHT OF WATER DISPLACED
... NOT THE WEIGHT OF THE OBJECT ITSELF!
... UNLESS IT IS FLOATING!

4. Repeat parts *a* through *f* in the previous question for a 5-kg rock that has the same volume (1000 cm³), as shown in Figure 3. Assume the rock is suspended by a string in the container of water.

a. 1000 cm³ (SAME)
b. 1 kg (SAME)
c. 9.8 N (SAME)
d. 9.8 N (SAME)
e. 39 N DOWNWARD*
f. DOWNWARD

1000 cm³
WHEN THE WEIGHT OF AN OBJECT IS GREATER THAN THE BUOYANT FORCE EXERTED ON IT, IT SINKS!

Figure 3

*NET FORCE = W − BF = 49 N − 9.8 N ≈ 39 N

Appendix E: Physics of Fluids

More on Archimedes' Principle

1. The water lines for the first three cases are shown. Sketch in the appropriate water lines for cases *d* and *e*, and make up your own for case *f*.

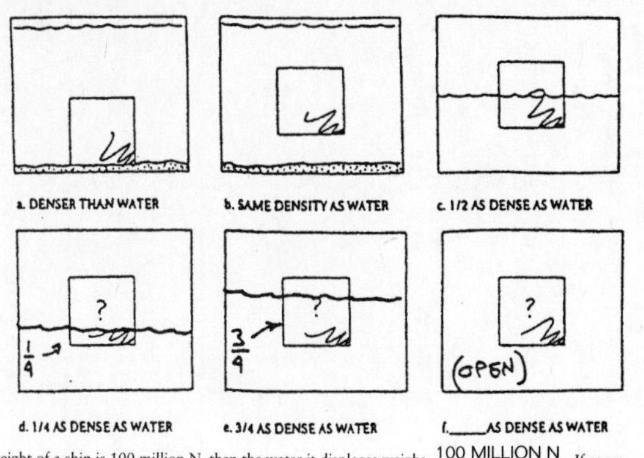

2. If the weight of a ship is 100 million N, then the water it displaces weighs 100 MILLION N. If cargo weighing 1000 N is put on board, then the ship will sink down until an extra 1000 N of water is displaced.

3. The first two sketches below show the water line for an empty and a loaded ship. Draw in the appropriate water line for the third sketch.

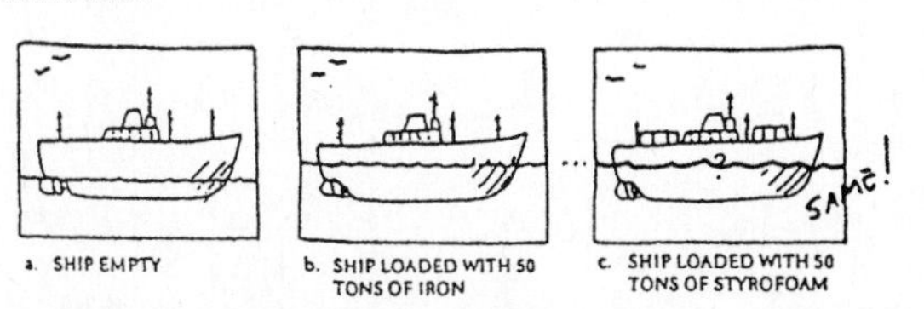

More on Archimedes' Principle—continued

4. Here is a glass of ice water with an ice cube floating in it. Draw the water line after the ice cube melts. (Will the water line rise, fall, or remain the same?)
REMAINS SAME. VOL OF WATER WITH SAME WT OF ICE CUBE EQUALS VOL OF SUBMERGED PORTION OF ICE CUBE. THIS IS ALSO THE VOL OF WATER FROM MELTED ICE CUBE.

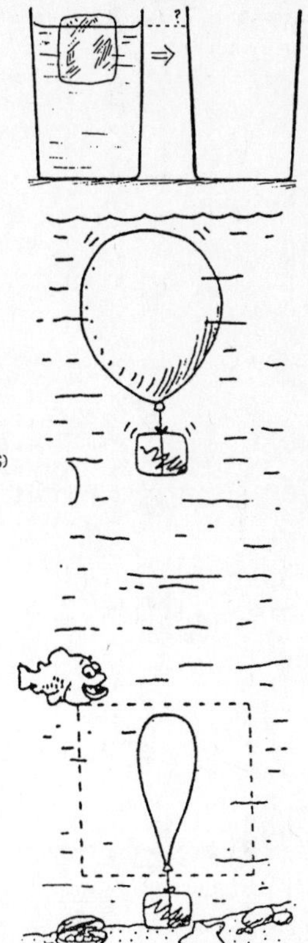

5. The air-filled balloon is weighted so it sinks in water. Near the surface, the balloon has a certain volume. Draw the balloon at the bottom (inside the dashed square) and show whether it is bigger, smaller, or the same size.

   a. Since the weighted balloon sinks, how does its overall density compare to the density of water?
   THE DENSITY OF BALLOON IS GREATER

   b. As the weighted balloon sinks, does its density increase, decrease, or remain the same?
   DENSITY INCREASES (BECAUSE VOL DECREASES)

   c. Since the weighted balloon sinks, how does the buoyant force on it compare to its weight?
   BF IS LESS THAN ITS WEIGHT

   d. As the weighted balloon sinks deeper, does the buoyant force on it increase, decrease, or remain the same?
   BF DECREASES (BECAUSE VOL DECREASES)

6. What would be your answers to Questions 5 *a*, *b*, *c*, and *d* for a rock instead of an air-filled balloon?

   a. DENSITY OF ROCK IS GREATER
   b. DENSITY REMAINS SAME (SAME VOL)
   c. BF IS LESS THAN ITS WEIGHT
   d. BF STAYS SAME (VOL STAYS SAME)

Appendix E: Physics of Fluids

Gases

1. A principle difference between a liquid and a gas is that when a liquid is under pressure, its volume
(increases) (decreases) (doesn't change noticeably) [circled: doesn't change noticeably]
and its density
(increases) (decreases) (doesn't change noticeably). [circled: doesn't change noticeably]
When a gas is under pressure, its volume
(increases) (decreases) (doesn't change noticeably) [circled: decreases]
and its density
(increases) (decreases) (doesn't change noticeably). [circled: increases]

PHYSICS SIGH

2. The sketch shows the launching of a weather balloon at sea level. Make a sketch of the same weather balloon when it is high in the atmosphere. In words, what is different about its size and why?
BALLOON GROWS AS IT RISES. ATM PRESSURE TENDS TO COMPRESS THINGS—EVEN BALLOONS. MORE PRESSURE AT GROUND LEVEL, & MORE COMPRESSION. LESS COMPRESSION AT HIGH ALTITUDES, & BIGGER BALLOON

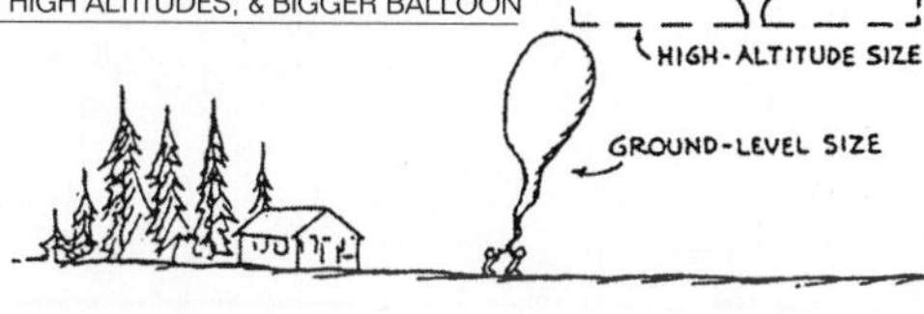

3. A hydrogen-filled balloon that weighs 10-N must displace 10 N of air in order to float in air. If it displaces less than 10 N it will be buoyed up with less than 10 N and sink. If it displaces more than 10 N of air it will move upward.

4. Why is the cartoon at right more humorous to physics types than to non-physics types? What physics has occurred?

RATS TO YOU TOO, DANIEL BERNOULLI!

IN ACCORD WITH BERNOULLI'S PRINCIPLE, MOVEMENT OF AIR OVER CURVED TOP OF UMBRELLA CAUSES A REDUCTION OF AIR PRESSURE (LIKE AIRPLANE WING). THIS LIKELY PRODUCED A NET UPWARD FORCE THAT TURNED THE UMBRELLA INSIDE OUT.

Answers and Solutions to Odd-Numbered Integrated Science Concepts, Exercises, and Problems from *Conceptual Integrated Science*

Chapter 1: About Science

Answers to Integrated Science Concepts

An Investigation of Sea Butterflies

1. The disciplines of biology and chemistry are needed to understand the behavior of the Antarctic amphipod.
3. McClintock and Baker's hypothesis was that amphipods carry sea butterflies because sea butterflies produce a chemical that deters a predator of the amphipod. This is a scientific hypothesis because it would be proven wrong if the chemical secreted by sea butterflies were found to not deter amphipod predators.

Answers to Exercises

1. The various branches of science overlap as we see by the existence of these hybrid fields: astrobiology; biochemistry; biophysics; ecology (biology and earth science); geochemistry, etc.
3. No; religion is a subjective area of study so that it cannot be wrong in the sense of being provably false. However, religions that do claim to be based on a factual knowledge of the physical world that is in fact provably false, can be said to be logically flawed. A person can certainly be wrong in their understanding of a scientific concept—experiments and observation often can correct such misunderstandings.
5. The printing press greatly accelerated the progress of science by facilitating communication—suddenly practitioners of science could collaborate across distance. The Internet takes communication to a new level because it is so fast, open, and accessible.

Answers to Chapter 1

Problems

1. Direct proportion
3. Sample answer: The more you practice shooting a basketball (P), the fewer shots you miss (M); $P = k/M$.

Part 1 Physics

Chapter 2: Describing Motion

Answers to Integrated Science Concepts

Friction Is Universal

1. Synovial fluid is a lubricant. It protects the bones against the wearing effects of friction—bones rub against the lubricating synovial fluid instead of against each other.
3. One might argue that friction prevents earthquakes in the sense that large blocks of rock are held still because of the friction between them. However, friction truly is implicated as a cause of earthquakes because if there were no friction, the blocks of rock could move along one another smoothly, never building up the strain that is released violently and suddenly in an earthquake.

Hang Time

1. Your speed is zero at the top of your jump.

Answers to Exercises

1. Aristotle would likely say the ball slows to reach its natural state. Galileo would say the ball is encountering friction, an unbalanced force that slows it.
3. mass
5. The density of lead is 11.3 g/cm^3. The density of aluminum is 5.4 g/2 cm^3 = 2.7 g/cm^3.
7. Maximum, 25 N + 15 N = 40 N. Minimum, 25 N − 15 N = 10 N.
9. From $\Sigma F = 0$, the upward forces are 400 N + tension in right scale. This sum must equal the downward forces of 250 N + 300 N + 300 N. Arithmetic shows the reading on the right scale is 450 N.
11. Each scale shows half her weight.
13. Yes, the forces are equal and opposite and cancel to zero, putting the person in equilibrium.
15. No, for the force of gravity acts on the object. Its motion is undergoing change, as a moment later should be evident.
17. Relative speed is 2 km/h.
19. More than 2 hours, for you cannot maintain an average speed of 60 miles/hour without exceeding the speed limit. You begin at zero, and end at zero, so even if there's no slowing down along the way you'll have to exceed 60 mi/h to average 60 mi/h. So it will take you more than 2 hours.
21. Distance increases as the square of the time, so each successive distance covered is greater than the preceding distance covered.
23. Both hit the ground with the same speed (but not in the same time).
25. The ball on B finishes first, for its average speed along the lower part as well as the down and up slopes is greater than the average speed of the ball along track A.

Solutions to Chapter 2 Problems

1. a. 30 N + 20 N = 50 N. b. 30 N − 20 N = 10 N.
3. From $\Sigma F = 0$, friction equals weight, mg, = (100 kg)(9.8 m/s^2) = 980 N.
5. $$a = \frac{\textit{change in velocity}}{\textit{time interval}} = \frac{-90 \text{ km/h}}{10 \text{ s}} = \frac{(-90 \text{ km/hr}) \times (1 \text{ hr}/3600 \text{ s}) \times (1000 \text{ m/km})}{10 \text{ s}}$$
   $= -2.5$ km/h · s
   (The vehicle decelerates at 2.5 km/h · s)
7.

Time (*in seconds*)	Velocity (*in meters/second*)	Distance (*in meters*)
0	0	0
1	10	5
2	20	20
3	30	45
4	40	80
5	50	125
6	60	180
7	70	245
8	80	320
9	90	405
10	100	500

Chapter 3: Newton's Laws of Motion

Answers to Integrated Science Concepts

Gliding

1. Gliding describes a mode of locomotion in which animals move through the air in a controlled fall.

3. "Flying" squirrels have large flaps of skin between their front and hind legs; draco lizards have long extendable ribs that support large gliding membranes; "flying" frogs have very long toes with extensive webbing between them; gliding geckos have skin flaps along their sides and tails in addition to webbed toes.

Animal Locomotion

1. In animal locomotion, an animal typically pushes against some medium (the ground, water, or air) that pushes back on it, providing the force needed for the animal to accelerate.

3. The force of friction between your back foot and the floor pushes you forward.

Answer to Exercises

1. Poke or kick the boxes. The one that more greatly resists a change in motion is the one with the greater mass—the one filled with sand.

3. Newton's first law again—when you jump, you tighten the disks. This is similar to how you can tighten a hammerhead by banging it against a surface. The greater inertia of the massive hammerhead makes it harder to stop than the less massive hammer handle. Similarly, when you jump you tighten your vertebrae. This effect also explains why you're shorter at the end of the day. At night, while lying prone, relaxation undoes the compression and you're taller!

5. You exert a force to overcome the force of friction. This makes the net force zero, which is why the wagon moves without acceleration. If you pull harder, then net force will be greater than zero and acceleration will occur.

7. The force that you exert on the ground is greater than your weight, for you momentarily accelerate upward. Then the ground simultaneously pushes upward on you with the same amount of force.

9. Only on hill B does the acceleration along the path decrease with time, for the hill becomes less steep as motion progresses. When the hill levels off, acceleration will be zero. On hill A, acceleration is constant. On hill C, acceleration increases as the hill becomes steeper. In all three cases, speed increases.

11. Air resistance on the thrown object decreases the net force on it ($mg - R$), making its acceleration less than that of free fall.

13. Neither a stick of dynamite nor anything else "contains" force. We will see later that a stick of dynamite contains *energy*, which is capable of producing forces when an interaction of some kind occurs.

15. 1000 N.

17. In accord with Newton's Third Law, the force on each will be of the same magnitude. But the effect of the force (acceleration) will be different for each because of the different mass. The more massive truck undergoes less change in motion than the motorcycle.

19. The person with twice the mass slides half as far as the twice-as-massive person. That means the lighter one slides 4 feet and the heavier one slides 8 feet (for a total of 12 feet).

21. In accord with Newton's Third Law, Steve and Gretchen are touching each other. One may initiate the touch, but the physical interaction can't occur without contact between both Steve and Gretchen. Indeed, you cannot touch without being touched!

23. The terminal speed attained by the falling cat is the same whether it falls from 50 stories or 20 stories. Once terminal speed is reached, falling extra distance does not affect the speed. (The low terminal velocities of small creatures enables them to fall without harm from heights that would kill larger creatures.)

25. Before reaching terminal velocity, weight is greater than air resistance. After reaching terminal velocity, both weight and air resistance are of the same magnitude. Then the net force and acceleration are both zero.

27. Air resistance is not really negligible for so high a drop, so the heavier ball does strike the ground first. But, although a twice-as-heavy ball strikes first, it falls only a little faster, not twice as fast, which is what followers of Aristotle believed. Galileo recognized that the small difference is due to friction and would not be present if there were no friction.

29. A hammock stretched tightly has more tension in the supporting ropes than one that sags. The tightly stretched ropes are more likely to break.

31. (a) As shown. (b) Yes. (c) Because the stone is in equilibrium.

33. It would be the same except that the upward vector would be absent. Only the downward *mg* vector acts.

35. (a) Weight and normal force only. (b) As shown.

Solutions to Chapter 3 Problems

1. Constant velocity means zero acceleration, and therefore zero net force. So the friction force must be equal to the bear's weight, *mg*.

3. Acceleration $a = F_{net}/m = (20\text{ N} - 12\text{ N})/2\text{ kg} = 8\text{ N}/2\text{ kg} = 4\text{ m/s}^2$.

5. Acceleration $a = F_{net}/m = 2\text{ N}/2\text{ kg} = 1\text{ m/s}^2$, the same.

7. Acceleration $a = F_{net}/m = (4 \times 30{,}000\text{ N})/30{,}000\text{ kg} = 4\text{ m/s}^2$.

9. $F = ma = m\,\Delta v/\Delta t = 0.003\text{ kg} \times [(25\text{ m/s})/0.05\text{ s}] = 1.5\text{ N}$.

11. By the Pythagorean theorem,
$V = \sqrt{[(3\text{ m/s})^2 + (4\text{ m/s})^2]} = 5\text{ m/s}$.

13. a. Force of air resistance will be equal to her weight, *mg*, or 500 N. b. She'll reach the same air resistance, but at a slower speed, 500 N. c. The answers are the same, but for different speeds. In each case she attains equilibrium (no acceleration).

Chapter 4: Momentum and Energy

Answers to Integrated Science Concepts

The Impulse–Momentum Relationship in Sports

1. a. An extended hand has room to move backward when the ball is caught. This stretches the time, resulting in less force; b. The force of impact will be less if momentum

changes over a long time. By making t long, F will be smaller; c. The shorter time is accompanied by a greater force when the momentum of the arm is reduced. The impulse will be greater if her hand bounces from the bricks. If the time of contact is not increased, a greater force is then exerted on the bricks (and on her hand).

3. The impulse will be greater if her hand bounces back from the bricks. If the time of contact is not increased, a greater force is then exerted on the bricks (and on her hand.)

Glucose: Energy for Life

1. The "burning" that goes on in cells differs from the burning or combustion of a log on a campfire in that the cellular process is much slower and more controlled.

Answer to Exercises

1. A steady collapse in a crash extends the time that the seat belt and air bags slow the passengers less violently.
3. The time during which momentum decreases is lengthened, thereby decreasing the force that brings the wine glass to rest. Less force means less chance of breaking.
5. Impact with a boxing glove extends the time during which momentum of the fist is reduced, and lessens the force. A punch with a bare fist involves less time and therefore more force.
7. Without this slack, a locomotive might simply sit still and spin its wheels. The loose coupling enables a longer time for the entire train to gain momentum, requiring less force of the locomotive wheels against the track. In this way, the overall required impulse is broken into a series of smaller impulses. (This loose coupling can be very important for braking as well.)
9. The momentum of both bug and bus change by the same amount because both the amount of force and the time, and therefore the amount of impulse, are the same on each. Momentum is conserved. Speed is another story. Because of the huge mass of the bus, its reduction of speed is very tiny—too small for the passengers to notice.
11. Oops, the conservation of momentum was overlooked. Your momentum forward equals (approximately) the momentum of the recoiling raft.
13. The magnitude of force, impulse, and change in momentum will be the same for each. The Civic undergoes the greater acceleration because its mass is less.
15. If the rocket and its exhaust gases are treated as a single system, the forces between rocket and exhaust gases are internal, and momentum in the rocket-gases system is conserved. So, any momentum given to the gases is equal and opposite to momentum given to the rocket. A rocket attains momentum by giving momentum to the exhaust gases.
17. They may fly in opposite directions wherein the momenta cancel to zero. But, if moving, there is no way kinetic energy can cancel. Hence, the difference between a vector quantity (momentum) and a scalar quantity (kinetic energy).
19. The KE of the tossed ball relative to occupants in the airplane does not depend on the speed of the airplane. The KE of the ball relative to observers on the ground below, however, is a different matter. KE, like velocity, is relative.
21. In the physical scientific sense, energy cannot be created or destroyed. When consuming energy, however, we can use more than we need and be wasteful. So we speak of saving energy, using it more wisely. Not in the science sense, conserving it.
23. Twenty-five times as much energy (as speed is squared for kinetic energy).
25. The KE of a pendulum bob is maximum where it moves fastest, at the lowest point; PE is maximum at the uppermost points. When the pendulum bob swings by the point that marks half its maximum height, it has half its maximum KE, and its PE is halfway between its minimum and maximum values. If we define PE = 0 at the bottom of the swing, the place where KE is half its maximum value is also the place where PE is half its maximum value, and KE = PE at this point. (In accordance with energy conservation: total energy = KE + PE).
27. The design is impractical. Note that the summit of each hill on the roller coaster is the same height, so the PE of the car at the top of each hill would be the same. If no energy were spent in overcoming friction, the car would get to the second summit with as much energy as it starts with. But in practice there is considerable friction, and the car would not roll to its initial height and have the same energy. So the maximum height of succeeding summits should be lower to compensate for friction.
29. Yes, a car burns more gasoline when its lights are on. The overall consumption of gasoline does not depend on whether or not the engine is running. Lights and other devices run off the battery, which "run down" the battery. The energy used to recharge the battery ultimately comes from the gasoline.

Solutions to Chapter 4 Problems

1. From $Ft = \Delta mv$, $F = \Delta mv/t = (1000\text{ kg})(20\text{ m/s})/10\text{ s} = 2000\text{ N}$.
   (Can you see this could also be solved by Newton's Second Law:

   $F = ma = (1000\text{ kg})(20\text{ m/s}/10\text{ s}) = 2000\text{ N}$.)

3. a. Momentum before lunch = momentum after lunch
   $(5\text{ kg})(1\text{ m/s}) + 0 = (5\text{ kg} + 1\text{ kg})v$
   $5\text{ kg}\cdot\text{m/s} = (6\text{ kg})\ v$
   $v = 5/6\text{ m/s}$

   b. Momentum before lunch = momentum after lunch
   $(5\text{ kg})\ (1\text{ m/s}) + 1\text{ kg}\ (-4\text{ m/s}) = (5\text{ kg} + 1\text{ kg})\ 3v$
   $5\text{ kg}\cdot\text{m/s} - 4\text{ kg}\cdot\text{m/s} = (6\text{ kg})\ v$
   $v = 1/6\text{ m/s}$

5. The freight cars have only half the KE possessed by the single car before collision. Here's how to figure it:
   $KE_{before} = 1/2\ mv^2$.
   $KE_{after} = 1/2\ (2\ m)(v/2)^2 = 1/2\ (2m)\ v^2/4 = 1/4\ mv^2$.
   What becomes of this energy? Most of it goes into nature's graveyard—thermal energy.

7. $(Fd)_{input} = (Fd)_{output}$
   $(100\text{ N} \times 10\text{ cm})_{input} = (? \times 1\text{ cm})_{output}$
   So, we see that the output force is 1000 N (or less if the efficiency is less than 100%).

9. Efficiency = (mechanical power output)/(power input) = 100 W/1000 W = 1/10, or 10%.

Chapter 5: Gravity

Answers to Integrated Science Concepts

Your Biological Gravity Detector

1. Yes, people have sense organs that allow them to sense gravity. They are called the vestibular organs and they are located in the inner ear.

Center of Gravity of People

1. Its center of gravity lies within its base.

Escape Speed

1. Minumum speed is 8 km/s; maximum speed is 11.2 km/s. An object projected from Earth at a speed greater than 11.2-km/s will escape Earth's gravitational field and no longer orbit Earth.

Answers to Exercises

1. In accord with the law of inertia, the Moon would move in a straight-line path instead of circling both the sun and Earth.
3. The force of gravity is the same on each because the masses are the same, as Newton's equation for gravitational force verifies. When dropped the crumpled paper falls faster only because it encounters less air drag than the sheet.
5. The Earth and Moon pull on each other equally in a single interaction. In accord with Newton's third law, the pull of Earth on the Moon is equal and opposite to the pull of the Moon on Earth.
7. The high-flying jet plane is not in free fall. It moves at approximately constant velocity so a passenger experiences no net force. The upward support force of the seat matches the downward pull of gravity, providing the sensation of weight. The orbiting space vehicle, on the other hand, is in a state of free fall. No support force is offered by a seat, for it falls at the same rate as the passenger. With no support force, the force of gravity on the passenger is not sensed as weight.
9. The pencil has the same state of motion that you have. The force of gravity on the pencil causes it to accelerate downward alongside of you. Although the pencil hovers relative to you, it and you are falling relative to Earth.
11. You'd follow a straight-line path; the force preventing this motion while you are riding is the centripetal force applied to you by the Merry-Go-Round platform.
13. When air drag is negligible, the vertical component of motion is identical to free fall.
15. Minimum speed occurs at the top, which is the same as the horizontal component of velocity anywhere along the path.
17. The bullet falls beneath the projected line of the barrel. To compensate for the bullet's fall, the barrel is elevated. How much elevation depends on the velocity and distance to the target. Correspondingly, the gunsight is raised so the line of sight from the gunsight to the end of the barrel extends to the target. If a scope is used, it is tilted downward to accomplish the same line of sight.
19. Neither the speed of a falling object (without air resistance) nor the speed of a satellite in orbit depends on its mass. In both cases, a greater mass (greater inertia) is balanced by a correspondingly greater gravitational force, so the acceleration remains the same ($a = F/m$, Newton's second law).
21. The Moon has no atmosphere (because escape velocity at the Moon's surface is less than the speeds of any atmospheric gases). A satellite 5 km above the Earth's surface is still in considerable atmosphere, as well as in range of some mountain peaks. Atmospheric drag is the factor that most determines orbiting altitude.

Solutions to Chapter 5 Problems

1. From $F = GmM/d^2$, three times d squared is $9\ d^2$, which means the force is one-ninth of surface weight.
3. From $F = GmM/d^2$, 10 d squared is $100\ d^2$, with a force 100 times smaller.
5. From $F = GmM/d^2$, five times d squared is $1/25\ d^2$, with a force 25 times greater.
7. From $F = G2m2M/(2\ d^2) = 4/4\ (GmM/d^2)$, with the same force of gravitation.
9. From $F = GmM/d^2 = [(6.67 \times 10^{-11}\ \text{N} \cdot \text{m}^2/\text{kg}^2)(6 \times 10^{24}\ \text{kg})(2 \times 10^{30}\ \text{kg})]/(1.5 \times 10^{11}\ \text{m})^2 = 3.6 \times 10^{22}$ N.
11. Use the equation: $v = distance/time$ where distance is the circumference of the Earth's orbit and time is 1 year. Then $$v = \frac{d}{t} = \frac{2\pi r}{1 \text{ year}} = \frac{2\pi(1.5 \times 10^{11} \text{ m})}{365 \text{ day} \times 24 \text{ h/day} \times 3600 \text{ s/h}} =$$ 3×10^4 m/s $=$ 30 km/s.

Chapter 6: Heat

Answers to Integrated Science Concepts

Entropy—The Universal Tendency Toward Disorder

1. When energy is transformed, some of it is converted to heat. Heat is the least useful form of energy because it is the least concentrated form.
3. No, it is not thermodynamically favored because the products have less entropy than the reactants. So, this reaction will not proceed without an energy input.

The Specific Heat Capacity of Water Affects Global Temperature

1. North Atlantic water cools, and releases energy to the air, which blows over Europe.
3. In winter months, when the water is warmer than the air, the air is warmed by the water to produce a seacoast climate warmer than inland. In summer months, when the air is warmer than the water, the air is cooled by the water to produce a seacoast climate cooler than inland. This is why seacoast communities and especially islands do not experience the high and low temperature extremes that characterize inland locations.

Answers to Exercises

1. Inanimate things such as tables, chairs, furniture, and so on, have the same temperature as the surrounding air (assuming they are in thermal equilibrium with the air—i.e., no sudden gush of different-temperature air or such). People and other mammals, however, generate their own heat and have body temperatures that are normally higher than air temperature.
3. Since Celsius degrees are larger than Fahrenheit degrees, an increase of 1°C is larger. It's 9/5 as large.
5. Only a small percentage of the electric energy that goes into lighting a lamp becomes light. The rest is thermal energy. But even the light is absorbed by the surround-

ings, and also ends up as thermal energy. So by the first law, all the electrical energy is ultimately converted to thermal energy. By the second law, organized electrical energy degenerates to the more disorganized form, thermal energy.

7. Gas pressure increases in the can when heated, and decreases when cooled. The pressure that a gas exerts depends on the average kinetic energy of its molecules, therefore, on its temperature.

9. The hot rock will cool and the cool water will warm, regardless of the relative amounts of each. The amount of temperature change, however, does depend in great part on the relative masses of the materials. For a hot rock dropped into the Atlantic Ocean, the change in temperature would be too small to measure. Keep increasing the mass of the rock or keep decreasing the mass of the ocean and the change will be evident.

11. Different substances have different thermal properties due to differences in the way energy is stored internally in the substances. When the same amount of heat produces different changes in temperatures in two substances of the same mass, we say they have different specific heat capacities. Each substance has its own characteristic specific heat capacity. Temperature measures the average kinetic energy of random motion, but not other kinds of energy.

13. Water is an exception. Below 4°C, it expands when cooled.

15. Gas is sold by volume. The gas meter that tallies your gas bill operates by measuring the number of volume units (such as cubic feet) that pass through it. Warm gas is expanded gas and occupies more space, and if it passes through your meter, it will be registered as more gas than if it were cooled and more compact. The gas company gains if gas is warm when it goes through your meter because the same amount of warmer gas has a greater volume.

17. Brass expands and contracts more than iron for the same changes in temperature. Since they are both good conductors and are in contact with each other, one cannot be heated or cooled without also heating or cooling the other. If the iron ring is heated, it expands—but the brass expands even more. Cooling the two will not result in separation either, for even at the lowest temperatures the shrinkage of brass over iron would not produce separation.

19. Water has the greatest density at 4°C; therefore, either cooling or heating at this temperature will result in an expansion of the water. A small rise in water level would be ambiguous and make a water thermometer impractical in this temperature region.

21. No, the coat is not a source of heat, but merely keeps the thermal energy of the wearer from leaving rapidly.

23. The main reason for serving potatoes wrapped in aluminum foil is to increase the time that the potatoes remain hot after being removed from the oven. Heat transfer by radiation is minimized as radiation from the potatoes is internally reflected, and heat transfer by convection is minimized as circulating air cannot make contact with the shielded potatoes. The foil also serves to retain moisture.

25. Heat from the relatively warm ground is conducted by the gravestone to melt the snow in contact with the gravestone. Likewise for trees or any materials that are better conductors of heat than snow, and that extend into the ground.

27. The smoke, like hot air, is less dense than the surroundings and is buoyed upward. It cools with contact with the surrounding air and becomes more dense. When its density matches that of the surrounding air, its buoyancy and weight balance and rising ceases.

29. Put the cream in right away for at least three reasons. Since black coffee radiates more heat than white coffee, make it whiter right away so it won't radiate and cool so quickly while you are waiting. Also, by Newton's Law of Cooling, the higher the temperature of the coffee above the surroundings, the greater will be the rate of cooling—so again add cream right away and lower the temperature to that of a reduced cooling rate, rather than allowing it to cool fast and then bring the temperature down still further by adding the cream later. Also—by adding the cream, you increase the total amount of liquid, which for the same surface area, cools more slowly.

Solutions to Chapter 6 Problems

1. Work the hammer does on the nail is given by $F \times d$, and the temperature change of the nail can be found from using $Q = cm\,\Delta T$. First, we get everything into more convenient units for calculating: 5 grams = 0.005 kg; 6 cm = 0.06 m. Then $F \times d = 500\text{ N} \times 0.06\text{ m} = 30\text{ J}$, and $30\text{ J} = (0.005\text{ kg})(450\text{ J/kg°C})(\Delta T)$ which we can solve to get $\Delta T = 30/(0.005 \times 450) = 13.3°\text{C}$. (You will notice a similar effect when you remove a nail from a piece of wood. The nail that you pull out is noticeably warm.)

3. Raising the temperature of 10 gm of copper by one degree takes $10 \times 0.092 = 0.92$ calories, and raising it through 100 degrees takes 100 times as much, or 92 calories. By the formula, $Q = mc\Delta T = (10\text{ g})(0.092\text{ cal/g°C})(100°\text{C}) = 92\text{ cal}$. Heating 10 grams of water through the same temperature difference takes 1,000 calories, more than ten times more than for the copper—another reminder that water has a large specific heat capacity.

5. $0.5\,mgh = cm\Delta T$
$\Delta T = 0.5\,mgh/cm = 0.5\,gh/c =$
$(0.5)(9.8\text{ m/s}^2)(100\text{ m})/450\text{ J/kg} = 1.1°\text{C}$.
Again, note that the mass cancels, so the same temperature would hold for any mass ball, assuming half the heat generated goes into warming the ball. As in the previous problem, the units check because $1\text{ J/kg} = 1\text{ m}^2/\text{s}^2$.

Chapter 7: Electricity and Magnetism

Answers to Integrated Science Concepts

Electric Shock

1. Electric current passing through the body produces electric shock.

3. In the first case, the current passes through your chest, in the second case, current passes though your arm only. You can cut off your arm and survive, but cannot survive without your heart.

Earth's Magnetic Field and the Ability of Organisms to Sense It

1. Animals sense Earth's magnetic field and orient themselves with respect to it.

Answers to Exercises

1. Charge and mass are alike in that both determine the strength of a force between objects. Both appear in an inverse-square law of force. They differ in that charge can be positive or negative while mass is always positive. They differ also in the strength of force they determine.

3. The leaves, like the rest of the electroscope, acquire charge from the charged object and repel each other because they both have the same sign of charge. The weight of the conducting gold foil is so small that even tiny forces are clearly evident.

5. The forces they exert on each other are still the same! Newton's third law applies to electrical forces as well as all forces.

7. For both electricity and heat, the conduction is via electrons, which in a metal are loosely bound, easy flowing, and easy to get moving. (Many fewer electrons in metals take part in heat conduction than in electric conduction, however.)

9. The forces on the electron and proton will be equal in magnitude, but opposite in direction. Because of the greater mass of the proton, its acceleration will be less than that of the electron, and be in the direction of the electric field. How much less? Since the mass of the proton is nearly 2000 times that of the electron, its acceleration will be about 1/2000 that of the electron. The greater acceleration of the electron will be in the direction opposite to the electric field. The electron and proton accelerate in opposite directions.

11. Only circuit number 5 is complete and will light the bulb. (Circuits 1 and 2 are "shortcircuits" and will quickly drain the cell of its energy. In circuit 3 both ends of the lamp filament are connected to the same terminal and are threfore at the same potential. Only one end of the lamp filament is connected to the cell in circuit 4.)

13. Current will be greater in the bulb connected to the 220-volt source. Twice the voltage would produce twice the current if the resistance of the filament remained the same. (In practice, the greater current produces a higher temperature and greater resistance in the lamp filament, so the current is greater than that produced by 110 volts, but appreciably less than twice as much for 220 volts. A bulb rated for 110 volts has a very short life when operated at 220 volts.)

15. There is less resistance in the higher wattage lamp. Since power = current × voltage, more power for the same voltage means more current. And by Ohm's law, more current for the same voltage means less resistance. (Algebraic manipulation of the equations $P = IV$ and $I = V/R$ leads to $P = V^2/R$.)

17. If the parallel wires are closer than the wingspan of birds, a bird could short circuit the wires by contact with its wings, be killed in the process, and possibly interrupt the delivery of power.

19. As more bulbs are connected in series, more resistance is added to the single circuit path and the resulting current produced by the battery is diminished. This is evident in the dimmer light from the bulbs. On the other hand, when more bulbs are connected to the battery in parallel, the brightness of the bulbs is practically unchanged. This is because each bulb in effect is connected directly to the battery with no other bulbs in its electrical path to add to its resistance. Each bulb has its own current path.

21. An electric field surrounds a stationary electric charge. An electric field and a magnetic field surround a moving electric charge. (And a gravitational field also surrounds both).

23. Yes, the physics is similar. The iron nails become magnetized as their domains are induced into alignment by the magnet's strong magnetic field. Similarly, the electric field of the balloon induces a charge on the surface of the wall.

25. Charged particles moving through a magnetic field are deflected most when they move at right angles to the field lines, and least when they move parallel to the field lines. If we consider cosmic rays heading toward the Earth from all directions and from great distance, those descending toward northern Canada will be moving nearly parallel to the magnetic field lines of the Earth. They will not be deflected very much, and secondary particles they create high in the atmosphere will also stream downward with little deflection. Over regions closer to the equator, like Mexico, the incoming cosmic rays move more nearly at right angles to the Earth's magnetic field, and many of them are deflected back out into space before they reach the atmosphere. The secondary particles they create are less intense at the Earth's surface. (This "latitude effect") provided the first evidence that cosmic rays from outer space consist of charged particles—mostly protons, as we now know.)

27. Agree with your friend, for light is electromagnetic radiation having a frequency that matches the frequency to which our eyes are sensitive.

Solutions to Chapter 7 Problems

1. By the inverse-square law, twice as far is 1/4 the force; 5-N. The solution involves relative distance only, so the magnitude of charges is irrelevant.

3. a. $\Delta V = \dfrac{\text{energy}}{\text{charge}} = \dfrac{12\text{ J}}{0.0001\text{ C}} = 120{,}000$ volts.

   b. ΔV for twice the charge is $\dfrac{24\text{ J}}{0.0002} =$ same 120 kV.

5. From power = current × voltage, current = $\dfrac{\text{power}}{\text{voltage}} = \dfrac{1200\text{ W}}{120\text{ V}} = 10$ A.

   From the formula derived above, resistance = $\dfrac{\text{voltage}}{\text{current}} = \dfrac{120\text{ V}}{10\text{ A}} = 12$ ohms.

Chapter 8: Waves-Sound and Light

Answers to Integrated Science Concepts

Sensing Pitch

1. The higher the pitch of a sound, the higher its frequency.

Mixing Colored Lights

1. All colors of light can be obtained from red, green, and blue light as described by the principles of additive color mixing. Since any color can be obtained by sensing red, green, and blue light, sensing these three primary colors of light is all that is needed.

The Doppler Shift and the Expanding Universe

1. The pitch of the buzzer increases as it gets closer to your ears and decreases as it moves away. The reason is that, as the Doppler effect explains, the frequency of the waves increases as the source of sound approaches, and decreases as the sound source gets farther away.

Answers to Exercises

1. To produce a transverse wave with a Slinky, shake it to and fro in a direction that is perpendicular to the length of the Slinky itself (as with the garden hose in the previous exercise). To produce a longitude wave, shake it to and fro along the direction of its length, so that a series of compressions and rarefactions is produced.
3. Light travels about a million times faster than sound in air, so you see a distant event a million times sooner than you hear it.
5. The lower strings are resonating with the upper strings.
7. The fundamental source of electromagnetism radiation is oscillating electric charges, which emit oscillating electric and magnetic fields.
11. Clouds are transparent to ultraviolet light, which is why clouds offer no protection from sunburn. Glass, however, is opaque to ultraviolet light, and will therefore shield you from sunburn.
13. They are most likely to be noticed if they are yellow-green. That is where the eye is most sensitive.
15. Only light from card number 2 reaches her eye.

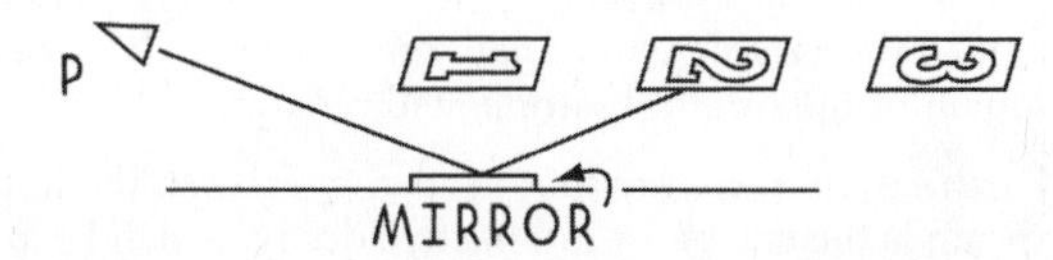

17. You would throw the spear below the apparent position of the fish, because the effect of refraction is to make the fish appear closer to the surface than it really is. But in zapping a fish with a laser, make no corrections and aim directly at the fish. This is because the light from the fish you see has been refracted in getting to you, and the laser light will refract along the same path in getting to the fish. A slight correction may be necessary, depending on the colors of the laser beam and the fish.
19. A cyan wave absorbs red light and reflects blue-green light.
21. If a single disturbance at some unknown distance away sends longitudinal waves at one known speed, and transverse waves at a lesser known speed, and you measure the difference in time of the waves as they arrive, you can calculate the distance. The wider the gap in time, the greater the distance—which could be in any direction. If you use this distance as the radius of the circle on a map, you know the disturbance occurred somewhere on that circle. If you telephone two friends who have made similar measurements of the same event from different locations, you can transfer their circles to your map, and the point where the three circles intersect is the location of the disturbance.
23. Radio waves are much larger and therefore diffract more than the shorter waves of light.
25. *Electric eye:* A beam of light is directed to a photosensitive surface that completes the path of an electric circuit. When the beam is interrupted, the circuit is broken, compromising a switch for another circuit. *Light meter:* the variation of photoelectric current with variations in light intensity. *Sound track:* An optical sound track on motion picture film is a strip of emulsion of variable density that transmits light onto a photoelectric surface, which in turn produces a variable current. This current is amplified and activates a speaker.
27. Letters to Grandpa should discuss the wave behavior of light as indicated by the wave properties it exhibits including interference. Letters should also discuss the particle behavior of light in its interaction with matter: the photoelectric effect. Taking into account all observations of light, scientists conclude it has properties of both waves and particles and therefore is both—"a wavicle."

Solutions to Chapter 8 Problems

1. a. $f = 1/T = 1/0.10s = 10$ Hz; b. $f = 1/5 = 0.2$ Hz; c. $f = 1/(1/60)$ s $= 60$ Hz.
3. From $c = f\lambda$, $\lambda = \frac{c}{f} = \frac{3 \times 10^8 \text{m/s}}{6 \times 10^{14}\text{Hz}} = 5 \times 10^{-7}$ m, or 500 nanometers. This is 5000 times larger than the size of an atom, which is 0.1 nanometer. (The nanometer is a common unit of length in atomic and optical physics.)
5. $\lambda_{air} = 600$ nm $= (6 \times 10^{-7}$ m). So $\lambda_{water} = (0.75)(6 \times 10^{-7}$ m) $= 4.5 \times 10^{-7}$ m $= 450$ nm. And $\lambda_{Plexiglas} = (0.67)(6 \times 10^{-7}$ m) $= 4.0 \times 10^{-7}$ m $= 400$ nm
7. You and your image are both walking at 2 m/s.

Chapter 9: The Atom

Answers to Integrated Science Concepts

Physical and Conceptual Models

1. When we use a scanning tunneling microscope, we see atoms indirectly because we are seeing a computer-generated diagram of the contours of atoms.
3. A physical model is tangible while a conceptual model is a mental image.

The Shell Model

1. The valence electrons are the electrons most responsible for the properties of an atom.
3. There are 8 orbitals present in the third shell.

Answers to Exercises

1. The cat leaves a trail of molecules on the grass. These in turn leave the grass and mix with the air, where they enter the dog's nose, activating its sense of smell.
3. You really are a part of every person around you in the sense that you are composed of atoms not only from every person around you, but from every person who ever lived on Earth!
5. With every breath of air you take, it is highly likely that you inhale one of the atoms exhaled during your very first breath. This is because the number of atoms of air in your lungs is about the same as the number of breaths of air in the atmosphere of the world.
7. The one on the far right where the nucleus is not visible.
9. The remaining nucleus is that of Carbon-12.
11. The outsides of the atoms of the chair are made of negatively charged electrons, as are the outsides of the atoms that make up your body. Atoms don't pass through one another because of the repulsive forces that occur between these electrons. When you sit on the chair these repulsive forces hold you up against the force of gravity, which is pulling you downward.
13. The iron atom is electrically neutral when it has 26 electrons to balance its 26 protons.

15. The carbon atoms that make up Leslie's hair or anything else in this world originated in the explosions of ancient stars.
17. Observe the atomic spectra of each using a spectroscope.
19. The blue frequency is a higher frequency and therefore corresponds to a higher energy level transition.
21. The first visible color will be red because this is the visible frequency with the lowest amount of energy per photon.
23. The spectral patterns emanating from the Sun indicate the spectral patterns of heated iron atoms.
25. Twice the frequency means twice the energy.
27. When a wave is confined, it is reinforced only at particular frequencies. The electron wave being confined to the atom, therefore, only exhibits particular frequencies where each frequency represents a discrete energy value.
29. Because of its wave nature, it would be better to say that the electron actually exists in both lobes at the same time.
31. The *s* orbital is a sphere which cannot be rotated without being in the same orientation . . . the sphere is perfectly symmetrical!
33. This would make the element 118.
35. Both the potassium and sodium atoms are in group 1 of the periodic table. The potassium atom, however, is larger than the sodium atoms because it contains an additional shell of electrons.
37. Open ended

Problems

1. $[(0.7553) \times (34.97)] + [(0.2447) \times (36.95)] = 35.4545$
3. A 50:50 mix of Br-80 and Br-81 would result in an atomic mass of about 80.5, while a 50:50 mix of Br-79 and Br-80 would result in an atomic mass of about 79.5. Neither of these is as close to the value reported in the periodic table as is a 50:50 mix of Br-79 and Br-81, which would result in an atomic mass of about 80.0. The answer is (c).
5. a. 10^4 atoms (length 10^{-6} m divided by size 10^{-10} m).
b. 10^8 atoms ($10^4 \times 10^4$).
c. 10^{12} atoms ($10^4 \times 10^4 \times 10^4$).
d. $10,000 buys a good used car, for instance. $100 million buys a few jet aircraft and an airport on which to store them, for instance. $1 trillion buys a medium-sized country, for instance. (Answers limited only by the imagination of the student.)

Chapter 10: Nuclear Physics

Answers to Integrated Science Concepts

Doses of Radiation

1. Since the biological effects of radiation exposure are cumulative, the radiation received from artificial sources increases the potential hazards of radiation.
3. The most significant artificial source of radioactivity is smoking, which can be eliminated by giving up this habit, which is hazardous in many ways.
5. A cell can survive a dose of radiation that would otherwise be lethal if the dose is spread over a long period of time to allow intervals for healing.

Isotopic Dating

1. The proportion of lead and uranium tells us that the oldest rocks are nearly 3.7 billion years old. Rock, for the Moon, dates 4.2 billion years.
3. Radioactive isotopes are produced by cosmic-ray-induced transmutation.

Answers to Exercises

1. Radioactivity is a part of nature, going back to the beginning of time.
3. It is impossible for a hydrogen atom to eject an alpha particle, for an alpha particle is composed of a pair of hydrogen isotopes (deuterium). It is equally impossible for a one-kilogram melon to spontaneously break into four one-kilogram melons.
5. The proton "bullets" need enough momentum to overcome the electric force of repulsion they experience once they get close to the atomic nucleus.
7. The strong nuclear force holds the nucleons of the nucleus together while the electric force pushes these nucleons apart.
9. No, it will not be entirely gone. Rather, after 1 day one-half of the sample will remain while after 2 days, one-fourth of the original sample will remain.
11. Earth's natural energy that heats the water in the hot spring is the energy of radioactive decay, which keeps Earth's interior molten. Radioactivity heats the water, but doesn't make the water itself radioactive. The warmth of hot springs is one of the "nicer effects" of radioactive decay. You'll most likely encounter more radioactivity from the granite outcroppings of the foothills than from a nearby nuclear power plant. Furthermore, at high altitude you'll be exposed to increased cosmic radiation. But these radiations are not appreciably different than the radiation one encounters in the "safest" of situations. The probability of dying from something or other is 100%, so in the meantime we all should enjoy life anyway!
13. Gamma radiation is generally the most harmful radiation because it is so penetrating. Alpha and beta radiation is dangerous if you ingest radioactive material, which is comparatively uncommon.
15. You can tell your friend who is fearful of the radiation measured by the Geiger counter that his attempt to avoid the radiation by avoiding the instrument that measures it, is useless. He might as well avoid thermometers on a hot day in effort to escape the heat. If it will console your fearful friend, tell him that he and his ancestors from time zero have endured about the same level of radiation he receives whether or not he stands near the Geiger counter. There are no better options.
17. Nuclear fission is a poor prospect for powering automobiles primarily because of the massive shielding that would be required to protect the occupants and others from the radioactivity, and the problem of radioactive waste disposal.
19. Because plutonium triggers more reactions per atom, a smaller mass will produce the same neutron flux as a somewhat larger mass of uranium. So, plutonium has a smaller critical mass than a smaller shape of uranium.
21. Plutonium has a short half-life (24,360 years), so any plutonium initially in the Earth's crust has long since decayed. The same is true for any heavier elements with even shorter half lives from which plutonium might orig-

inate. Trace amounts of plutonium can occur naturally in U-238 concentrations, however, as a result of neutron capture, where U-238 becomes U-239 and after beta emission becomes Np-239, and further beta emission to Pu-239. (There are elements in Earth's crust with half-lives even shorter than plutonium's, but these are the products of uranium decay; between uranium and lead in the periodic table of elements.)

23. A nucleus undergoes fission because the electric force of repulsion overcomes the strong nuclear force of attraction. This electric force of repulsions is of the very same nature as static electricity. So, in a way, your friend's claim that the explosive power of a nuclear bomb is due to static electricity is valid.
25. Less in the fission fragments.
27. To predict the energy release of a nuclear reaction, simply find the difference in the mass of the beginning nucleus and the mass of its configuration after the reaction (either fission or fusion). This mass difference (called the "mass defect") can be found from the curve of Figure 10.26 or from a table of nuclear masses. Multiply this mass difference by the speed of light squared: $E = mc^2$. That's the energy released!
29. If uranium were split into three parts, the segments would be nuclei of smaller atomic numbers, more toward iron on the graph of Figure 10.27. The resulting mass per nucleon would be less, and there would be more mass converted to energy in such a fissioning.
31. Letters to Grandma should discuss natural sources of radiation, including the radioactive decay that occurs in Earth's interior, which is as old as Earth itself. Letters also should relate the public's fear of radioactivity to the general lack of knowledge of this subject.

Solutions to Chapter 10 Problems

1. In accord with the inverse-square law, at 2 m, double the distance, the count rate will be 1/4 of 360, or 90 counts/minute. At 3 m, the count rate will be 1/9 of 360, or 40 counts/min.
3. At 3:00 PM (after 3 half-lives) there will be 1/8 of the original remaining, 0.125 grams. At 6:00 PM, after 3 more half-lives, there are 1/8 of 1/8 left, 0.016 grams. At 10:00 PM the amount remaining has halved ten times, which leaves (1/2)10, or about 1/1000 of the original. So, the remaining amount will be 0.001 g, or 1 mg.
5. It will take four half-lives to decrease to one-sixteenth the original amount. Four half-lives of cesium-137 corresponds to 120 years.
7. Your count is 10/5000 for the gallon you remove. That's a ratio of 1/500, which means the tank must hold 500 gallons of gasoline.
9. The atom of uranium has an initial mass m = 232.03174 amu. The products of radioactive decay have a combined mass of 228.02873 amu + 4.00260 amu = 232.03133 amu. Thus, a mass equal to 232.03174 amu − 232.03133 amu = 0.00041 amu has been "lost" or converted to energy in this reaction. Since 1 amu = 1.6605×10^{-27} kg, the converted mass = (0.00041 amu) $\times$ (1.6605×10^{-27} kg/1 amu) = 6.81×10^{-31} kg. By $E = mc^2$, the energy released in the decay of uranium to thorium and helium is: E = $(6.81 \times 10^{-31}\text{ kg})(3.0 \times 10^8\text{ m/s})^2 = 6.1 \times 10^{-14}\text{ J}$.

Part 2 Chemistry

Chapter 11: Investigating Matter

Answers to Integrated Science Concepts

Origin of the Moon

1. Summaries should discuss the chemical analysis of moon rocks, specifically the similarity of the chemical composition of moon rocks to Earth's mantle and the lack of water and other volatile compounds from moon rocks.

Evaporation Cools You Off, Condensation Warms You Up

1. Evaporation is a change of phase from liquid to gas at the surface of a liquid. Faster molecules are the ones that evaporate, leaving slower ones.
3. Steam gives up considerable energy when it changes phase, condensing to a liquid and wetting the skin.

Answers to Exercises

1. When looked at macroscopically, matter appears continuous. On the submicroscopic level, however, we find that matter is made of extremely small particles, such as atoms or molecules. Similarly, a TV screen looked at from a distance appears as a smooth continuous flow of images. Up close, however, we see this is an illusion. What really exists are a series of tiny dots (pixels) that change color in a coordinated way to produce the series of images.
3. At 25°C there is a certain amount of thermal energy available to all the submicroscopic particles of a material. If the attractions between the particles are not strong enough, the particles may separate from each other to form a gaseous phase. If the attractions are strong, however, the particles may be held together in the solid phase. We can assume, therefore, that the attractions among the submicroscopic particles of a material in its solid phase at 25°C are stronger than they are within a material that is a gas at this temperature.
5. At the cold temperatures of your kitchen freezer, water molecules in the vapor phase are moving relatively slowly, which makes it easier for them to stick to inner surfaces within the freezer or to other water molecules.
7. As the alcohol evaporates it soaks up energy from the table top which is thus cooled. This transfer of energy that occurs during a change in phase is discussed in more detail in Chapter 8.
9. As each kernel is heated, the water within each kernel is also heated to the point that it would turn into water vapor. The shell of the kernel, however, is air tight and this keeps the water as a superheated liquid. Eventually the pressure exerted by the superheated water exceeds the holding power of the kernel and the water bursts out as a vapor, which causes the kernel to pop. These are physical changes. The starches within the kernel, however, are also cooked by the high temperatures, and this is an example of a chemical change.
11. The gas particles take time to cross a room because they bump into each other as well as other particles in the air.
13. Across any period (horizontal row), the properties of elements gradually change until the end of the period. The next element in the next period has properties that are abruptly different.

15. The change from A to B represents a physical change because no new types of molecules are formed. The collection of blue/red molecules on the bottom of B represents these molecules in the liquid or solid phase after having been in the gaseous phase in A. This must occur with a decrease in temperature. At this lower temperature the purely red molecules are still in the gaseous phase which means that they have a lower boiling point, while the blue/red molecules have a higher boiling point.
17. Based upon its location in the periodic table we find that gallium, Ga, is more metallic in character than germanium, Ge. This means that gallium should be a better conductor of electricity. Computer chips manufactured from gallium, therefore, operate faster than chips manufactured from germanium. (Gallium has a low melting point of 30°C, which makes it impractical for use in the manufacture of computer chips. Mixtures of gallium and arsenic, however, have found great use in the manufacture of ultra fast, though relatively expensive, computer chips.)
19. Calcium is readily absorbed by the body for the building of bones. Since calcium and strontium are in the same atomic group they have similar physical and chemical properties. The body, therefore, has a hard time distinguishing between the two and strontium is absorbed just as though it were calcium.
21. At higher temperatures the molecules of the material have sufficient kinetic energy to break loose of the attractions to neighboring molecules to form new attractions to other neighboring molecules and so on. In this way, the molecules are able to tumble around one another much like marbles in a bag, which describes the liquid phase.
23. At lower temperatures the molecules of the material have insufficient kinetic energy to prevent the electrical attractions between them from holding them within fixed positions.

Solution to Chapter 11 Problems

1. $mgh = mL$, so $gh = L$ and $h = L/g$. $h = (334{,}000 \text{ J/kg})/(9.8 \text{ m/s}^2) = 34{,}000 \text{ m} = 34 \text{ km}$. Note that the mass cancels and the unit J/kg is the same as the unit m^2/s^2. So in the ideal case of no energy losses along the way, any piece of ice that freely falls 34 km would completely melt upon impact. Taking air resistance into account, only partial melting would occur.

Chapter 12: The Nature of Chemical Bonds

Answers to Integrated Science Concepts

How Geckos Walk on Walls—The Adhesive Force

1. The adhesive force is an interparticle force that acts between two different substances while the cohesive force is an interparticle force that acts between molecules of a single substance. The force between a wall and molecules in a gecko's spatulae is an adhesive force; cohesive forces between water molecules pull water into spherical drops.

Mixtures

1. A skin cell and Earth's layered atmosphere are examples of heterogeneous mixtures.

Fish Depend on Dissolved Oxygen

1. The solubilities of gases in liquids *decrease* with increasing temperature.
3. In warmer water, there is less dissolved oxygen—causing fish to occasionally suffocate during hot summer months.

Answers to Exercises

1. Hydrogen's electron joins the valence shell of the fluorine atom. Meanwhile, fluorine's unpaired valence electron joins the valence shell of hydrogen.
3. The neon atom tends not to gain electrons because there is no more room available in its outermost occupied shell; it doesn't lose electrons because its outermost electrons are held tightly to the atom by a relatively strong effective nuclear charge.
5. The charges on the aluminum and oxide ions of aluminum oxide are greater than the charges on the sodium and chloride ions of sodium chloride. The network of aluminum and oxide ions within aluminum oxide, therefore, are more strongly held together, which gives the aluminum oxide a much higher melting point (more thermal energy is required to allow these ions to roll past one another within a liquid phase.)
7. The hydrogen atom has only one electron to share.
9. To form a covalent bond, an atom must have a fairly strong attraction for at least one additional electron. Metals atoms, however, tend not to have such an attraction. Instead, they tend to lose electrons to form positively charged metal ions.
11. The most polar molecule is the least symmetrical molecule (c) O=C=S.
13. Water is a polar molecule because in its structure the dipoles do not cancel. Polar molecules tend to stick to one another, which gives rise to relatively high boiling points. Methane, on the other hand, is nonpolar because of its symmetrical structure, which results in no net dipole and a relatively low boiling point. The boiling points of water and methane are less a consequence the masses of their molecules and more a consequence of the attractions that occur among their molecules.
15. The covalent bonds within a molecule are many times stronger than the attractions occurring between neighboring molecules. We know this because while two molecules can move away from each other (as occurs in the liquid or gaseous phase) the atoms within a molecule remain stuck together as a single unit. To pull the atoms apart requires some form of chemical change.
17. The charges in sodium chloride are balanced, but they are not neutralized. As a water molecule gets close to the sodium chloride it can distinguish the various ions and it is thus attracted to an individual ion by ion–dipole forces. This works because sodium and chloride ions and water molecules are of the same scale. We, on the other hand, are much too big to be able distinguish individual ions within a crystal of sodium chloride. From our point of view the individual charges are not apparent.
19. Salt is composed of ions that are too attracted to themselves. Gasoline is nonpolar, so salt and gasoline will not interact very well.
21. When an ionic compound melts, the ionic bonds between the ions are overcome. When a covalent compound melts, the molecular attractions between molecules are overcome. Because ionic bonds are so much stronger than molecular attractions, the melting points of ionic compounds are typically much higher.

Solutions to Chapter 12 Problems

1. Multiply concentration by volume: (0.5 g/L)(5 L) = 2.5 g.

3. a. $\frac{1 \text{ mole}}{1 \text{ Liter}} = 1$ Molar (1 M)

   b. $\frac{2 \text{ moles}}{0.5 \text{ Liters}} = 4$ Molar (4 M)

Chapter 13: Chemical Reactions

Answers to Integrated Science Concepts

Acid Rain

1. It is the alkaline character of limestone (also known as calcium carbonate) that serves to neutralize waters that might be acidified in the Midwestern United States.

3. The burning of fossil fuels produces sulfur dioxide, which reacts with water to produce sulfuric acid, acidifying rain.

Fuel Cells

1. As long as fuel is supplied, fuel cells don't run down, but batteries die when the electron-producing chemicals are consumed.

The Effect of Temperature on Reaction Rate

1. Lightning results in the formation of nitrogen monoxide in the atmosphere, which reacts further to produce nitrates, which are chemicals plants need to survive.

Answers to Exercises

1. a. 4, 3, 2 b. 3, 1, 2 c. 1, 2, 1, 2 d. 1, 2, 1, 2 (Remember that, by convention, 1's are not shown in the balanced equation.)

3. There is the same number of the same types of atoms on both sides of the arrow, which means this reaction is balanced: atoms are neither created nor destroyed.

5. Equation "d" best describes the reacting chemicals.

7. The oxygen atom.

9. The corrosive properties are no longer present because the acid and base no longer exist. Instead, they have chemically reacted with each other to form completely new substances—salt and water—that are not so corrosive.

11. This solution would have a hydronium ion concentration of 10^3 M, or 1000 moles per liter. The solution would be impossible to prepare because only so much acid can dissolve in water before the solution is saturated and no more will dissolve. The greatest concentration possible for hydrochloric acid, for example, is 12 M. Beyond this concentration, any additional HCl, which is a gas, added to the water simply bubbles back out into the atmosphere.

13. The hydrochloric acid solution becomes more dilute with hydronium ions as the weak acid is added to it. The pH of the hydrochloric acid solution therefore increases. Conversely, the pH of the weak acid solution has a relative decrease in its pH as the many hydronium ions from the hydrochloric acid solution are mixed in.

15. The tin ion, Sn^{2+}, gains electrons and is reduced, while the silver atom, Ag, loses electrons and is oxidized. The iodine atoms, I, gain electrons and are reduced, while the bromine ions, Br-, lose electrons and are oxidized.

17. Within carbon dioxide there are two oxygen atoms for every one carbon. With the product of photosynthesis (glucose), however, there is only one oxygen for every one carbon. Furthermore, the carbon now has more hydrogens around it. This all tells us that carbon is getting reduced. The oxygen of the water molecule winds up with fewer hydrogen atoms whether it ends up being an oxygen within the carbon based product ($C_6H_{12}O_6$, glucose, where it needs to share these hydrogens with carbon) or within the oxygen molecule, O_2. This tells us that the oxygen of the water molecule is getting oxidized, which is not an easy thing for oxygen. It takes the energy of sunlight to make this happen.

19. We exhale carbon dioxide and water vapor, which are the products of the oxidation of the food we eat.

21. As an exothermic reaction proceeds from reactants to products, the result is a release (dispersion) of thermal energy, which is favorable. Typically, this amount of energy dispersion is significantly larger than the difference in chemical entropies of the products and reactants.

Solutions to Chapter 13 Problems

1. The pH of this solution is 0, which is very acidic.

3. a.

Energy to break bonds:	Energy released from bond formation:
H—H = 436 kJ	H—Cl = 431 kJ
Cl—Cl = 243 kJ	H—Cl = 431 kJ
Total = 679 kJ absorbed	Total = 862 kJ released

NET = 679 kJ absorbed − 862 kJ released = 83 kJ released (exothermic)

b.

Energy to break bonds:	Energy released from bond formation:
C≡C = 837 kJ	4 × O═C = 3212 kJ
H—C = 414 kJ	4 × C═O = 3212 kJ
C—H = 414 kJ	H—O = 464 kJ
O═O = 498 kJ	H—O = 464 kJ
O═O = 498 kJ	O—H = 464 kJ
O═O = 498 kJ	O—H = 464 kJ
O═O = 498 kJ	Total = 8280 kJ released
O═O = 498 kJ	
Total = 4155 kJ absorbed	

NET = 4155 kJ absorbed − 8280 kJ released = −4125 kJ released (very exothermic)

Chapter 14: Organic Chemistry

Answers to Integrated Science Concepts

Drug Action and Discovery

1. The drug is viewed as the key.

3. Whether a drug is isolated from nature or synthesized in the laboratory makes no difference as to "how good it may be for you." There are a multitude of natural products that are downright harmful, just as there are many synthetic drugs that are also harmful. The effectiveness of a drug depends on its chemical structure, not the source of this chemical structure.

Answers to Exercises

1. To make it to the top of the fractionating column, a substance must remain in the gaseous phase. Only substances with very low boiling points, such as methane (−160°C) are able to make it to the top. According to Figure 14.3, gasoline travels higher than kerosene and so it must have a lower boiling point. Kerosene, therefore, has the higher boiling point.

3. The percent carbon increases as the hydrocarbon gets bigger. Methane's percent carbon is 20%; ethane, 25%; propane, 27%; butane, 29%.

5. C_4H_8O

7. In order of least to most oxidized b < a < d < c, whereas c is the most oxidized. Note how this was the order of their presentation within the chapter. The most reduced hydrocarbons were introduced first, followed by the alcohols, followed by the aldehydes, followed by the carboxylic acids.

9. The second and the fourth structures are the same. In all, there are three different structures shown.

11. The long, nonpolar hydrocarbon tail embeds itself in a person's oily skin, where the molecule initiates an allergic response. Scratching the itch spreads tetrahydrourushiol molecules over a greater surface area, causing the zone of irritation to grow.

13. Aspirin's chemical name is acetyl salicylic acid. It is the acidic nature of aspirin that gives rise to its sour taste.

15. The transformation of benzaldehyde to benzoic acid is an oxidation.

17. Ultimately, this is the energy that was captured from the sun by photosynthetic plants that turned into fossil fuels after decaying under anaerobic conditions.

19. Note the similarities between the structure of SBR and polyethylene and polystyrene, all of which possess no heteroatoms. SBR is an addition polymer made from the monomers 1,3-butadiene and styrene mixed together in a 3:1 ratio. Notably, SBR is the key ingredient that allows the formation of bubbles within bubble gum.

21. Today we take polymers and their remarkable properties for granted. Not so back during the time of World War II when their remarkable properties had a significant impact on how the war was won—Nylon for parachutes, synthetic rubber for tires, polyethylene for RADAR, Plexiglas for airplane gunner turets, and Teflon to help in the development of the nuclear bomb.

Solutions to Organic Chemistry Problems

1.

3.

Chapter 15: The Basic Unit of Life—The Cell

Answers to Integrated Science Concepts

Macromolecules Needed for Life

1. Proteins perform a wide range of functions in living organisms. The protein keratin provides structure in the form of skin, hair, and feathers. Insulin is a protein that acts as a hormone, allowing cells to communicate with one another. Actin and myosin are proteins that allow muscles to contract. Hemoglobin, a protein found in red blood cells, transports oxygen to body tissues. Antibodies are proteins that protect the body from disease. Proteins known as digestive enzymes break down food during digestion.

3. Four nitrogenous bases are used in DNA—adenine, cytosine, guanine, and thymine.

The Microscope

5. Electron microscopes are able to resolve objects about a nanometer (10^{-9} meter) in size, which allows cellular structures to be viewed in fine detail.

Chemical Reactions in Cells

7. No, it costs more energy to make ATP than cells eventually get out of it. This is consistent with the second law of thermodynamics.

Answers to Exercises

1. Like other living things, humans use energy when we move, speak, or perform any activity. Like other animals, we obtain this energy from the food we eat. Humans certainly develop and grow—think how much a newborn differs from a two-year-old, and how much a two-year-old differs from you! Humans maintain themselves by repairing injuries to skin and bone and other organs, and we maintain a stable internal environment where body temperature, oxygen, water content, and numerous other variables are carefully controlled. (This will be discussed further in Chapters 19 and 20). Humans have the capacity to reproduce through sexual reproduction. Human populations also evolve. In Chapter 16, for example, we will learn why the sickle cell anemia allele is comparatively common in people of African descent, but not in other human groups.
3. Eukaryotic cells also have some DNA in their mitochondria (and in their chloroplasts, if they are plant cells). Both mitochondria and chloroplasts have their own cell membrane and their own circular chromosome of DNA. Because of this resemblance to prokaryotes, it has been hypothesized that these organelles evolved from prokaryotes living inside early eukaryotes.
5. Cells with lots of mitochondria require lots of energy. The mitochondria are there to make lots of ATP for these busy cells.
7. Proteins are made of carefully folded chains of amino acids. The varying amino acid sequences combined with the complex folding allows proteins to take on the complex shapes required to create many different types of "locks" and "keys."
9. Glucose is quickly broken down by cells to make ATP (through the process of cellular respiration). As a result, there usually isn't much glucose inside cells.
11. Highly branched roots increase the amount of available surface area for absorption of water and other soil nutrients.
13. Both gap junctions and plasmodesmata allow messages to pass directly from one cell to an adjacent cell. So, they have similar functions. However, they are different structurally. Gap junctions are tiny channels in the cell membrane surrounded by specialized proteins. Plasmodesmata are slender threads of cytoplasm that link adjacent plant cells.
15. Mitosis.
17. Plants remove carbon dioxide from the atmosphere during photosynthesis.
19. The bubbles come from carbon dioxide released during alcoholic fermentation.
21. Prokaryotes don't have or require telomeres because they have circular, rather than linear, chromosomes.

Solutions to Chapter 15 Problems

1. The amount of energy the bear stores for the winter is:

   (50 kilograms of fat) (1000 grams /1 kilogram) (9 kilocalories/gram) = 450,000 kilocalories.

   In order to store the same amount of energy as carbohydrate:

   (x kilograms of carbohydrate) (1000 grams /1 kilogram) (4 kilocalories/gram) = 450,000 kilocalories

   $x = 450{,}000/(4)(1000) = 112.5$ kilograms

3. Surface area of a sphere is $4\pi r^2$
   For 1 micrometer cell, surface area = $4\pi(1)^2 = 4\pi$
   For 5 micrometer cell, surface area = $4\pi(5)^2 = 100\pi$

   Volume of a sphere is $4/3\pi r^3$
   For 1 micrometer cell, volume = $4/3\pi(1)^3 = 4/3\pi$
   For 5 micrometer cell, volume = $4/3\pi(5)^3 = 500/3\pi$

   So, the surface area to volume ratio is
   For 1 micrometer cell, surface area/volume = $4\pi/(4/3\pi) = 3$
   For 5 micrometer cell, surface area/volume = $100\pi/(500/3\pi) = 3/5$

   The larger cell is able to obtain more molecules through diffusion because it has a larger surface area. It is nonetheless more challenging for the larger cell to meet its needs through diffusion because it has a smaller surface area to volume ratio. This ratio measures how easy it is for a cell to meet its needs through diffusion because a cell's need for molecules depends on its volume, and its ability to obtain these molecules depends on its surface area.

5. A chain of two amino acids can include 20 amino acids for the first position and 20 amino acids for the second position, so the number of possibilities is:

   $$(20)(20) = 400$$

   A chain of three amino acids can have any of 20 amino acids in the first position, any of 20 amino acids in the second position, and any of 20 amino acids in the third position—so the number of possibilities is:

   $$(20)(20)(20) = 8000$$

   For ten amino acids, the number of possibilities is:

   $$(20)(20)(20)(20)(20)(20)(20)(20)(20)(20) = 10{,}240{,}000{,}000{,}000$$

Chapter 16: Genetics

Answers to Integrated Science Concepts

The Structure of DNA

1. DNA is often described as a double helix because it consists of two strands twisted into a spiral or helix.
3. The four nitrogenous bases in DNA are adenine (A), guanine (G), cytosine (C), and thymine (T). A pairs with T and G pairs with C.

How Radioactivity Causes Genetic Mutations

5. Frequently dividing cells are particularly vulnerable to radiation damage. These include cells in the bone marrow (where blood cells are made), in the lining of the gastrointestinal tract, in the testes, and in the developing fetus.

Environmental Causes of Cancer

7. The most important cancer-related environmental risk factors include tobacco, diet, ionizing radiation, UV light, disease-causing viruses and bacteria, and mutagens present in air, water, and soil.
9. Radon is a radioactive gas produced by the decay of uranium. Because minute amounts of uranium are found in many rocks, radon is present in many areas. When radon decays, it releases small radioactive particles that can be inhaled. These radioactive particles damage DNA in lung cells, making lung cancer more likely to develop.

Answers to Exercises

1. Your finger is made of diploid cells, like most of your body except for your sex cells (and, in women, eggs do not actually complete meiosis and become haploid until they are fertilized!).
3. Every new DNA molecule has one old strand and one new strand because during DNA replication, the old molecule is unzipped and each strand is used as a template for putting together a new strand.
5. "aggfr" is an intron. "not" is an exon.
7. Point mutations in the different positions of a codon are not equally likely to change the amino acid sequence of a protein. Changes in the third position are least likely to change amino acid sequence.
9. This person has Down syndrome, which is a result of trisomy 21 (having three copies of chromosome 21 rather than two). Down syndrome is characterized by mental retardation and defects of the heart and respiratory system. Chromosomal abnormalities such as trisomies usually result from mistakes that occur during meiosis, resulting in an egg or sperm having two copies of a chromosome rather than the usual single copy. (The third copy is added at fertilization).
11. If you breed two pink-flowered snapdragons, blending inheritance predicts you get all pink-flowered snapdragons. However, what you get are pink, red, and white snapdragons. Specifically, breeding two pink-flowered snapdragons RW x RW yields a quarter RR (red-flowered), half RW (pink-flowered), and a quarter WW (white-flowered) offspring.
13. It is possible for two parents with widow's peaks to have a child that has a straight hairline if both parents are heterozygotes (Ss) and the child inherits a recessive (s) allele from each parent. Two parents with straight hairlines must be ss, so all their children will only inherit s alleles and will also have straight hairlines. Thus, two parents with straight hairlines cannot have a child with a widow's peak.
15. People with type O blood are universal donors because their blood cells have neither A nor B molecules that could cause the cells to be rejected and attacked. People with type AB blood are universal receivers—their bodies accept both A and B molecules, so they can receive type O, type A, type B, or type AB blood.
17. Because it's a condition that affects more males than females, it is likely to be (and, in fact, is) a sex-linked trait found on the X chromosome.
19. A nonsense mutation produces a stop codon in the middle of a gene-coding sequence. The one that occurs near the beginning of the gene (and therefore ends amino acid assembly earlier) is more likely to affect protein function than the one that occurs near the end of the gene.
21. First, with a trait that shows incomplete dominance, the child would resemble neither parent, but have a phenotype in between. Second, the child could inherit a recessive allele from each heterozygous parent; thus, the parents would have the dominant phenotype and the child would have the recessive phenotype. Third, with codominance, the child could have a phenotype different from both parents—a bloodtype A mother and bloodtype B father could have a child with bloodtype AB.
23. Human height and weight are both determined partially by genetic factors and partially by environmental factors. The environment has little effect on traits such as eye color, dimples versus no dimples, straight versus widow's peak hairline, etc.

Solutions to Chapter 16 Problems

1. $64/2 = 32$ chromosomes in its haploid cells.
3. For RNA, G, C, A, and T pair with C, G, U, and A, respectively. (Recall that RNA uses U instead of T). So, the RNA molecule will have nucleotides GACUCCAGUCCU.
5. We can break the sequence into triplet codons and consult the genetic code table. (AGU)(CGU)(UGG)(CAG)(GAA)(GUA) = serine-arginine-tryptophan-glutamine-glutamic acid-valine
7. Many answers are possible. ACATGTCCAGACTAATTGCAA and ACCTGTCCAGACTAATTGCAA are two possible answers—point mutations in the first codon still code for the amino acid threonine, so the amino acid sequence (and protein produced) are unaffected.
9. The red-green colorblindness allele is recessive, and the gene is found on the X chromosome. Because the woman's husband is not red-green colorblind, his allele must be normal. All the daughters will inherit this normal allele (since they all inherit an X chromosome from their father), so they will not be red-green colorblind no matter which allele they inherit from their mother. The sons, however, only receive an X chromosome from their mother, so they have a $\frac{1}{2}$ chance of being red-green color blind (since she has one red-green colorblindness allele and one normal allele).

Chapter 17: The Evolution of Life

Answers to Integrated Science Concepts

Did Life on Earth Originate on Mars?

1. Because in 1996, NASA scientists found what could be fossils of tiny bacteria in a Martian meteorite. Moreover, the potential fossils were found very close to complex organic molecules and carbonate minerals that, on Earth, are associated with living organisms.
3. Perhaps, scientists proposed, life found its way to Earth in Martian dust set adrift in space when a comet collided with Mars.

Animal Adaptations to Heat and Cold

5. The heat an animal generates is proportional to its volume. The heat an animal dissipates is proportional to its surface area, since heat is lost to the environment through its body surface. Consequently, animals are better able to lose heat if they have a high surface area-to-volume ratio and better able to retain heat if they have a low surface area-to-volume ratio.
7. Allen's Rule says that desert species typically have long legs and large ears that increase the surface area available for heat dissipation, whereas Arctic species typically have short appendages and small ears that help conserve heat. Desert and Arctic rabbit species provide an example of Allen's Rule.

Earth's Tangible Evidence of Evolution

9. Fossils of now-extinct relatives of the horse show that species grew larger in size over time, as well as more specialized for eating grass and running. Some fossil whales exhibit some of the characteristics of the hoofed animals they evolved from—hind limbs, nostrils on their noses rather than blowholes, and different types of teeth. *Archaeopteryx,* the famous 150-million-year-old fossil bird, has many birdlike features—feathers, wings, a wishbone—but also has dinosaur-like features absent in modern birds, including claws on its wings, bones in its tail, and teeth.

Answers to Exercises

1. The types of environments where living organisms were thought to spontaneously appear—rotting carcasses or meat broths—had to be isolated from living organisms so that they would not be contaminated by life that already existed. This isolation proved difficult to achieve, and contamination often did occur.
3. Liposomes have double membranes and behave in ways that are eerily cell-like, growing and shrinking, even budding and dividing. Liposomes also control the absorption of materials and run chemical reactions within their membranes, like cells. However, they do not have genetic material like real cells.
5. Lamarck would say that over a lifetime of swimming, the shapes of fish became more streamlined as they fought their way through the water. They then pass this more streamlined shape to their offspring. Darwin would say that fish vary in their body shape, and that more streamlined individuals were more effective swimmers, so survived and reproduced better, leaving more streamlined individuals in the population.
7. Answers will vary. Traits that don't show variation include having a four-chambered heart, five fingers on each hand, two arms, two legs, one nose, etc. Traits that show variation include height, weight, arm length, foot length, eye color, hair color, etc.
9. No, color band color is not a heritable trait.
11. Alternative explanations are genetic drift, migration into or out of the population, and mutation pressure. To determine whether natural selection is responsible for the shift, you could compare the fitness (number of offspring left) of red individuals versus yellow individuals. If this turned out to be difficult, you could also compare their survival or ability to acquire mates in an attempt to identify underlying causes of potential fitness differences.
13. It could be, but it could also be the result of better nutrition.
15. The kit fox is the one with pale brownish fur, large ears, and long limbs. The fur color helps it reflect heat and stay camouflaged in its environment. The large ears and long limbs help it increase surface area available for heat dissipation. The Arctic fox has white fur, which helps it stay camouflaged, and small ears and short limbs that help it decrease the surface area from which heat is lost.
17. No, you cannot conclude they are distinct species merely because they are distinguishable. You can determine whether they are distinct species by figuring out whether they interbreed.
19. Hawaii is extremely isolated from all mainlands, so organisms that arrived there then had plenty of time to evolve in isolation and speciate from mainland species.
21. Both these biogeographical patterns suggest that organisms dispersed where they could, not that they were purposefully distributed across the globe.

Solutions to Chapter 17 Problems

1. Yes, this is natural selection because color is a variable, heritable trait and brown and green individuals have different fitness. What is happening here is that, over time, the population is shifting toward a greater and greater proportion of brown individuals.

Generation	Brown	Green	Proportion Brown
1	2	2	0.50
2	4	2	0.67
3	8	2	0.80
4	16	2	0.89
5	32	2	0.94
6	64	2	0.97
7	128	2	0.98
8	256	2	0.992
9	512	2	0.996
10	1024	2	0.998

3. Since there is a RR organism and a RW organism, the frequency of the R allele in the population is $\frac{3}{4} = 0.75$. The frequency of the W allele is $\frac{1}{4} = 0.25$.

Chapter 18: Biological Diversity

Answers to Integrated Science Concepts

Coral Bleaching

1. Increases in seawater temperature that last for an extended period of time.
3. Because of continued global warming due to human greenhouse gas emissions.

How Birds Fly

5. Birds move forward through the air by flapping their wings. During the downstroke, the wings push against the air and the air pushes back. This propels them forward.

Answers to Exercises

1. Because archaea are more closely related to eukaryotes than either is to bacteria, classifying archaea and bacteria together to the exclusion of eukaryotes obscures the evolutionary history of the three groups. It is like the example in the text of grouping humans and daisies together to the exclusion of elephants. In terms of a cladistic classification, the fact that archaea and bacteria are both prokaryotes is not relevant–only the evolutionary relationships among the three groups matters in constructing biological groups.
3. Spores are very hardy, able to survive for long periods under tough conditions. This allows bacteria, fungi, and other organisms capable of generating spores to produce descendants that can survive tough periods and then grow into mature individuals when conditions improve.
5. No, life would do just fine without eukaryotes, as it did for billions of years before the first eukaryotes evolved.

7. Mosses are most dependent on living in a moist environment. This is because they have no vascular systems—instead, every part of a moss plant receives water directly from the environment. In addition, mosses have swimming sperm that require moisture in the environment in order to travel to and fertilize moss eggs. Seed plants are least dependent on living in a moist environment because they use pollen rather than swimming sperm during sexual reproduction, and they have vascular systems. (The third group, ferns, have vascular systems, but also use swimming sperm, so are more moisture-dependent than seed plants.)
9. Seed plants produce pollen. Pollen is carried to the female flower/cone by wind or by animals. Most flowering plants use animal pollinators, particularly insects.
11. Wind-pollinated plants are most likely to cause allergies because they make larger quantities of pollen due to the haphazard nature of wind pollination.
13. Many cnidarians, including jellyfish and sea anemones, catch prey using tentacles armed with barbed stinging cells. Corals, however, house dinoflagellates in their bodies and obtain the bulk of their nutrients from these photosynthesizers.
15. The muscles of roundworms all run longitudinally (from head to tail) down the body. As a result, roundworms move like flailing whips as muscles on alternate sides of the body contract. The muscles of annelids are arranged in both circular (around the body) and longitudinal (head-to-tail) orientations, allowing for great flexibility of motion. Unlike roundworms, for example, annelids are able to contract one part of the body while keeping the rest of the body still.
17. Amphibians have a skin made of living cells that is vulnerable to drying out. In addition, their eggs are unshelled and also vulnerable to drying out.
19. Because birds are descended from the last common ancestor of all reptiles. What they do or do not have in common with mammals is not a factor in how they are classified.
21. The advantage of genetic exchange is genetic diversity among the offspring. This way, you don't put all your eggs in one (genetic) basket, and at least some of your offspring are likely to do well under a wide array of potential environmental conditions.
23. Viruses are small pieces of genetic material wrapped in a protein coat. Many viruses have normal, double-stranded DNA genomes, but others use single-stranded DNA, single-stranded RNA, or double-stranded RNA. Viruses reproduce by infecting a host cell and then using the cell's enzymes and ribosomes to copy their genetic material and build viral proteins. These are then assembled to form new viruses. One feature of viruses that makes them hard to deal with from the point of view of disease control is that they mutate very quickly. This is particularly true of viruses with RNA genomes, since there is no error-checking and repair system for copying RNA, as there is with DNA. Bird flu, which has devastated populations of domesticated birds, is caused by a virus that occurs naturally among wild birds. In 1997, the first case of a human infected by bird flu was reported in Hong Kong, and dozens of additional cases have been seen since then. So far, however, the virus cannot be transmitted easily from person to person. The evolution of this capability is the event scientists await with trepidation. This fear turns out to be more than justified. Scientists recently discovered that the infamous "Spanish flu" epidemic of 1918—which killed more people than any other disease over a similar length of time—was a bird flu that became easily transmissible among humans.

Solutions to Chapter 18 Problems

1. The population doubles every 20 minutes, so:

8:00	1
8:20	2
8:40	4
9:00	8
9:20	16
9:40	32
10:00	64
10:20	128
10:40	256
11:00	512
11:20	1024
11:40	2048
12:00	4096

3.

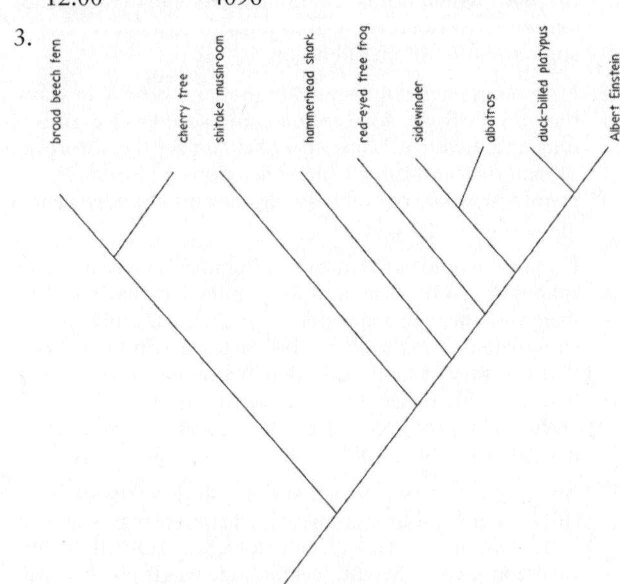

Chapter 19: Human Biology I: Control and Development

Answers to Integrated Science Concepts

How Fast Can Action Potentials Travel?

1. An action potential's speed depends in part on how quickly successive parts of the axon's cell membrane (that is, parts further down the axon) can be induced to increase to threshold. How quickly the membrane reaches threshold is in turn dependent on how fast sodium ions flow downstream to increase the membrane potential there. And how fast sodium ions flow down an axon depends on Ohm's Law. Ohm's Law tells us that current = voltage/resistance, so the lower the resistance, the more current (that is, ions) flows and the faster the action potential travels. Like any other material, an axon has lower resistance if it is thicker around—a thick axon resists current less than a thin axon the same way a wide pipe resists water flow less than a thin pipe.
3. The myelin sheath that surrounds an axon insulates it so that ions cannot escape out the cell membrane, but must flow down the axon. The end result is the same as for a giant axon—sodium ions are able to travel more efficiently down the axon. Moreover, in myelinated axons, the action potential is not regenerated at every point

along the axon; instead, it "jumps" from one gap in the sheath to the next. An action potential at one gap causes sodium ions to move into the axon, flow down to the next gap, generate a new action potential there, and so on. This jumping propagation makes for extremely rapid signal transmission.

Endorphins

5. Because they bind to the same receptors (opiate receptors).

Answers to Exercises

1. The temporal lobes interpret sound, including language comprehension. The frontal lobes control the voluntary movements required to produce speech.
3. A motor neuron goes to your biceps muscle and tells you to bend your elbow. A sensory neuron transmits information from your feet as to whether they feel cold.
5. The action potential would travel both forward and backward along the axon away from the spot where you artificially excited it. It travels backward in addition to forward, unlike a "real" action potential, because the area behind where the axon is stimulated has not just experienced an action potential, and so is able to be stimulated.
7. Neurotransmitters do not enter the target cell. Instead, they have their effect by binding to receptors on the target cell, starting a sequence of events that results in the target cell's response to the neurotransmitter.
9. Rods are more light-sensitive than cones, and are responsible for vision in dim light.
11. If the middle ear bones stiffen, they are less effective at transferring and amplifying sound vibrations from the outer ear to the inner ear, where "hearing" actually occurs.
13. Antidiuretic hormone helps regulate the amount of water in the body. Specifically, it helps the body conserve water by producing a more concentrated urine. Parathyroid hormone, which raises calcium levels in the blood, and calcitonin, which lowers calcium levels in the blood, help regulate calcium levels. Insulin and glucagon regulate the amount of glucose in the blood. Insulin lowers blood glucose levels by directing muscle and other cells to take in glucose and by stimulating the liver to convert glucose to the storage substance glycogen. Glucagon increases blood glucose levels by causing the liver to break down glycogen. (Other answers are possible).
15. Parathyroid hormone raises calcium levels in the blood, partly by causing calcium to be released from bones. Calcitonin has the opposite effect—it lowers blood calcium levels by causing bones to take up calcium. This is why it is useful for treating osteoporosis—calcitonin helps increase bone strength by increasing bone calcium.
17. Meiosis in women is unequal meiosis, with one of the four daughter cells—the future egg—getting the bulk of cytoplasm and nutrients. So, meiosis in women produces only a single egg. Meiosis in men is not unequal, and produces four functional sperm.
19. Sperm are unable to reach the unfertilized egg, which also cannot reach the uterus.
21. In each of our muscles, there are many sarcomeres lined up end to end—if each of these contracts and shortens a tiny amount, the entire muscle shortens by a significant distance.
23. Both curare and sarin affect the nerve to muscle connection, and both cause death through asphyxiation. However, their precise mechanisms differ. Curare, an arrow poison used in the South American tropics for hunting, binds to acetylcholine receptors on muscle cells, preventing acetylcholine itself from binding. Curare causes paralysis and then death as the respiratory muscles become paralyzed. The powerful nerve gas sarin prevents acetylcholine from being broken down after muscles contract. Muscles are stimulated continuously and soon become exhausted. Again, death occurs through asphyxiation as the respiratory muscles stop working.

Solutions to Chapter 19 Problems

1. For the fast type, time = distance/speed = 1/25 = 0.04 seconds.

   For the slow type, time = distance/speed = 1/0.5 = 2 seconds.
3. 1000/30,000 = 0.033 or 3.3%

Chapter 20: Human Biology II: Care and Maintenance

Answers to Integrated Science Concepts

Hemoglobin

1. A molecule of hemoglobin consists of four subunits, each of which contains a component known as a heme group that includes an iron atom at its center. It is this iron atom that binds oxygen.
3. An active, working tissue makes and uses more ATP and so releases more carbon dioxide during cellular respiration. Because carbon dioxide reacts with water in the blood to form carbonic acid, the presence of high carbon dioxide levels decreases blood pH. This acidity decreases the oxygen affinity of local hemoglobin molecules, making it easier for them to unload oxygen to the working tissue.

Low-Carb Versus Low-Cal Diets

5. Studies have confirmed that, for many people, low-carb diets do produce weight loss more quickly and more consistently than low-calorie diets. This appears to be because many people find low-carb diets easier to stick to because of their permissive attitude towards fats. People on low-carb diets lose weight for the same reason that people on low-calorie diets lose weight—they consume fewer total calories. In addition, low-carb diets cause you to retain less water in the body—this water is used during excretion to flush out the extra proteins consumed.

Answers to Exercises

1. Hearing requires the functioning of our skin (skin makes up part of the eardrum) and skeleton (the cartilaginous outer ear as well as the middle ear bones), in addition to sensory cells. Smelling requires the functioning of the respiratory system, which brings air in to be "sampled" by our sensory cells for smell. Tasting requires saliva, part of the digestive system, since molecules must be dissolved in liquid in order to be sensed.
3. The contractions of our voluntary muscles help move blood back toward the heart.
5. Near the source of oxygen—near the alveoli.

7. It contributes to our sense of smell by bringing chemical molecules in the air to the cells responsible for smelling. It also is involved in speech, which depends on air vibrating our vocal cords as we exhale.

9. Carbon dioxide and nitrogen-containing wastes (in the form of ammonia, which is then quickly converted to urea) are the waste products produced during cellular respiration. Carbon dioxide is removed from the body by the respiratory system. Nitrogenous wastes are removed by the excretory system.

11. Antidiuretic hormone causes more water to be reabsorbed from the filtrate during excretion. Parathyroid hormone decreases calcium excretion. Mineralocorticoids help regulate water and salt balance in the body by affecting excretion of these substances. (Other answers are possible.)

13. Elimination is associated with the digestive system, excretion with the excretory system. Elimination eliminates substances in food that cannot be absorbed or used by the body. (Feces are composed primarily of living and dead bacteria and indigestible materials, such as plant cellulose.) Excretion helps control the amount of many substances in the body, but it helps us get rid of urea in particular, which is a product of breaking down proteins during cellular respiration.

15. The innate immune system is described as nonspecific because its defenses work against a wide variety of potential pathogens. The acquired immune system is described as specific because the cells of this system recognize very specific features of specific pathogens and take action only when these features are encountered.

17. Many of the symptoms of allergies result from an excessive inflammatory response to irritants. In order to trigger the inflammatory response of the innate immune system, injured tissues produce histamines. Antihistamines help to counter these histamines and reduce the redness and swelling that make up the inflammatory response.

19. Perhaps it does, via the placebo effect.

Solutions to Chapter 20 Problems

1. (25,000,000,000,000) (300,000,000) (4) = 30,000,000,000,000,000,000,000 molecules of oxygen.

3. For heartbeat:

   70 beats per minute (60 minutes per hour) = 4200 beats per hour

   4200 beats per hour (24 hours per day) = 100,800 beats per day

   100,800 beats per day (365 days per year) = 36.8 million beats per year

   For breathings:

   12 breaths per minute (60 minutes per hour) = 720 breaths per hour

   720 breaths per hour (24 hours per day) = 17,280 breaths per day

   17,280 breaths per day (365 days per year) = 6.3 million breaths per year

Chapter 21: Ecosystems and Environment

Answers to Integrated Science Concepts

Energy Leaks Where Trophic Levels Meet

1. The second law of thermodynamics states that natural systems tend to move from organized energy states to disorganized energy states; that is, useful energy dissipates to unusable energy. Specifically, any time energy is converted from one form to another—including in any chemical reaction—some energy is lost to the environment as heat. Moving energy from one trophic level to another—such as by breaking down plant matter in the digestive tract of a rabbit, and then using the molecules to build more rabbit muscle—involves a long series of chemical reactions, one after another. So, the second law of thermodynamics explains the reason for energy loss between trophic levels.

Materials Cycling in Ecosystems

3. The word "biogeochemical" emphasizes that substances cycle between living organisms ("bio") and Earth ("geo")—in particular, Earth's atmosphere, crust, and waters.

5. Legumes such as peas, beans, clover, and alfalfa have evolved a mutualistic symbiotic relationship with nitrogen-fixing bacteria. These bacteria live in nodules on legume roots and provide them with nitrogen.

Answers to Exercises

1. This is a population-level study because it considers a group of individuals of a single species that occupies a given area.

3. Yes, all producers are autotrophs—they make their own food from inorganic substances. Not all producers photosynthesize; some are chemoautotrophs.

5. Mutualism—both species benefit. The insects receive food and the plants are pollinated, a key step in their reproduction.

7. Because, in all ecosystems, a huge amount of energy is lost as you go up the food chain. On average, only about 10 percent of the energy at one trophic level becomes available to the next level. First of all, not every organism at one trophic level is exploited by the next level—for example, not every plant gets eaten by a herbivore. Second, when a consumer eats, the energy it receives from food goes into things other than building biomass—feces and maintenance, to be specific. Feces contains organic materials that the consumer is unable to digest. Maintenance is the energy the consumer requires to live—the energy it takes to find and eat food, run, mate, breathe, and so on. During these activities, a lot of energy is also lost to the environment as heat. So, by the time feces and maintenance have taken their share, only about 10 percent is left for growth and reproduction—for building new biomass.

9. We heterotrophs get all our carbon from the food we eat. Ultimately, all organic carbon comes from producers such as photosynthesizing plants. As to whether every carbon atom in our body came ultimately from a plant, the answer is that *most* of the carbon in our bodies came to us via a plant. However, *some* of the carbon atoms may have come to us through diatoms or other oceanic plankton, seaweeds, etc.

11. Because of the tremendous density and diversity of life, most of the nutrients present in tropical forests are being used by one or another living organism—as a result, the soil tends to be poor.

13. Many answers are possible. Marine plankton include diatoms, dinoflagellates, and the larvae of many animals. Marine nekton include most fishes, whales, seals, penguins, etc. Benthic species include mussels, clams, marine worms, lobsters, etc.

15. Secondary succession occurs more frequently—it is much more common for soil to remain intact than for bare rock to be revealed. Examples of the latter, which lead to primary succession, include when volcanism creates new land, or when glaciers retreat.

17. Yes. According to the intermediate disturbance hypothesis, regular disturbances, if not too extreme, actually contribute to biodiversity because different species make use of different habitats, and periodic disturbances guarantee that there will always be habitat at varying stages of recovery. However, a habitat that received regular, extreme disturbances would probably always be found in the early stages of succession and would probably be less diverse.

19. Humans are K-selected. We are described by Type I survivorship. Like other K-selected species, we have large body size, relatively few offspring, high investment in offspring (lots of parental care, for example), we reach sexual maturity slowly, and we have long life expectancy.

21. Most people are young. The population is growing rapidly.

23. Human population has grown exponentially for thousands of years. Exponential growth occurs when a population grows at a rate that is proportional to its size. Logistic growth occurs when population growth slows as it reaches the habitat's carrying capacity; that is, the maximum number of individuals or maximum population density the habitat can support. Interestingly, it has become evident in the last several decades that, although world population continues to grow rapidly, the rate of growth has slowed, so that growth is no longer exponential. In fact, scientists now believe that if present trends continue, the global human population will peak at around 10 billion soon after 2050.

Solutions to Chapter 21 Problems

1. 25 two-kilogram carnivores = 50 kilograms of carnivores
   This means there were 500 kilograms of herbivores (since 10% of 500 = 50)
   And 5000 kilograms of grass and other producers (since 10% of 5000 = 500)

3. This is the table:

Year	Individuals Alive
2000	1,000,000
2001	50,000
2002	2,500
2003	125
2004	6
2005	0

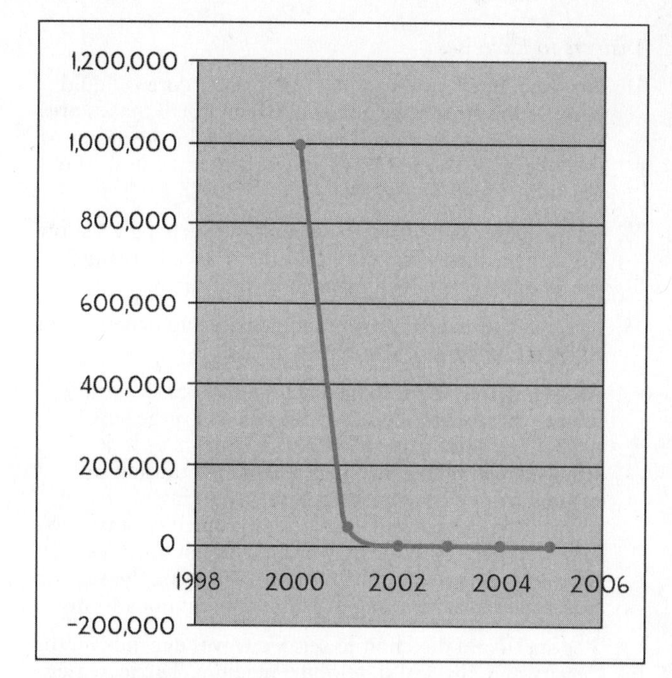

This is Type III survivorship.

Part 4 Earth Science

Chapter 22: Plate Tectonics

Answers to Integrated Science Concepts

Isostasy

1. Continental crust is less dense than oceanic crust so it must sink further to achieve sufficient buoyant force to counter gravity.

What Forces Drive the Plates?

1. Heat transfer away from Earth's extremely hot core toward its surface, as required by the Second Law of Thermodynamics, is the reason there is convection in the mantle and thus the reason there are moving plates.

3. The core is like a spherical set of burners. There is convection in Earth's interior just as there would be convection in the water heating on the stove.

Life in the Trenches

1. Zero light, extreme cold, and crushing pressure.

3. Hydrothermal vents are fissures created by the grinding action of the plates.

Anatomy of a Tsunami

1. Both are waves, but a ripple's energy moves across a surface as it travels, whereas a tsunami's energy moves inside the body of water as it goes.
3. A tsunami can be generated by subduction at a convergent boundary as the subducting plate "sticks" beneath the overriding plate. When the compressive force on the plates exceeds the force of friction between them, the plates jerk, which causes an earthquake. As rock snaps upward, it acts like a piston by pushing the water above it, generating a wave.

Answers to Exercises

1. No—the inner core is solid and the outer core is liquid because the greater weight of overlying earth makes pressure greater at the core. The pressure packs the atoms of the inner core too tightly to allow them to be in the liquid state.
3. As the plates pull apart, the weight of overlying crust on the asthenosphere is reduced. Under reduced pressure, the asthenosphere rock melts to form magma.
5. oceanic–oceanic convergent boundaries and oceanic–continental convergent boundaries
7. An island arc, or arc-shaped region of volcanic islands, forms where there is oceanic–oceanic convergence of plates. The subducting plate partially melts as it descends, creating magma. This magma in addition to magma from partial melting of the overlying plate rise toward the surface and erupts. The erupted magma cools and accumulates to form a volcanic mountain. This process is repeated in various locations along the subduction zone so that a string of volcanic mountains forms.
9. The speed and direction of seismic waves depends on the density and physical state of the medium. The increases and decreases in speed of P waves and the inability of S waves to penetrate the outer core led scientists to deduce the presence of Earth's layers.
11. Fossils of single species were found on facing edges of continents. It would be highly improbable for an organism to evolve in exactly the same way in two separate locations. And such organisms would have no way of swimming or otherwise transporting themselves across the intervening ocean.
13. The evidence comes largely from sea floor spreading, which is suggested by the ages of rocks, magnetic striping, and the existence of mid-ocean ridges, trenches, and other signs of underwater geological activity. Correlation of intense geologic activity along plate boundaries also supports the movement of plates. Finally, evidence for continental drift includes the migration of ancient ice sheets, and the matching of fossils and rock formations.
15. Sea floor spreading is faster in the Pacific.
17. It cannot. It is traveling northward with respect to the rest of California, not westward, into the ocean. Besides, isostasy would prevent this.
19. Either the climate has changed or the regions themselves have migrated southward, away from the poles. As it turns out, explanation #2 is correct.
21. The circulation of charges in the liquid outer core.
23. The Cascade mountain range is a coastal volcanic mountain range. This type of mountain range is formed when an oceanic plate subducts beneath a continental plate.
25. Tiny quakes do not release enough strain to prevent major quakes. The Richter scale is logarithmic—quakes with low magnitude have millions of times less energy than large magnitude quakes, so they do not help significantly in releasing built-up strain in the rocks.

Solutions to Chapter 22 Problems

1. $t = d/r = (600 \text{ km})/(3.5 \text{ cm/yr}) = (6 \times 10^7 \text{ cm})/(3.5 \text{ cm/yr}) = 1.7 \times 10^7$ years = about 17 million years
3. Calculate this based on the distance from the center of the rise to the edge of the rise, which equals 200 km/2 = 100 km. Then: $t = d/r = (1 \times 10^2 \text{ km})/(14.2 \text{ cm/yr}) = 7.04 \times 10^5$ yr or about 700,000 years.

Chapter 23: Rocks and Minerals

Answers to Integrated Science Concepts

The Silicate Tetrahedron

1. silicon and oxygen
3. It has four unpaired electrons available for bonding.

Coal

1. Photosynthesis
3. It is combusted.
5. It is made of plant remains rather than minerals.

Igneous Rock Texture

1. The magma that produced the rock crystallized over a broad range of temperatures—the large crystals cooled slowly, but the small ones cooled quickly; porphyritic.
3. Large numbers of small crystals form quickly in a fast cooling magma as the atoms lose energy. They cannot grow large without bumping into neighboring crystals. Possible example: rhyolite.
5. Large

Answers to Exercises

1. They have been weathered.
3. In crystallization, ions, molecules, or atoms solidify into orderly arrangements characterized by a repeating basic structural unit. It is characterized by two steps: nucleation and crystal growth.
5. No, it is actually very common, making up limestone, etc. It is the most abundant nonsilicate mineral.
7. All fossils are contained in sedimentary rocks.
9. (a) Decreasing temperature and pressure. (b) Retrograde metamorphism occurs when temperature and/or pressure are reduced. Chemical processes occur more slowly as temperature is decreased.
11. Yes; added thermal energy can break apart bands in a mineral, while added pressure can alter a crystal's geometry by squeezing poitive and negative ions.
13. Granite is exposed as layers of surface rock are eroded.
15. Metamorphic rocks are formed at high temperatures and/or pressures or from solutions associated with magma.

17. Rock is not conserved in the sense that energy is conserved. Rock can be created (by crystallization of magma) and destroyed (by melting)—although the atoms rock is composed of cannot be created or destroyed (except by nuclear means).

19. Minerals with high melting points crystallize out of a magma first; the remaining magma therefore has a different composition. Rocks form successively as different mixtures of minerals are produced from different magmas at different temperatures.

21. Yes; limestone is a sedimentary rock formed from calcium carbonate which can be obtained from the hard body parts of animals.

23. Possible problems are high consumption of fossil fuels in supplying the energy needed for processing minerals; disruption of ecosystems; erosion and landslides; toxic waste.

25. Dikes and plutons are igneous rock intrusions, forming only below Earth's surface. They are visible only when overlying rock erodes.

27. Asbestos is primarily harmful to lung tissue when inhaled.

Solutions to Chapter 23 Problems

(a) Approximating the curve as a linear graph, the slope is about 30°C per kilometer. (b) Temperature changes faster near the surface. Deeper in the earth, rock is surrounded by more insulating rock, which slows heat loss.

Chapter 24: Earth's Surface—Land and Water

Answers to Integrated Science Concepts

Ocean Waves

1. They become too tall from the bunching of many waves.
3. Energy.

Ocean Water

1. Only a few elements and compounds are present in abundance.
3. The composition of sea water remains relatively constant.

Groundwater Contamination

1. Sewage, agricultural chemicals such as nitrate fertilizers, and pesticides
3. Gas stations typically store MTBE in underground tanks. If a tank leaks, MTBE can infiltrate to underground and pollute wells.

Answers to Exercises

1. Colliding tectonic plates produce compressive forces that deform rock plastically into folds. Folds on a large scale produce folded mountains and upwarped mountains. Volcanoes form where magma erupts then accumulates—often where tectonic plates diverge. Fault-block mountains result where huge blocks of rock are subject to tension that occurs as tectonic plates move apart.

3. Where it is below sea level—for example on the shores of the Dead Sea, 400 meters below sea level.

5. Extraction of groundwater caused subsidence.

7. Some will evaporate, some will become surface run-off; the rest will percolate underground.

9. A well must be drilled past the level of the water table.

11. Convergent plate boundaries

13. Ice. Glaciers moving across a landscape loosen and lift up blocks of rock and incorporate them into the ice. They literally pick up everything in their path. As the ice melts, the rock debris is deposited.

15. No. Water flows downhill from higher elevations to the sea. So, it can't carry sediment up to sea level from where the land is below sea level.

17. Underground rivers occur, but they are very rare. Water typically moves through pore spaces, faults, and fractures and does not have large channels to move through.

19. Freshwater provides the sustenance for life. This includes water for drinking, agricultural uses, sanitation, and transportation.

21. Infiltration is greater on gentle sandy slopes because sandy materials have a high porosity and because runoff is greater on steeper slopes.

Solutions to Chapter 24 Problems

1. Porosity = (volume of open space)/(volume of open space + volume of solids) = $325\ cm^3/325\ cm^3 + 975\ cm^3 = 0.25$; the volume of open space is one-fourth of the total volume of the rock.

3. Volume of the mountain = $4\ km \times 3\ km \times 4\ km = (4000\ m) \times (3000\ m) \times (4000\ m) = 4.8 \times 10^{10}\ m^3$. Assume the rate of erosion is about $146\ m^3$ per year as stated in the *Math Connection* example. Then the duration of the mountain = (volume of mountain)/(rate of erosion) = $3.29 = 10^8$ years or 329 million years.

Chapter 25: Weather

Answers to Integrated Science Concepts

The "Ozone Hole" in the Atmosphere

1. Good ozone is the O_3 molecule when it is present in the stratosphere; bad ozone is O_3 found in the air we breathe.

The Coriolis Effect

1. The Coriolis Effect influences winds as well as surface currents by causing them to rotate with respect to Earth from left to right (as viewed from above the north pole.)

The Greenhouse Effect

1. Whitewash is sometimes applied to greenhouses in order to better reflect light thus reducing the amount of incoming solar radiation and subsequent indoor temperature.

Answers to Exercises

1. A sea breeze blows from the sea toward the shore and occurs mostly during the day. The reason is that land cools off faster than the ocean, so during the day cooler, high-pressure air blows toward land. At night, the cooler, high-pressure air forms over land, so it flows toward the area of lower pressure out to sea.

3. Friction
5. The change in environment from cold to warm. As we leave the cold outdoors the warm air inside comes into contact with the cold surface of the eyeglasses. As the air touching the glasses cools to its dew point, water vapor condenses onto the eyeglasses.
7. The layer of cold air only allows minimal convection currents to occur.
9. Seasonal temperatures are caused by solar intensity, solar radiation per area. In the winter in the Northern Hemisphere, the tilt of Earth's axis leads to solar radiation at the widest angle, reducing solar intensity to a minimum.
11. It is not accurate. Atmospheric circulation is broken up into six convection cells due to the influence of the Coriolis Effect.
13. Because cool air has slower moving molecules, and warm air can hold more water vapor than cold air.
15. The low cloud cover acts as an insulation blanket inhibiting the outflow of terrestrial radiation.
17. Warm air is able to hold more water vapor before becoming saturated than can cold air. As warm moist air blows over cold water it cools, which causes the water vapor to condense into tiny droplets of fog.
19. The Sun heats the ocean unevenly; equatorial waters are warmed more than parts of the ocean nearer the poles. Currents redistribute heat so that it is dispersed more evenly.
21. At high altitudes, there is more UV radiation due to a decrease in the concentration of UV absorbing atmospheric gases.
23. Since pressure decreases in a regular way with altitude, a device for measuring pressure can be used to measure altitude by recalibrating the scale.
25. People who prefer definite seasons prefer inland areas; those who like more moderate climates would prefer to live in coastal areas, due to the moderating effect of large bodies of water on climate (due to water's high specific heat capacity.)
27. The leeward side of a mountain has a much drier climate than the windward side, so leeward vegetation would consist of plants adapted for dry climates—cacti and succulents, rather than ferns and conifers, for example.

Solutions to Chapter 25 Problems

1. Standard atmospheric pressure = 14.7 lb/in^2. Convert this to units of feet: (14.7 lb/in^2) × (144 in^{2}ft^2) = 2,100 lb/ft^2. Pressure (P) is defined as force (F) per area (A). Rearrange: $F = P \times A$. Then the total force that the air inside the house exerts upward on the ceiling: $P \times A$ = (2,100 lb/ft^2) × (2,000 ft^2) = 4.2×10^6 lb of force.
3. Relative humidity = [(water vapor content)/(water vapor capacity)] × 100%. Rearrange to solve for the mass of water in 1 m^3 of air: water vapor content = [(relative humidity)(water vapor capacity)]/100%. Then: water vapor content at 50°C = [(40%)(9 g/m^3)]/100% = 3.6 g/m^3

Chapter 26: Earth's History

Answers to Integrated Science Concepts

Radiometric Dating of Rock

1. The half-life of uranium-238 is 4.5 billion years; potassium-40 has a half-life of 1.3 billion years; carbon-14 has a half-life of 5730 years.
3. Carbon-14

The Great Transformation of Earth's Atmosphere

1. a. The early Earth atmosphere had a much higher concentration of carbon dioxide (perhaps 80%) than it does today (only 1/10 of 1%); b. Free oxygen was not present in appreciable amounts in the Earth's early atmosphere, but is plentiful enough now to support air-breathing organisms.
3. Earth's atmosphere was drastically transformed by cyanobacteria that absorbed carbon dioxide and produced free oxygen through photosynthesis.

The Permian Extinction

1. The Permian extinction caused the demise of about 90 percent of species living at the time.

The Cretaceous Extinction

1. The iridium layer in the rock record corresponding to the Cretaceous.

Answers to Exercises

1. We can assume that the ancient rocks represent the location of an ancient stream. This is an application of the Principle of Uniformitarianism.
3. Fossils are used to establish the relative ages of rocks because geologists have been able to arrange different groups of fossils—and the time periods they were associated with—in a chronological sequence.
5. The collision between the westward moving North American Plate and the Pacific Ridge System occurred about 30 million years ago, creating the San Andreas fault.
7. Abundant microscopic life appeared at the end of the Precambrian.
9. a. Uranium-238 (or possibly Potassium-40, depending on how early); b. Uranium-235 or Potassium-40 are best, but Uranium-238 will work too (but is not as precise); c. Carbon-14.
11. The fossil leaves must be older than the shale, but the shale is on top of the formation. Most of the rocks in the sequence are older than the fossil leaves. So, we can say that the average age of the formation is younger than the trilobite and older than the fossil leaves. By average we mean that the formation was deposited over some finite time period, and the age we get from the fossil brackets the beginning and end of that formation.
13. At the time of deposition, the climate of Antarctica was mild enough to support swamps.
15. Melting of the polar ice caps, caused by global warming, for example. An increase in the rate of sea floor spreading could also make sea levels rise.

Solutions to Chapter 26 Problems

1. By ratio and proportion, a 1-km thick sequence would accumulate in 100 million years.

3. Pie graphs should depict this data: 23% mammals, 12% birds, 61% reptiles, 31% amphibians, 46% fishes, 73% insects, 45% mollusks, 86% mosses, 67% ferns, and 73% flowering plants.

Part 5 Astronomy

Chapter 27: The Solar System

Answers to Integrated Science Concepts

The Solar Nebula Heats Up, Spins Faster, and Flattens

1. It spun faster, and became flatter and hotter.

The Chemical Composition of the Solar System

1. Astronomers call the elements heavier than hydrogen the "heavy elements." Elements from lithium through iron are formed through thermonuclear fusion in stars. Elements heavier than iron are formed by large-mass stars that supernovate.

What Makes a Planet Suitable for Life?

1. Scientists believe that most, perhaps all, life in the universe is based on carbon because carbon can form four covalent bonds and therefore serve as the backbone for large biomolecules.

Why One Side of the Moon Always Faces Us

1. The rates are the same.

3. The fact we see one side is evidence that it rotates. If it didn't rotate, we'd only need to wait until it completed a half orbit to see its opposite side.

Answers to Exercises

1. Planets should be labeled in order from the Sun: Mercury, Venus, Earth, Mars, Jupiter, Saturn, Uranus, Neptune, Pluto.

3. Comet tails point away from the Sun because they are blown away from the Sun by the solar wind.

5. Yes; the Martian poles are covered with ice and the surface exhibits dry river beds or flood plains—surface features produced by flowing water.

7. The Giant Impact Theory is a subset of the nebular theory, so there is no contradiction between the nebular theory and Giant Impact Theory.

9. Observations are made during the new moon part of the month, when the sky is moonless. It makes a difference because moonlight is not there to be scattered and obscure a good view.

11. Extend the bite to complete a circle, and the patch of Earth's shadow appears to be a circle with a diameter 2.5 moon diameters. Does this mean Earth's diameter is 2.5 moon diameters? No, because Earth's shadow at the distance of the Moon has tapered. How much? According to the tapering that is evident during a solar eclipse, by 1 moon diameter. So, add that to the 2.5 and we find Earth is 3.5 times wider than the Moon.

13. The Sun's output of energy is that of thermonuclear fusion. Because fusion in the sun is the result of gravitational pressure, we can say the prime source of solar energy is gravity. Without the strong gravity, fusion wouldn't occur.

15. It has more surface area in the disk shape which allows it to radiate more energy not reradiated.

17. In star interiors.

19. The Jovian planets are large gaseous low-density worlds, and have rings. The terrestrial planets are rocky and have no rings.

21. Using the following mnemonic to state the order of the planets: *My very excellent mother just served us nine pizzas* (Mercury, Venus, Earth, Mars, Jupiter, Saturn, Uranus, Neptune, Pluto).

23. Meteorites are more easily found in Antarctica, because so many are imbedded in ice. On regular ground, they are not so obvious. Found on the surface of ice indicates they came from above.

25. A comet continually orbits the Sun.

27. The essential reason is the relative proximity of the inner planets to the Sun. Distance from the sun determines the temperature differences of the planets which in turn accounts for differences in the planets' chemical composition. The inner planets consist of materials that stay solid at the higher temperatures found nearest the Sun. The outer planets consist largely of hydrogen and helium gases that cohere under the influence of gravity in regions distant from the Sun, where gas pressure is lowered because the thermal energy of the gases is lower.

Solutions to Chapter 27 Problems

1. The mean distance from Earth to the Sun is defined as 1 AU = 1.5×10^8 km. From Table 27.1, the diameter of Earth is 12,760 km. So the number of Earth diameters between Earth and the Sun is 1.5×10^8 km/1.276×10^4 km = 11,800. About 12,000 Earths would fit in the distance between Earth and the Sun.

3. Radio signals travel at the speed of light, 3×10^8 m/s = 3×10^5 km/s. From Table 27.1, the distance from the Sun to Saturn is 9.54 AU, so the distance from Saturn to Earth is 9.54 AU − 1 AU = 8.54 AU. Convert this to kilometers: 8.54 AU × (1.5×10^8 km/AU) = 1.28×10^9 km. Using the distance formula, we have $t = d/r$ = (1.28×10^9 km)/(3×10^5 km/s) = 0.4267×10^4 s = 1.19 h or about 71 minutes. A similar calculation shows that the time required for the radio signal to reach Pluto is 5.4 h or about 321 minutes.

Chapter 28: The Universe

Answers to Integrated Science Concepts

Radiation Curves of Stars

1. The hotter a star is, the shorter the wavelength of its peak frequency.

The Search for Extraterrestrial Life

1. The SETI (Search for Extraterrestrial Intelligence) program is an effort to locate evidence of past or present communicative civilizations in the universe, particularly within our own galaxy.

3. Open-ended.

Answers to Exercises

1. He didn't know that the constellations are not always overhead in the sky, but vary with Earth's motion around the Sun.
3. The figure shows that the background of a solar eclipse is the nighttime sky normally viewed 6 months earlier or later.
5. The nuclei of atoms that compose our bodies were once parts of stars. All nuclei beyond iron in atomic number were in fact manufactured in supernovae.
7. Since all the heavy elements are manufactured in supernovae, the newer the star, the greater percentage of heavy elements available for its construction. Very old stars were made when heavy elements were less abundant.
9. Thermonuclear fusion reactions produce an outward pressure that counteracts the inward pressure that would lead to collapse due to gravity.
11. Thermonuclear fusion is caused by gravitational pressure, wherein hydrogen nuclei are squashed together. Gravitational pressures in the outer layers are insufficient to produce fusion.
13. Bigger stars live faster, and collapse more energetically when they burn out.
15. Stars with fewer heavier elements formed at an earlier time than the Sun.
17. You are simply closer to the center of gravity of the star, in accord with Newton's law of gravitation.
19. Yes, the central bulge of the Andromeda Galaxy, which covers an area about five times that of the full moon, can be seen with the naked eye on a clear night. The Magellanic clouds are two galaxies visible to the naked eye in the Southern Hemisphere.
21. Space exists in the universe, not the other way around.
23. Both you and the Earth don't have to occupy a central location to be special. The Earth is certainly special among planets in the solar system in that it is the only one with abundant water and an atmosphere—and us.
25. Polaris lies on the axis of Earth's rotation, other stars do not. As Earth revolves on its axis, the celestial sphere seems to rotate.
27. Elements heavier than iron are created by high-mass stars that supernova while elements lighter than iron are created by fusion in low- and medium-mass stars.
29. Stars vary in color because they vary in temperature; they vary in brightness because they are different distances from Earth and because they produce different amounts of radiant energy.
31. For some cultures, study of the constellations involved storytelling; to other cultures they served as navigational aids for travelers and sailors; to other cultures, the constellations provided a guide for the planting and harvesting of crops because constellations were seen to move in the sky in concert with the seasons.

Solutions to Chapter 28 Problems

1. If these stars were the same distance from Earth, apparent brightness would depend just on luminosity and Star A appears four times as bright as Star B. However, if Star A were twice as far away as Star B, the stars would have equal apparent brightness because apparent brightness is related to distance through the inverse-square law.

